Searching for ORDER IN the COMPLEXITY of Evolving Worlds

ACKNOWLEDGMENTS

*The SFI Press would not exist without the support of William H. Miller
and the Miller Omega Program, Andrew Feldstein and the Feldstein
Program on History, Law, and Regulation, and Alana Levinson-Labrosse.*

LAW AS DATA

Computation, Text, & the Future of Legal Analysis

MICHAEL A. LIVERMORE
DANIEL N. ROCKMORE

editors

⁵ᶠᴵPR✦SS

THE SANTA FE INSTITUTE PRESS

1399 Hyde Park Road
Santa Fe, New Mexico 87501

The Energetics of Computing in Life & Machines
ISBN (HARDCOVER): 978-1-947864-13-9
Library of Congress Control Number: 2019941906

The SFI Press is supported by the
Feldstein Program on History, Regulation, & Law,
the Miller Omega Program, and Alana Levinson-LaBrosse.

This volume emerged from workshops supported by the
National Science Foundation under Grant Nos. 1741021
and 1749019. Any opinions, findings, and conclusions or
recommendations expressed in this material are those of
the author(s) and do not necessarily reflect the views
of the National Science Foundation.

IT IS A CAPITAL MISTAKE to theorize before one has data. Insensibly one begins to twist facts to suit theories, instead of theories to suit facts.

SIR ARTHUR CONAN DOYLE
"A Scandal in Bohemia,"
The Adventures of Sherlock Holmes (1892)

CONTRIBUTORS

Charlotte S. Alexander, *Georgia State University*

Daniel Argyle, *FiscalNote*

Elliott Ash, *ETH Zurich*

Jon Ashley, *University of Virginia*

Adam B. Badawi, *University of California, Berkeley*

Keith Carlson, *Dartmouth College*

Daniel L. Chen, *University of Toulouse*

Ryan Copus, *Harvard University*

Giuseppe Dari-Mattiacci, *University of Amsterdam*

Marion Dumas, *London School of Economics and Political Science*

Vlad Eidelman, *FiscalNote*

Mohammad Javad Feizollahi, *Georgia State University*

Adam Feldman, *SCOTUSBlog*

Jens Frankenreiter, *Max Planck Institute for Research on Collective Goods*

Brian Grom, *FiscalNote*

Yosh Halberstam, *University of Toronto*

Ryan Hübert, *University of California, Davis*

Khalifeh al Jadda, *Georgia State University*

Anastassia Kornilova, *FiscalNote*

Manoj Kumar, *NYU Center for Data Science*

Hannah Laqueur, *University of California, Davis*

Michael A. Livermore, *University of Virginia*

Allen Riddell, *Indiana University Bloomington*

Daniel N. Rockmore, *Dartmouth College*

Anne M. Tucker, *Georgia State University*

Anna Venancio, *Primer AI*

Alan C. L. Yu, *University of Chicago*

TABLE OF CONTENTS

Introduction: From Analogue to Digital Legal Scholarship
The Editors xiii

Frameworks

1: Distant Reading the Law
Michael A. Livermore and Daniel N. Rockmore 3

2: Big Data, Machine Learning, and the Credibility
Revolution in Empirical Legal Studies
Ryan Copus, Ryan Hübert, and Hannah Laqueur 21

3: Text as Observational Data
Marion Dumas and Jens Frankenreiter 59

4: Prediction Before Inference
Allen Riddell 73

Investigation

5: Style and Substance on the US Supreme Court
*Keith Carlson, Daniel N. Rockmore, Allen Riddell, Jon Ashley, and
Michael A. Livermore.* 83

6: Predicting Legislative Floor Action
Vlad Eidelman, Anastassia Kornilova, and Daniel Argyle 117

7: Writing Style and Legal Traditions
Jens Frankenreiter 153

8: A Computational Analysis of California
Parole Suitability Hearings
Hannah Laqueur and Anna Venancio 193

9: Analyzing Public Comments
Vlad Eidelman, Brian Grom, and Michael A. Livermore. 233

Exploration

10: Using Text Analytics to Predict Litigation Outcomes
Charlotte S. Alexander, Khalifeh al Jadda, Mohammad Javad Feizollahi, and Anne M. Tucker 275

11: Case Vectors: Spatial Representations of the Law Using Document Embeddings
Elliott Ash and Daniel L. Chen 313

12: Reference Networks and Civil Codes
Adam B. Badawi and Giuseppe Dari-Mattiacci 339

13: Attorney Voice and the US Supreme Court
Daniel L. Chen, Yosh Halberstam, Manoj Kumar, and Alan C. L. Yu 367

14: Detecting Ideology in Judicial Language
Marion Dumas 383

15: Opinion Clarity in State and Federal Trial Courts
Adam Feldman 407

Conjecture

16: Machine Learning and the Rule of Law
Daniel L. Chen 433

17: The Law Search Turing Test
Michael A. Livermore and Daniel N. Rockmore 443

Reference

Bibliography 455

INTRODUCTION: FROM ANALOGUE TO DIGITAL LEGAL SCHOLARSHIP

The Editors

In 1978, legal philosopher Ronald Dworkin proposed a thought experiment to argue that difficult questions of legal interpretation have unique correct answers. Dworkin acknowledged that normal mortal human judges, flawed and incomplete as they are, might not know these answers. But he proposed a kind of super-judge, a "lawyer of superhuman skill, learning, patience, and acumen," able to read and understand every possible legal document (Dworkin 1978, 105). Dubbed "Hercules" by Dworkin, this super-judge—possessing memory and cognitive capacities far outstripping those of any human—would be able to decide new cases in ways that conform to the best interpretation of all relevant legal sources.

Four decades later, the idea of the super-judge "of superhuman skill, learning, patience, and acumen" has taken on new meaning. In recent years, artificial intelligence platforms have steadily continued to beat the best humans at increasingly sophisticated games of strategy and memory: checkers in 1995, chess in 1997, Jeopardy! in 2011, Go in 2016, and StarCraft in 2019. Legal contestation isn't play, but nonetheless requires similar skills, such as the application of strategic thinking and logic and the ability to draw on vast stores of information. In other professional areas, advances in computational technology have hastened the spread of automation to cognitive tasks once performed by white collar workers. Some commentators have speculated that if current trends continue, a day may come when the legal workforce becomes almost completely automated (Susskind 1998), with lawyers finding themselves replaced by blinking computerized Herculi, operating in server farms rather than law offices.

This dramatic vision is provocative, but there is another possibility that is both more realistic (at least for the foreseeable future) and in some ways even more arresting. After losing to the IBM-created chess program Deep Blue in 1997, world champion Gary Kasparov proposed a new game called "advanced chess" in which human–computer teams compete against each other (Kasparov 2017). The idea is that humans and machines have different cognitive styles that balance the relative strengths and weakness of each other, leading to a game that is more technically complex and strategically sophisticated—in some ways, at least, a better form of chess. Similarly, perhaps as computers come to augment the cognitive capacities of lawyers in more profound ways, computer-assisted human lawyers could begin to practice a kind of "advanced law," ultimately approaching "superhuman skill, learning, patience, and acumen."

It is hard to know whether advanced law of this variety will be socially beneficial. The spread of ever-more-sophisticated computational tools within law practice presents both potential and peril, and whether businesses and firms competing in the marketplace and legal forums use these techniques for good or ill is likely to depend largely on how their incentives line up with social well-being.

This dynamic is well illustrated by the transition from analog to digital in the practice of discovery in the United States. The pretrial information disclosure procedure known as discovery is famously contentious, expensive, and time-intensive. Lawyers in the midst of discovery face a number of challenges. On the receiving end of a document request, lawyers must figure out which documents are responsive, which documents are irrelevant, and which might be relevant but need not be released (such as those protected as attorney work product). On the receiving end of a request, lawyers must comb through a large number of documents to look for the smoking gun, if there is one.

Analog discovery involves an army of associates sitting in document repositories combing through documents. Entry-level associates may not covet the assignment, but, with documents totaling some thousands of pages, analog discovery is a manageable undertaking. Document digitization, followed by the falling price of electronic storage, massively expanded the number of documents that parties to a litigation might have in their possession. In many cases, this information explosion moved analog discovery—that is, passing human eyes over every document—from the realm of the expensive and difficult to the impossible. Law firms started to adopt computational tools to adapt, first relying on simple keyword searches, and then moving on to more sophisticated machine learning approaches that use human attorneys to train algorithms that are then unleashed on the document pile. These algorithms now routinely beat their human trainers in retrieving information, so much so that a landmark 2012 discovery ruling in the Southern District of New York concluded that, at least some of the time, computer-assisted review was "better than the available alternatives, and thus should be used for appropriate cases."[1]

Today's "e-discovery" represents a kind of advanced law practice that is somewhat akin to Kasparov's advanced chess. Once again, machine intelligence is used to augment human cognition in a competitive strategic environment. Whether this form of advanced law is better than the system it replaced

---

[1] Monique Da Silva Moore, et al. v. Publicis Groupe & MSL Group, 11 Civ. 1279 (ALC) (AJP) (SDNY 2012).

is a difficult question. Given their competitive environment, lawyers engaged in discovery have no choice but to rely on the most advanced computational tools that they can access. That does not necessarily mean that advanced law discovery will promote social well-being. It may be that lawyers are simply engaged in an e-discovery arms race that drives up legal costs with no appreciable benefit. Defenders of the adversarial system, on the other hand, might argue that if computational tools allow lawyers to better serve their clients, a hotter crucible will ultimately lead to better outcomes. Under this account, advanced e-discovery might better serve the purposes of unearthing relevant documents, resulting in earlier settlement (avoiding litigation costs) or better-informed adjudication of legal rights. How one views the growth of advanced e-discovery, then, largely turns on one's views about how well the incentives of the underlying actors in the adversarial system line up with social well-being.

Of course, computation in one form or another has been central to legal practice for decades. Examples include word processing, search, document retrieval, email, and even the Internet. Although these tools are now so quotidian that they almost fade into the background, they clearly have had a major influence on legal practice as well as the study of the law. In the workplace, developments range from the increased length of a workday (e.g., 3:00 a.m. client emails) to more substantial changes such as computerized search. Advanced law—in the Kasparovian sense—is already well underway.

Scholars have different goals and face different incentives than legal practitioners, but information technology has changed life in academia as well. Teaching, of course, has been affected by technology: professors and students communicate on web forums, guest lecturers pop in via Skype, and new teaching tools (like classroom response systems, or "clickers") are deployed. But the work of scholarship and knowledge production has also been affected. Research once done in libraries is now done online; work in progress is shared instantly through portals like the Social Science Research Network rather than awaiting publication in paper journals; empirical scholarship facilitated by advances in data storage and computer-assisted statistics is now regularly featured in law journals.

For legal scholars, advances in computational text analysis also empower new ways to read and think about the law, culminating in a new approach that we will call *computational legal analysis*. This new approach has important consequences for scholars who study the law, but will also likely ultimately filter into legal practice. Computational legal analysis sits at the intersection of traditional doctrinal analysis and quantitative empirical legal studies. With doctrinal analysis, which has been a mainstay of legal practice and commentary for centuries, computational legal analysis shares its emphasis on the text of legal documents—on what is said by legal actors like courts and legislatures. With quantitative empirical legal studies, a comparatively recent intellectual movement primarily within US law schools, computational legal analysis shares its emphasis on the use of mathematical methods to describe and understand legal phenomena, especially through the collection and analysis of data.

This volume explores the new field of computational legal analysis. The hallmark of this approach is its use of *legal texts as data*. Advances in computational text analysis, natural language processing, and machine learning are in the process of transforming research in a variety of disciplines; a new movement to use "text as data" has begun reshaping fields across the social and behavioral sciences (Grimmer and Stewart 2013). Human life is dominated by language and, since the advent of written writing, textual representations of language have served as a core means of human interaction. Advances in technology have created ever more opportunities for people to communicate via text (indeed, "texting" is now a basic and common feature of everyday life). As the price of digital storage has fallen, massive new textual datasets are continually created, covering everything from scientific publications to Wikipedia to microblogs. These datasets have dramatically increased the number and kinds of raw materials now available for quantitative social and behavioral science research. As noted by Grimmer and Stewart (2013), the question for researchers now is not whether to use these new data, but how best to integrate textual data into scientific research programs that push forward the frontier of knowledge and understanding.

If text as data is transforming the social and behavioral sciences generally, using the specific text of law as data will transform legal scholarship specifically. The origin of this book was a working group held at the Santa Fe Institute (SFI) and sponsored by the Feldstein Program on Law, History, and Regulation in December 2017. At that meeting, a group of scholars in law, mathematics, computer science, machine learning, and the social sciences gathered for two days in Santa Fe to discuss how the study of law will be transformed as researchers grab hold of new text-as-data techniques. The explosion in availability of law as data opens up important new vistas for researchers, enabling the construction of new conceptual models of law and legal development; the invention of new methods and analysis; and the curation and organization of new sources of data. Together, these concepts, methods, and data can help form the basis for new research programs built around a "standard package" (Fujimura 1988) that can galvanize and coordinate a concerted multidisciplinary inquiry with the potential to generate new scientific insights into law as a social and political phenomenon.

Our goals with this volume are to provide an introduction for the legal world of the broad range of computational tools that are already proving themselves relevant to law scholarship and practice, and to highlight the early steps in what promises to be an exciting new approach to studying the law. The following chapters collect and discuss some of the best and most innovative scholarship happening in this new field of computational legal analysis. The overarching theme of the volume is scholarship that uses the text and related underlying data of legal documents as the direct object of quantitative statistical analysis. These documents can include texts that are traditionally recognized as law, such as judicial opinions, constitutions, and statutes, as well as a broader set of legal documents, such as regulatory filings and administrative proceedings. The eclectic mix of subject matter presents these diverse textual corpora as objects of study that are the target of a range of advanced forms of text analysis and related statistical tools. These include topic modeling, stylometry, network analysis, and various machine learning approaches. The emphasis across the volume is work that pushes methodological boundaries, either by using new tools to study longstanding questions within legal studies or by identifying new questions revealed by recent developments in data availability and analysis.

The chapters in this book are likely to be of interest to legal scholars; scholars in other fields interested in empirical legal studies, such as political science and economics; and researchers in fields related to applied quantitative text analysis, such as computer science and digital humanities. As the first general effort to collect work across this new field, this volume is intended to help map and define a research agenda that will continue to flourish and ultimately take up a central place within law scholarship and legal studies.

The volume is divided into four sections: Frameworks, Investigation, Exploration, and Conjecture. *Frameworks* contains four chapters that discuss cross-cutting issues faced by law-as-data researchers, especially concerning how quantitative analyses of legal texts should be interpreted.

In the first chapter, "Distant Reading the Law," we discuss the notion of distant reading, which serves as the conceptual leap at the heart of computational legal analysis. This idea was first developed within the field of literary criticism (Moretti 2013). Loosely characterized, distant reading involves reducing text to numeric data that can be quantitatively analyzed to identify and characterize patterns. The transformation enables huge and sometimes exhaustive corpora of text to be analyzed at a coarse scale (the "distance" in distant reading). Traditional reading, which is really a cluster of related activities that range from skimming (reading quickly or cursorily) to perusing (reading slowly and carefully, or "close reading"), has been around since the advent of the written word. Although early distant reading can be found at least as far back as the nineteenth century (Hughes et al. 2012), the extensive and increasingly all-inclusive digitization of text and advances in computational hardware and sophisticated natural language–processing software have radically transformed the practice to position distant reading as a legitimate—and sometimes necessary—alternative method of textual comprehension. Distant reading will transform (and to some extent already has transformed) the relationship of lawyers and society to legal texts. In this chapter, we delve into the nature of distant reading and how it differs from traditional ways of comprehending textual information. We also discuss how distant reading can fit into what Justice Oliver Wendell Holmes described as the "rational study of law," how distant reading expands the criteria for what counts as a legal text, and how the relative success of different quantitative approaches to distant reading sheds light on the nature of law and legal development.

In the second chapter, "Big Data, Machine Learning, and the Credibility Revolution in Empirical Legal Studies," Ryan Copus, Ryan Hübert, and Hannah Laqueur discuss the intersection of two intellectual trends that fundamentally interact with the law-as-data movement: the "credibility revolution" that re-emphasized the importance of causal inference and experimental design in social science research, and the growth of machine learning techniques that, merged with big data, are capable of highly accurate prediction. Drawing from several examples from their own research on law and legal decision-making, the authors argue that machine learning and experimental design should be understood as complements, rather than substitutes, with each bringing epistemic strengths and weaknesses. Researchers in the law-as-data movement must be mindful of both the limitations and promise of these two types of analysis.

In "Text as Observational Data" by Marion Dumas and Jens Frankenreiter, the authors return to thorny epistemic terrain, exploring the distinction between "causal inference, the description of processes, and the discovery of mechanisms." While the first has more traditionally been the domain of quantitative social science scholarship, the latter two have been primarily the subject of qualitative research. Dumas and Frankenreiter argue that computational legal analysis creates new opportunities for researchers to explore the processes and mechanisms of legal change through quantitative tools. Nevertheless, there may often be trade-offs to be made between detailed description of process and mechanism and robust causal inference concerning counterfactuals or the likely effects of interventions. This trade-off requires researchers to attend carefully to the limits of their analyses, but also favors a multiple-methods approach that values different analytic methods and the unique perspectives they offer.

The final chapter in this section is a short intervention by Allen Riddell, "Prediction Before Inference," that challenges the social sciences' emphasis on causal inference. For Riddell, useful and accurate out-of-sample prediction rather than experimental design should be the calling card, and focusing overmuch on the latter risks stunting researchers' ability to address important questions or identify tantalizing patterns in the world that can then serve as the basis for future work.

The next section of the book, *Investigation*, collects chapters that discuss the methods and results from several sustained law-as-data research projects. The first chapter, "Style and Substance on the US Supreme Court," discusses some of our work with collaborators Keith Carlson, Allen Riddell, and Jon Ashley on the United States Supreme Court. SCOTUS, as it's known to its friends, is a singular institution in American public life. It is a court, but it is also the subject of high politics and near-constant partisan contestation. The judicial opinions produced by SCOTUS are among the most important pieces of legal "data" in understanding legal thought and sociopolitical dynamics in the United States. In this chapter, we discuss two related projects. First, we examine trends in writing style on the court through sentiment analysis and a form of stylometry based on the frequency of function words in a text. Second, we examine whether the Supreme Court has begun to carve out a unique judicial genre by examining how the content of Supreme

Court opinions differs from the federal appellate courts that it supervises.

The second chapter in this section, "Predicting Legislative Floor Action" by Vladimir Eidelman, Anastassia Kornilova, and Daniel Argyle, discusses a project that use machine learning and a large dataset of state legislative actions to study the dynamics of lawmaking. Drawing from prior political science literature, the authors identify several political features—such as the identity of bill sponsors—that have predictive potential, and pair those features with information extracted from bill text concerning the subject matter of legislation. Feeding these data into a machine learning model, the authors generate strikingly accurate predictions concerning the likelihood that a given bill will make it to the floor for a vote. The authors also examine the features that have the most predictive power, including various political dynamics (such as whether a bill is sponsored by a member of the majority party), the semantic content of proposed legislation, and legislative actions taken subsequent to introduction but prior to floor actions (such as the introduction of amendments).

In the following chapter, "Writing Style and Legal Traditions," author Jens Frankenreiter looks outside the United States to the European Court of Justice (ECJ). As a transnational court with power over many different national jurisdictions, the judicial culture on the ECJ and its relationship to national legal cultures is a consistent source of controversy. This controversy covers substantive issues, of course, but also generates criticism over seemingly more mundane issues, such as the writing style of the judges on the Court. Leveraging rounds of EU enlargement, Frankenreiter uses stylometric tools to examine time trends in writing style, and particularly whether jurists from French language jurisdictions—who dominated the early ECJ—continue to exhibit stylistic influence. In general, this analysis finds that over time, jurists from non-French jurisdictions moved closer to their French jurisdiction peers, while the opposite is less true, evidencing continued influence by French-speaking jurisdictions over the writing style of the ECJ.

In "A Computational Analysis of California Parole Suitability Hearings," authors Hannah Laqueur and Anna Venancio focus on parole decisions, a process with profound consequences for the lives of incarcerated individuals as well as the penal system more generally. Drawing on data from eight thousand transcripts from California state parole hearings held between 2011 and 2014, the authors examine the factors that may affect parole decisions, including expert evaluation of inmate risk, cognitive biases among parole officials, and narratives of remorse and redemption given by inmates themselves. At a time of growing attention to the problem of mass incarceration, the authors' analysis of this crucial but understudied legal phenomenon are relevant to a range of potential proposals currently in play that aim to reform the criminal justice system.

The final chapter in the section, "Analyzing Public Comments" by Livermore, Vladimir Eidelman, and Brian Grom, explores the use of text analysis tools in the investigation of one of the largest unstructured textual datasets intentionally accumulated by the federal government: public comments received in response to proposed regulations. Every year, federal agencies propose hundreds of major rulemakings, and under the Administrative Procedure Act, they must have a process in place to solicit, collect, analyze, and respond to comments submitted by the public. With the transition to online platforms in the early 2000s, participation in notice-and-comment rulemaking (as it is typically described) has increased exponentially, and "mega-rulemakings," which receive more than a million comments, are now common. In this chapter, the authors examine how computational text analysis tools can be used to manage this new phenomenon by analyzing several million public comments received in response to rules proposed during the presidential administration of Barack Obama.

The next section of the book, *Exploration*, collects chapters at the experimental edge of computational legal analysis. The first, "Using Text Analytics to Predict Litigation Outcomes," is by a group of researchers at Georgia State University: Charlotte S. Alexander, Khalifeh al Jadda, Mohammad Javad Feizollahi, and Anne M. Tucker. They tackle one of the most pressing questions in this area, at least from the practitioner's perspective, which is whether computational tools can be used to predict the outcomes of legal disputes. The chapter discusses the goals, methodologies, and preliminary results of an ongoing litigation outcomes–prediction project that focuses on a "small data" population of several thousand employment law cases filed and closed in the US District Court for the Northern District of Georgia in the period 2010–2017. The authors describe the challenges of collecting and curating the relevant data, and discuss some promising early returns in applying predictive analytic tools to this constrained set, with the goal of learning lessons that can be applied at scale.

In "Case Vectors: Spatial Representations of the Law Using Document Embeddings," authors Elliott Ash and Daniel L. Chen apply the natural language–processing tool of vector space models to legal language, specifically judicial opinions from the US Supreme Court and US appellate courts. The goal of vector space models is to construct a space in which location and distance carry some semantic meaning. In the well known "word-to-vec" model, words serve as the unit of interest (Mikolov, Chen, et al. 2013) but higher-level semantic objects, including documents, can also be captured this way. Ash and Chen develop the first (to our knowledge) application of document embeddings to a corpus of judicial opinions and engage in exploratory data analysis to estimate observed relationship between vector representations and variables of interest, such as legal topic and political party affiliation.

The following chapter, by Adam Badawi and Giuseppe Dari-Mattiacci, widens its focus to statutory law around the globe. In "Reference Networks and Civil Codes," the authors explore the question of whether history leaves its imprint on the internal structure of a nation's laws. Analyzing the network structure of the cross references in statutory law, the authors find marked differences in contemporary statutes based on their historical origins. This research has the potential to update and improve scholarship that examines how a legal system's origins affects contemporary development—a heated subject of dispute in empirical legal studies.

In "Attorney Voice and the US Supreme Court," authors Daniel L. Chen, Yosh Halberstam, Manoj Kumar, and Alan C. L. Yu examine an alternative channel for legal language: oral communication. While much of the law is textual, there are important forums where oral communication remains vital. At the same time, the widespread recording of these proceedings is breaking down the barriers between oral and written communication. In their study, the authors use a surprising new source of data for empirical legal study: voice recordings of attorneys presenting oral arguments before the US Supreme Court. Using both human-coded data as well as machine learning tools, they examine how vocal features of attorneys affect justices' voting, even when the attorneys are saying the same words. The authors find that gendered characteristics appear to have some affect on how justices perceive the strength of a legal argument.

One of the most persistent questions in empirical legal studies is the influence of judicial ideology on legal outcomes, and, in her chapter, "Detecting Ideology in Judicial Language," Marion Dumas takes a new look at this question. Rather than coding for case outcomes—which is standard in the literature—Dumas examines the language of judicial opinions using natural language-processing tools that were developed to detect ideological bias in scholarly writing (specifically economics research). Dumas finds that, using the same methodology, judges' writings do not appear to differ on the basis of their party affiliation. Dumas concludes that, in some relevant senses, the judicial genre is more neutral that other studied corpora, but she also notes that future technological developments may be able to recover ideological leanings in legal language that current tools fail to recognize.

Text-analysis tools can also provide new leverage on another question of great practical importance to the legal profession: the written clarity of judicial opinions. Well-written opinions are easy to understand and apply, while poor writing leads to confusion and misunderstanding. In "Opinion Clarity in State and Federal Trial Courts," author Adam Feldman compares writing clarity in state and federal trial courts to explore how writing differs in these two venues. Feldman finds that there are identifiable differences, with federal courts issuing longer and less clearly written opinions. This finding raises an important set of questions for future researchers concerning the features of judges and jurisdictions that might account for these differences.

The final section, *Conjecture*, comprises two short chapters that offer some speculation on what might be over the horizon for the field of computational legal analysis. In "Machine Learning and the Rule of Law," Daniel L. Chen argues that, given adequate data, machine learning tools can be used to reduce bias in the application of the law. Accurate models of judging can help identify those cases where the legitimate factors that bear on outcomes—that is, relevant law and facts—leave legal decision-makers in equipoise, as situations open the door for psychological biases to creep in. At a time when machine learning and predictive analytics have been criticized for their potential to perpetuate bias in the legal system (Starr 2014), Chen offers a more hopeful perspective on how these tools can be put to beneficial use.

In the final chapter, "Legal Data for Turing-Like AI Challenges," the two of us offer some speculation on how legal data can help push forward the field of artificial intelligence research. For as long as the field has existed, progress on AI has been measured through benchmark tasks, typically compared to humans. These include gameplay, but also tasks such as machine translation and image recognition and categorization. We argue that certain types of legal data have the characteristics that AI researchers crave: mountains of data, and a human standard that can be used to measure progress. We offer law search as an example, and discuss some of our early successes in using computational algorithms to simulate human-level law search.

This book is, quite obviously, not meant to be the final word on computational legal analysis. Rather, it is a tentative and preliminary offering in what we hope becomes a long and fruitful conversation. We expect that many of the techniques and applications discussed in these pages will continue to develop and yield interesting results and avenues for research. Others may end up as dead ends or false paths. Much of the work described herein is exploratory and experimental, and future refinements will sharpen both the questions asked and the methods used. Regardless, we hope that the chapters that we have collected here convey some of the excitement and potential for this new field. It is impossible to know how computational legal studies will reshape how the law is understood, by both scholars and practitioners. But, the process of transformation has already begun, with a growing intellectual community interested in putting the newest technologies to use to study and improve the law. ✒

Acknowledgments

This book represents the labor of many. We are grateful to all of the contributors to this volume as well as the participants at the 2017 working group at the Santa Fe Institute. SFI and SFI Press have provided tremendous support for this project; our special thanks to the staff at SFI Press whose work made this volume possible: Sienna Latham, Laura Egley Taylor, Katherine Mast, and Lucy Fleming (who is now pursuing graduate studies at Oxford). We thank SFI President David Krakauer for his long-standing support of this project. Databrew worked with the SFI Press and authors to generate the beautiful data visualizations throughout the book. In the past several years, we have worked with many outstanding collaborators whose ideas and contributions have shaped this volume. They include Jonathan Ashley, Peter A. Beling, Sarunas Burdulis, Keith Carlson, Faraz Dadgostari, Vlad Eidelman, Chen Fang, Nick Foti, Tom Ginsburg, Brian Grom, Mauricio Guim, Reed Harder, Greg Leibon, Jason Linehan, and Allen Riddell. For many interesting conversations on themes related to this volume, we thank Jenna Bednar, Gillian Hadfield, and Daria Roithmayer as well as our colleagues at the University of Virginia School of Law and Dartmouth College Departments of Computer Science and Mathematics. We also appreciate the efforts of the Free Law Project and CourtListener to provide public access to legal documents; many of our projects have relied on this source for data. We've received financial support for our work from MITRE,

LAW AS DATA

the Neukom Institute for Computational Science at Dartmouth College, and the University of Virginia's Presidential Fellowships in Data Science program. Finally, we are especially grateful for the generous support for this project provided by the Feldstein Program on History, Law, & Regulation.

FRAMEWORKS

॰ͻ

DISTANT READING THE LAW

Michael A. Livermore, University of Virginia
Daniel N. Rockmore, Dartmouth College

The law is a textual enterprise—it is through the written word that legislatures, executive bodies, and judges make the law. The study of the law requires interacting with that text by reading statutes, regulations, judicial opinions, or other legal sources. For hundreds of years, the process of reading the law began by picking up a book. With the digitization of massive corpora of legal documents, advances in computer hardware, machine learning, and natural language processing techniques have made legal documents a form of "big data." As a consequence, a new form of "reading"—grounded in quantitative analysis and mathematics—has taken hold. This new form of reading is poised to change how the law is studied and understood by providing fresh perspectives on old questions and spurring entirely new research agendas.

Distant Reading the Law

It is now common for cultural artifacts—be they texts, images, objects, or even cities—to be represented in digital form (Jones 2013). Some artifacts are generated natively as digital, such as online literary magazines, newspapers, or videos. Others are transformed from the physical to the digital, as occurs when a paper text is scanned, and still others exist as representations of an underlying, perhaps lost, physical artifact in the world, such as a virtual reality reconstruction of portions of Ancient Rome. Regardless, when realized in digital form, cultural objects become a form of data that can be subject to analysis through the tools and techniques of computer science, statistics, and mathematics.[1] This change in form allows for quantitative analysis of phenomena that have, to date, largely been the exclusive province of qualitative methods. The analysis of large bodies of text is one such realm.

Document organization and retrieval is perhaps among the most familiar of the success stories of such work (Salton, Wong, and Yang 1975; Berry 2004; Berry and Castellanos 2008). Less familiar has been the use of related ideas to understand broad trends in textual corpora where large-scale temporal patterns in a collection of documents are of interest. Literary scholar Franco Moretti (2013) invented the term "distant reading" to characterize the growing use of quantitative tools to study literary texts.

Distant reading is the consideration of literature from afar, using statistics to support large-scale claims about phenomena such as genre. Distant reading is juxtaposed with the classic "close reading" performed by a traditional scholar of literature—or the law—wherein analysis proceeds by the detailed analysis of a relative handful of texts. Moretti invokes something of an analogy: computer analysis is to literature as surveying is to landscapes, with the derived statistics enabling the invention of a cartography of literary works (Moretti 2013).

Distant reading can pose significant interpretive challenges of both qualitative and quantitative varieties. Easily accessible

~ 4 ~

---

[1]Whenever something physical is "captured" via a digital representation, questions arise concerning what is lost. The issue of "lossy capture" is well worth considering and is posed even in the distinctions between words in their native form on a page compared to data on a machine. For the most part, we proceed under the assumption that much, and perhaps most, of the information contained in physical form can be translated with reasonable fidelity to machine format.

software packages may put powerful statistical tools within reach, but it can be difficult to understand how these sophisticated tools work. To use these tools in a meaningful way, one needs a firm grasp of their strengths, their limitations, and how the analyses they generate relate to questions of interest. Even if one understands the underlying qualitative models, interpreting their meaning can pose additional difficulties. For example, it is far from obvious how one ought to understand the normative significance of a trend toward more negative language on the US Supreme Court (Chapter 5) or the influence of partisan affiliation on word choices in US appellate courts (Chapter 14), even if the underlying empirical reality of the distinctions can be discerned fairly clearly.

The growth of distant reading in the humanities, spurred by Moretti and others, has not been uncontroversial. A lively debate has broken out in the field about the relative merits and demerits of distant and traditional reading. These debates are similar to the disagreements that occurred over the past several decades within the legal community as the methods and approaches of other disciplines—most notably, economics, philosophy, and history—have taken on increasing importance for legal scholars. At the heart of these debates are not only classical conflicts between quantitative and qualitative intellectual cultures, but also disagreements about what counts as knowledge within a discipline and, consequently, the type of scholarly work that should be valued within the disciplinary community. Although to outsiders the stakes in these debates can seem small, to participants they can have profound intellectual importance. Moreover, over time, decisions within the community of legal scholars about what to study and how to study it can affect the law, bringing unanticipated but substantial political and social consequences.

Legal Scholarship

Readers bring different purposes to their tasks. One might read a novel for pleasure at the beach, or the back of a package of breakfast cereal for nutrition information, or a newspaper article to stay informed for the conversation at the workplace water cooler. The details that a reader attends to within a text are determined by the reason the reader is engaging with the text. What counts as reading comprehension for some purposes would not count for others.

Some law professors are fond of saying that the goal of law school is to teach students how to "think like a lawyer." Whatever one thinks of that, a major emphasis of legal education, especially in the first year, is to train students to read like a lawyer. The archetypal example of this is in the first-year classroom where students are asked to recite certain information from the cases in their assigned readings, such as "the facts" or "the rule" from a case. To prepare, students are taught a skill called case briefing, which is essentially a structured form of reading in which students learn to extract the information that is most likely to be responsive to the types of questions that law professors like to ask. In particular, students learn to focus on the legal rules and their application— the abstract principles discussed by a judge in justifying his or her decision. Students are taught that legal rules are important because they might be applicable to future analogous cases. At the same time, students learn to avoid focusing on specific factual details discussed in judicial opinions that are unlikely to be presented in other contexts.

This basic set of skills is at the heart of doctrinal analysis, in which lawyers attempt to extract "blackletter law"—general legal propositions that have been applied over many specific cases to different circumstances—from individual opinions.

Law students learn the basics of doctrinal analysis, and it is an important part of legal practice. In addition, for many years, this same basic methodology served as the basis for legal scholarship. Judge Richard Posner describes the situation within legal academia from the mid-nineteenth century through the mid-twentieth century:

> The task of the legal scholar was seen as being to extract a doctrine from a line of cases or from statutory text and history, restate it, perhaps criticize it or seek to extend it, all the while striving for "sensible" results in light of legal principles and common sense. Logic, analogy, judicial decisions, a handful of principles such as *stare decisis*, and common sense were the tools of analysis (Posner 2002, 1316).

But there have long been signs of dissatisfaction with the limits of academic doctrinal analysis as a scholarly discipline. Writing in the late nineteenth century, Oliver Wendell Holmes Jr. proposed that, "For the rational study of the law the blackletter man may be the man of the present, but the man of the future is the man of statistics and the master of economics" (Holmes 1897). A contemporary observer might view this prediction as almost prophetic, presaging, as it does, the growth of interdisciplinary legal scholarship and various "law and" movements, the most successful of which likely is law and economics, championed by Judge Posner, among many others.

Holmes was quite critical of the doctrinal scholarship of his day. Yet, although methodologies from a range of disciplines are now routinely applied to the law, the doctrinal analysis of the "blackletter [person]" retains a central place in legal practice and legal scholarship. Doctrinal analysis plays a central role in legal argumentation, law professors continue to write and publish analyses of judicial decisions of the kind described by Judge Posner above, and a glance through the leading law journals reveals that doctrinal analysis is typically incorporated in some form or another in most published law scholarship.

We might distinguish between Holmes's blackletter person and the "[person] of statistics" that he predicted for the future along two dimensions. The first dimension measures the degree to which quantitative or qualitative methods are used. Blackletter analysis is, traditionally, qualitative—it involves carefully reading cases and reasoning about how those cases fit together in light of underlying goals of the legal system. Statistics, on the other hand, is, by definition, a quantitative discipline—it involves the collection and analysis of numerical information for the purposes of drawing valid inferences about relations contained among the data.

The second dimension involves whether an internal or an external perspective is taken (Barzun 2015). In rough outline, the internal perspective takes the point of view of participants in the legal system and grapples with the law on its own terms by taking seriously the reasons provided by legal actors for their decisions. Scholarship in this vein can be thought of as an interpretive exercise in which the authors of and audience for legal texts are assumed to genuinely recognize the authoritative normative force of the law as good-faith actors within the legal order. From this vantage point, the purpose of legal interpretation is to provide some kind of best account of a set of legal sources, such as a line of cases or statutory text and history. What goes into this "best" account will itself depend on a theory of legal interpretation of which there are many, including originalism (in which the meaning of a text is fixed at the time of enactment), natural law (theories that depend, at least in part, on the law's moral content), and pragmatism (which places priority on the practical effects of legal decisions).

The external perspective, by contrast, examines the causes or consequences of legal phenomena. Research along these lines might examine how individual characteristics of immigration judges affect their disposition toward petitions for asylum (Ramji-Nogales, Schoenholtz, and Schrag 2007). Such a causal account might ignore the facts of the individual case or the relevant law to focus on whether, as a statistical matter, the gender of the assigned immigration judge affects the rate of asylum grants. When focusing on consequences, a researcher might examine the effects of state death penalties on crime (Chalfin, Haviland, and Raphael 2013)—regardless of whether legislatures specifically attempt to justify death penalty statutes on deterrence grounds.

~9~

It should be noted that the quantitative/qualitative divide between disciplines and differences between an external and internal perspective on the law can be overstated. Although it is probably fair to characterize some fields as more quantitatively oriented than others, the quantitative/qualitative divide also exists within, and not just between, disciplines. Many classic works in law and economics, typically thought of as on the quantitative side of the divide, involve neither data analysis nor formal mathematical modeling (Calabresi and Melamed 1972). Many historians, working in a traditionally qualitative discipline, will collect relevant numerical data about their object of study (Shammas, Salmon, and Dahlin 1987).

Similarly, although there is often a relatively clear distinction between the types of legal arguments that are welcome in a courtroom and the historical accounts of legal development or highly technical policy analyses that one might find in an academic journal, the internal/external boundary can be porous. Policy arguments of an informal variety are often thought to influence judges and regularly find their way into legal briefs—pragmatic theories of legal interpretation provide an explicit place for such policy considerations (Posner 1996). In addition, lawyers and judges sometimes reach for social or political facts surrounding lawmaking to provide insight into the intended purposes of constitutional provisions, statutes, or judicial decisions.

With these caveats in place, for expository purposes we can construct a simple two-by-two matrix that describes four potential "modes" for the "rational study of law" along the dimensions of internal–external and quantitative–qualitative (see fig. 1.1).

	Quantitative	Qualitative
External	"[person] of the future"	historians, anthropologists
Internal	???	"blackletter [person]"

Figure 1.1.

In the upper left box we have Holmes's person of the future, using quantitative tools and taking an external perspective on the law. Economic analysis of the law will frequently fall into this category. Less emphasized by Holmes, but also important, are approaches in the upper right box, which use qualitative methods to take an external perspective such as history and anthropology. Moving clockwise we find traditional doctrinal analysis, which Holmes believed would someday be obsolete, but which continues to have a vibrant place within the legal community. These are the skills taught to first-year law students in their common law classes and practiced by attorneys and judges when arguing and deciding legal questions. The final box is largely a conjectural new category that uses quantitative tools from an internal perspective.

The Place of Distant Reading

Distant reading has something to contribute to all four modes of scholarship. For external-quantitative scholarship, computational tools allow a whole new universe of data to be collected and analyzed. For some time, scholars have used quantitative tools to study legal texts. To date, these approaches have relied on the actual reading of texts in a traditional "close" manner. Often, this task is carried out by an army of student research assistants alongside mentoring and supervising senior scholars. The idea is for research assistants to read a large number of texts (often judicial opinions) and then "code" the texts according to whatever variables the mentoring researcher was interested in. For example, this is the methodology that produced the very important database put together by a group led by Harold Spaeth wherein the US Supreme Court decisions are coded according to a variety of variables, including the legal issues discussed in the decisions, whether each justice joined the majority or dissented, and whether the outcome tilted liberal or conservative (Spaeth et al. 2016). Through this coding procedure, the text in the decisions is rendered into (and summarized by) data that can be easily subjected to statistical analysis. Many hundreds of papers have been published based on the Spaeth dataset alone, and the general methodology is a mainstay of empirical legal scholarship (Hall and Wright 2008). Distant reading takes the natural further step of treating text more directly as data and effectively—although not always—removing the

~11~

human reader/coder.[2] The result is a new vantage point that enables the quantitative estimation of macro-level textual characteristics of the data (corpus) that are impossible for even the most dedicated group of human readers to perceive.

The value of distant reading for the external-qualitative scholar is less obvious. Indeed, there is a justifiable worry that quantitative analysis can veer toward reductive explanations that fail to convey important nuance. As with other quantitative methods, translating the law into data capable of tractable analysis generally requires some loss of information. Some text analysis methods are particularly "lossy," especially those that rely on "bag of words" representations of texts in which word order is ignored altogether and documents are treated as frequency lists of terms. This kind of reduction can interfere with sensible interpretation of the underlying phenomenon. There also may be some who fear that an expanded toolkit of quantitative approaches will hasten the displacement of qualitative methods. Both methodological concerns as well as issues arising from taste, inclination, and career incentives may help explain why scholars in traditionally qualitative disciplines have sometimes viewed distant reading as a threat.

Nevertheless, we (and others) are inclined to believe that distant reading offers something useful to disciplines like history, literature, and law that (to date) rely primarily on qualitative tools. First and foremost, the best use of distant reading, as with any approach or technique, is in a collaborative form: one more tool for the scholar's workbench. It is, of course, crucial that the techniques of distant reading be informed by

---

[2] Consider, for example, a standard classification text on a document corpus. Human coding might serve as a way to label a "training set" that the machine will use as input data for a classification algorithm. Conversely, human coding can also serve the purpose of validating a classification algorithm.

qualitative knowledge, both prior to research to inform what analyses should be carried out, and after the fact in the form of qualitative interpretation. In an appropriately informed collaborative setting, distant reading can provide a new lens for old questions and can also open entirely new lines of inquiry that are best addressed through qualitative techniques.

An example of the kind of profitable interaction that can occur between qualitative and quantitative scholars in a legal distant reading project is provided by Klingenstein, Hitchcock, and DeDeo (2014), also discussed in Chapter 3 of this volume. In that paper, the authors take an external perspective and investigate whether known cultural and social trends that lessened the social acceptability of interpersonal violence had an effect on how criminal cases were tried in London during the early modern period. They examined the influence of a hypothesized set of causes (cultural/social attitudes) on a legal phenomenon (criminal trial testimony). To test this hypothesis, more than twenty million words of testimony were run through a text analysis algorithm to measure its information content *vis-à-vis* the labelling of the case as violent or nonviolent. The authors found that, according to their measure, the testimony in violent cases became much more distinctive starting in the late 1700s, with the distinction growing until the end of their data in the early twentieth century. They attributed this gradual long-term shift not to any specific legislative or bureaucratic policy change, but to "a gradual process driven by evolving social attitudes" (Klingenstein, Hitchcock, and DeDeo 2014, 9423).

In addition to nicely illustrating the ways that qualitative and quantitative forms of analysis can work together in distant reading projects, the paper also helps demonstrate the unique perspective distant reading can bring to the law. The trends identified by the authors likely are not the kinds that a human reader would be able to identify. They are very subtle shifts in the frequency of certain terms and would likely pass unnoticed by even the most careful reader. They only become apparent when millions of words, spanning tens of thousands of documents spread over a century and a half, are analyzed. Computational analysis put to use on large digitized collections of legal sources allows new ways to examine legal and cultural phenomenon that previously would have been impossible. With a "distant" view, the collection of trees becomes a forest.

Distant reading also has a place within scholarship that takes an internal perspective on the law, the unlabeled leftmost lower box in figure 1.1. There is a nascent movement in support of using tools from the field of corpus linguistics to help judges interpret legal language (Lee and Mouritsen 2017). The idea is that frequently, legal disputes boil down to alternative interpretations of a word or phrase in a statute or contract. The field of corpus linguistics focuses on collecting a large body of texts for purposes of understanding the semantic content of words. Rather than relying on dictionaries—a common interpretive practice for judges and lawyers—judges could draw from curated corpora of documents that can be analyzed statistically, for example to see whether some usages for a term are common or extremely rare (Mouritsen 2010). This form of distant reading has the potential to help guide real-world decisions, and has already been deployed in some cases.[3]

Scholars who focus on doctrinal questions can use similar distant reading tools to aid interpretation, and can also play a role in pushing methodological boundaries. In an earlier time, the leading torts scholar of the mid-twentieth century,

---

[3]See, e.g., In re Adoption of Baby E.Z., 266 P.3d (Utah 2011)

William Prosser, was famous for his copious capacity to read and digest thousands of judicial opinions (Abraham and White 2013). Prosser's voracious appetite for the law gave him the unique authority to make general statements about how courts in general approached various legal questions in the area of torts—he was the only one who had read all of the cases and undertaken the effort to synthesize them into broader doctrinal themes. But even Prosser would be hard-pressed to keep up with the contemporary production of court cases, especially in an area like tort law that is primarily the domain of state courts.

Distant reading tools can expand the ability of contemporary scholars working in the "Prosser model" to cover more legal terrain. For example, the American Law Institute periodically publishes Restatements of the Law, meant to summarize and characterize the existing state of legal doctrine across various fields. Restatements are canonical examples of internal legal scholarship; their goal is to distill the meaning of legal texts specifically for use by participants in the legal system. But they could also benefit from distant reading tools, given the vast textual space that they frequently attempt to cover. For example, what if committees assigned to the Restatements of an area of law were staffed with an expert in computational legal analysis? That person could provide quantitative descriptive analysis of the law—for example, estimating the proportion of jurisdictions that adopt different variants of a legal rule—that could sit alongside more qualitative approaches.

Distant reading can also provide a means to add rigor to doctrinal analysis. Legal scholars William Baude, Adam Chilton, and Anup Malai have argued that standard forms of doctrinal analysis suffer from a lack of "systematic demonstration of supporting evidence" for their claims, which "makes it difficult for the reader to evaluate [their] validity," and may "impede future legal analysis and allow for either conscious or unconscious bias" (Baude, Chilton, and Malani 2017).

These authors propose that doctrinal analysis take up a form of systematic review, adopted from other disciplines, such as medicine, that requires researchers to explicitly state the research question and then develop and deploy a specific methodology for obtaining, weighing, and analyzing the relevant universe of cases.

As noted by Baude, Chilton, and Malani (2017), some distant reading techniques are already common in systematic review. For example, researchers can state exactly how they arrive at some set of documents to review by disclosing the databases searched and the exact search procedures used. This is a form of explicit quantitative preprocessing of textual data. Additionally, distant reading techniques allow for the use of statistical analyses that formalize the inferences that can be drawn from the textual data. General impressions about the prevalence or rarity of one or the other doctrinal development can give way to rigorous analyses able to separate out genuine trends from background noise.

Further integration of distant reading techniques into more systematic doctrinal analysis could spur an interesting interplay of quantitative and qualitative interpretive techniques. Two examples from the law literature provide a sense of where such a project may be heading. In each, law professors take advantage of topic models to summarize areas of law, but then rely on their expertise within their relevant fields to interpret the outputs of those models. Topic models are a family of mathematical and computational tools that represent documents in terms of probability distributions defined over the vocabulary in a corpus (the "topics") based on assumptions concerning patterns of word co-occurrence in documents. The foundational topic model is the latent Dirichlet allocation (LDA) introduced in the 2000s (Blei, Ng, and Jordan 2003; Blei and Lafferty 2007). Topic models have become especially important in social science and humanities analysis of texts. Blei (2012) provides an accessible overview, and the technique is discussed in more detail in subsequent chapters herein.

In a 2014 article, Jonathan Macey and Joshua Mitts examine cases that involve "piercing the corporate veil," which is the judicial doctrine covering exceptions to the normal limitations on liability for corporate shareholders (Macey and Mitts 2014). The authors develop a theoretical taxonomy of justifications for veil-piercing, based largely on notions of economic efficiency. They then statistically analyze over nine thousand cases using a topic model to detect patterns within the texts, finding that their taxonomy well predicts outcomes in veil-piercing cases, while earlier doctrinal categories do not. A project by David S. Law uses topic models to examine constitutional preambles (Law 2016). Law relies on existing theories within the comparative constitutional law discourse to associate the statistically derived topics with categories of thematic content. In both papers, the authors' deep knowledge of their respective legal areas is married with distant reading techniques to support new forms of doctrinal scholarship.

The topic models used by Macey and Mitts (2014) and Law (2016) are "unsupervised" forms of text analysis that seek to identify very general patterns in data with relatively little human intervention. More supervised approaches to text analysis can also prove useful for legal scholars. For example, in Rauterberg and Talley (2017) the authors train a machine learning model to identify "corporate opportunity waivers" to test the effects of a change in Delaware law on the contracting behavior of public corporations and analyze the types of firms that tend to contract out of fiduciary duties.

The articles mentioned above reflect a slowly growing trend toward integrating computational techniques into the study of law. The Supreme Court remains an important and oft-used target of such methods. Studies have examined time trends in the length of Supreme Court opinions (Black and Spriggs II 2008); others have estimated how justices modulate the clarity of their writing style depending on their audience (Black et al. 2016b); still others have quantified how Supreme Court attention to an issue affects litigation activity in the federal courts (Rice 2014). A National Science Foundation grant awarded in 2006 supported research that, among other things, applied machine learning techniques to classify the ideological direction of *amicus curiae* briefs submitted to the Supreme Court (Evans et al. 2007). Several scholars have developed computational approaches to examining the level of influence of briefs on judicial opinions (Oldfather, Bockhorst, and Dimmer 2012; Corley 2008). In another direction, automated content analysis and machine learning have also been used to distinguish fact-intensive from law-intensive cases (Smith 2014) and estimate the level of controversy over regulations issued by federal agencies (Stiglitz 2014).

With its focus on text and its enthusiasm for quantitative analysis, distant reading sits at the intersection of alternative forms of legal scholarship that have sometimes been at odds. For scholars in the empirical legal studies tradition, distant reading opens up vast new fields of data to explore, a kind of undiscovered country that offers exciting new opportunities. For more traditional legal scholars, distant reading offers a new approach to understanding the meaning of the law— those powerful words issues by judges and legislatures that can shape the world. Indeed, over time, distant reading may serve as a methodological bridge between competing approaches to studying the law, especially for those who are predisposed to crossing disciplinary lines.

Although it has a short past, the potential utility of distant reading for many different types of legal scholarship is great. For as long as advances in data availability and analytic techniques continue to pair with the need to better understand the law and its consequences, distant reading the law is likely to have a future. 🪶

6

BIG DATA, MACHINE LEARNING, AND THE CREDIBILITY REVOLUTION IN EMPIRICAL LEGAL STUDIES

Ryan Copus, Harvard University
Ryan Hübert, University of California, Davis
Hannah Laqueur, University of California, Davis

The so-called credibility revolution changed empirical research (see Angrist and Pischke 2010). Before the revolution, researchers frequently relied on attempts to statistically model the world to make causal inferences from observational data. They would control for confounders, make functional form assumptions about the relationships between variables, and read regression coefficients on variables of interest as causal estimates. In essence, they would rely heavily on ex post *statistical analysis* to make causal inferences. The revolution centered around the idea that the only way to truly account for possible sources of bias is to remove the influence of all confounders *ex ante* through better *research design*. Thus, since the revolution, researchers have attempted to design studies around sources of random or as-if random variation, either with experiments or what have become known as "quasi-experimental" designs. This credibility revolution has increasingly brought quantitative researchers into agreement that, in the words of Donald Rubin, "design trumps analysis" (Rubin 2008).

However, the research landscape has changed dramatically in recent years. We are now in an era of "big data." At the same time as the internet vastly expanded the number of available data sources, sophisticated computational resources became widely accessible. This has opened up a whole new frontier for social scientists and empirical legal scholars: textual data. Indeed, most of the information we have about law, politics, and society is contained in texts of one kind or another, almost all of which are now digitized and available online. For example, in the 1990s, federal courts began to adopt online case records management—known as CM/ECF—where attorneys, clerks, and judges file and access documents related to each case.[1] Using the federal government's PACER database (available at pacer.gov), researchers (both academic and professional) can now easily access the dockets and filings for each case that is filed in a federal court. LexisNexis, Westlaw, and other companies have further improved access by providing raw text versions of a wide range of legal documents, along with expert-coded metadata to help researchers more easily find what they are looking for. And yet, despite the potential of these newly available resources, the sheer volume presents challenges for researchers. A core problem is how to draw substantively important inferences from a mountain of often unstructured digitized text. To deal with this challenge, researchers are turning their attention back toward the tools of statistical analysis. As many of the essays in this volume demonstrate, there is now a surging interest among researchers in one particularly powerful tool of statistical analysis: machine learning.

This chapter addresses the place of machine learning in a post–"credibility revolution" landscape. We begin with an

---

[1] See United States Courts, "25 Years Later, PACER, Electronic Filing Continue to Change Courts" (2013), https://www.uscourts.gov/news/2013/12/09/25-years-later-pacer-electronic-filing-continue-change-courts.

overview of machine learning and then make four main points. First, design still trumps analysis. The lessons of the credibility revolution should not be forgotten in the excitement around machine learning; machine learning does nothing to address the problem of omitted variable bias. Nonetheless, machine learning can improve a researcher's data analysis. Indeed, with growing concerns about the reliability of even design-based research, perhaps we should be aiming for triangulation rather than design purism. Further, for some questions, we do not have the luxury of waiting for a strong design, and we need a best approximation of answer in the meantime. Second, even design-committed researchers should not ignore machine learning: it can be used in service of design-based studies to make causal estimates less variable, less biased, and more heterogeneous. Third, there are important policy-relevant prediction problems for which machine learning is particularly valuable (e.g., predicting recidivism in the criminal justice system). Yet even with research questions centered around prediction, a focus on design is still essential. As with causal inference, researchers cannot simply rely on statistical models but must also carefully consider threats to the validity of predictions. We briefly review some of these threats: GIGO ("garbage in, garbage out"), selective labels, and Campbell's law. Fourth, the predictive power of machine learning can be leveraged for descriptive research. Where possible, we illustrate these points using examples drawn from real-world research.

Learning with Machines

Machine learning is becoming increasingly popular among researchers. And yet, there is a great deal of ambiguity about what exactly it is. This is for good reason. Machine learning is not a specific research tool; it is a catch-all term that refers to any method that features *learning* by a *machine* about quantitative data. A unifying feature of these methods is that they leverage (ever increasing) computational power to apply (ever more complicated) techniques of statistical inference to (ever larger) datasets. As a result, one main distinction between "traditional" methods of statistical inference and machine learning methods is *scale*. In some sense, machine learning is statistical analysis on steroids.

And yet, this distinction is too simplistic. The point of machine learning techniques is to delegate the statistical learning process to complex algorithms that are designed to generate the best predictions possible using a given dataset. To illustrate, consider a simple example: A state parole board may wish to have high-quality predictions about the likelihood that a convicted felon will reoffend if released. Whether a potential parolee will reoffend is the *outcome* of interest to the board members. Outcome variables are also called "responses" or "dependent variables" and are usually denoted mathematically by the letter Y.

Outcomes are determined by specific constellations of other factors. Variables that help predict outcomes are appropriately called *predictors* (also known as "independent variables"), which are usually denoted concisely by a vector $\mathbf{X} = [X_1, X_2, ..., X_k]$, where k is the number of predictors. The parole board members could rely on their intuitions about the factors that increase the likelihood of reoffending, which might generate decent predictions to guide their decision-making. For example, a member may assume, based on their past experiences, that a potential parolee is much more likely to reoffend if that person had been convicted of a very serious offense. But these kinds of intuitions could also yield bad predictions. First of all, predictions based on these intuitions might be inaccurate on average and lead to many high-risk prisoners being released or many low-risk prisoners being denied parole. The problem here is that the decisions exhibit high *bias*. Second, the predictions the members make might be very noisy: an individual member's assessments across many cases might be inconsistent, or members may disagree with one another. The problem here is that the decisions exhibit high *variance*.

Alternatively, the members of the parole board could assemble the data on all past parolees and delegate to a machine the task of predicting a reoffense. Statistical analyses of relationships like this—where predictor variables are used to predict outcomes—are known as *supervised learning*.[2] The machine will perform the task agnostically. It will search over all the possible ways to make statistical predictions using the data and will return the best predictions it finds. Of course, while the machine may perform this task objectively, the quality of the predictions it produces depends critically on the data being used. In the section "Designing Good Predictions" below, we specifically address a set of concerns about the validity of predictive models when used in real-life applications.

Y-Hats, Not Beta-Hats

The machine doesn't care how or why it generates accurate predictions. This may be somewhat unsatisfying to a human who is unable to make sense of the machine's learning process. As a result, and especially if she uses a sophisticated machine

---

[2]Supervised learning can take many forms, but it is always "supervised" by the researcher's expectation that outcomes can be predicted by the predictors. There is an entire class of statistical problems where there is no obvious relationship between an outcome variable and a set of predictors. In these cases, a researcher is often interested in making sense of the relationship among the independent variables. This is called *unsupervised learning*. A common example is clustering.

learning method, the analyst will usually be unable to make inferences about *why* certain variables were more or less helpful in generating accurate predictions. For the members of a parole board, this may not be such a problem. Their central concern is whether or not a potential parolee will reoffend. And with enough historical data on reoffending to guide the learning process, a machine may be able to give them very accurate predictions about this outcome.[3]

But social scientists are not like parole board members. Whereas a parole board member might not need to probe into the determinants of the machine's predictions so long as they are confident in their accuracy, a social scientist would almost always approach the question of a parolee reoffending with an eye toward explaining *causes*. At the very least, the social scientist would seek to understand which of the variables in the dataset "explain" most of the reoffending behavior of past parolees. Indeed, social scientists don't generally want to predict what has happened or will happen in the world, except when prediction is useful for some auxiliary purpose on the way to explaining a larger phenomenon (e.g., estimating propensity scores for a matching analysis). This somewhat subtle point is fundamental to understanding the role of machine learning in social science, but it is often obscured by the fact that many of the tools social scientists use for explaining things from a causal perspective are also useful for predicting things.

To illustrate, consider an example drawn from a dataset of all Ninth Circuit civil cases from 1995 to 2013, used in Copus and Hübert (2017). A perennial question among scholars who

---

[3] An obvious caveat is that measures of reoffending (e.g., arrest and conviction) are themselves imperfect, which means even if predictions are very accurate, they may not be substantively useful. Below, we discuss a number of concerns that arise with machine prediction, such as embedded racial biases, in outcome measures.

study US courts is, What explains differential decision-making by federal judges? A researcher studying the Ninth Circuit might speculate that there is a linear relationship between the outcome—whether, on appeal, a lower court decision is reversed by the Ninth Circuit—and some case-specific and judge-specific predictors.[4] Accordingly, they might estimate an ordinary least squares (OLS) regression model, such as the following:[5]

$$Y_i = \beta_0 + \beta_1 X_{1i} + \cdots + \beta_k X_{ki} + \varepsilon_i$$

For concreteness, suppose the researcher is specifically interested in whether panels with majority Republican appointees are more likely to reverse lower court decisions. Accordingly, she regresses reversal on a dummy variable indicating whether the appellate panel is majority Republican appointees. Moreover, she expects this might also depend on whether the plaintiff won in the trial court, and so she includes that variable along with an interaction term. Using the data from Copus and Hübert (2017), she would get the following estimated model (with all coefficients statistically significant at 0.001 level):

$$\hat{Y}_i = 0.305 + 0.131 \cdot \text{Pltf}_i - 0.074 \cdot \text{MajRep}_i$$
$$+0.084 \cdot \text{Pltf}_i \cdot \text{MajRep}_i \tag{2.1}$$

The researcher ran this regression to try to detect whether politics affected case outcomes. Setting aside concerns about

---

[4] More specifically, we examine whether cases are "treated negatively" by the Ninth Circuit, which encompasses reversals, remands, and vacated cases. With some abuse of language, we will refer to the outcome variable from Copus and Hübert (2017) as "reversal."

[5] For expositional purposes, we present a variant of OLS with binary outcome variables, known as a linear probability model. Since models like this can generate predicted probabilities greater than 1 or less than zero, an analyst may instead opt to impose additional functional form assumptions to constrain the predicted probabilities. This can be achieved with a logistic or probit regression.

measurements and omitted variable bias, she would conclude that majority Republican panels are more deferential than majority Democratic panels to cases won by the defendant, but slightly less deferential than majority Democratic panels to cases won by the plaintiff. All else being equal, this regression shows that majority Republican panels reverse about 23% of cases when the defendant wins at trial (as opposed to 31% for majority Democratic panels), but about 45% of cases when the plaintiff wins at trial (as opposed to 44% for majority Democratic panels).

In this example, the researcher is interested in the $\hat{\beta}$s—the marginal effects of *specific* variables on the outcome of interest. However, the regression in model (2.1) doesn't just generate $\hat{\beta}$s, it also generates $\hat{Y}$s. That is, the regression can be used to generate a *prediction* about how a hypothetical case with certain characteristics would be decided. For example, suppose that there is a hypothetical case won by the plaintiff in the lower court and heard by a panel made up of mostly Republican appointees. Then, model (2.1) generates a prediction about the probability that case will be reversed. Specifically, if we do the math, we see that

$$\hat{Y}_i = 0.305 + 0.131 \cdot 1 - 0.074 \cdot 1 + 0.084 \cdot 1 \cdot 1 \approx 0.446$$

The model produced a prediction that this hypothetical case would be reversed with 44.6% probability.

OLS regression provides an analytically tractable way to recover both marginal effects of predictors on outcomes ($\hat{\beta}$s) as well as predictions about outcomes themselves ($\hat{Y}$s). That said, most of the time, OLS regression is used by social scientists to explore specific relationships between variables, *not* to generate predictions. A social scientist might therefore ask, Why do I care that the hypothetical case above would be reversed with 44.6% probability? We are sympathetic to this concern, because much of social science is about understanding *why*, in a causal sense, the world works the way it does. But this concern is also too quick to dismiss the myriad ways that prediction exercises do strengthen social science research, even research that is causally oriented. In the sections below, we will provide specific examples to demonstrate how prediction alone can aid in social scientific research.

In the meantime, we need to fix ideas. Machine learning is not (yet) widely used in the social sciences, and thus many readers will be unfamiliar with the ideas and jargon that motivate its applications. Moreover, understanding the promise of machine learning requires a relatively dramatic paradigm shift. As a methodological tool, analysts use machine learning precisely when they care about the $\hat{Y}$s, but not the $\hat{\beta}$s. This is because machine learning techniques generate very good $\hat{Y}$s, often *much* better $\hat{Y}$s than one can recover from simple regression-based methods familiar to most quantitative researchers. In the following section, we introduce the fundamental ideas motivating machine learning methods in a relatively nontechnical way, focusing attention on explaining why it is so good at prediction.[6]

What Are We Predicting?

Suppose that we are interested in generating high-quality predictions from an existing dataset. There are many potential ways to get predictions from data, and we have already seen one: OLS regression. Consider the specific prediction we generated from the Ninth Circuit data in Copus and Hübert (2017): a hypothetical case won by the plaintiff in the lower court and heard by an appellate panel of mostly Republican appointees has a 44.6% chance of being reversed by that panel. An obvious question

---

[6]There are several excellent primers on machine learning, which we highly recommend to interested readers: Hastie, Tibshirani, and Friedman (2008), James et al. (2013), and Grus (2015).

emerges: Is this a "good" prediction of how the hypothetical case will be resolved?

Answering this question is more complicated than it initially appears. First, what exactly are we predicting? Sometimes our goal is to make predictions within the dataset we're analyzing. This is what traditional OLS regression does: it generates *in-sample predictions*. An analyst may wish to do this for the express purpose of summarizing existing data more clearly. For example, consider figure 2.1, which presents a scatterplot of some simulated (i.e., fake) data. A researcher looking at the plotted data may struggle to see the relationship between the variables. To get a sense for the data, she runs a regression of

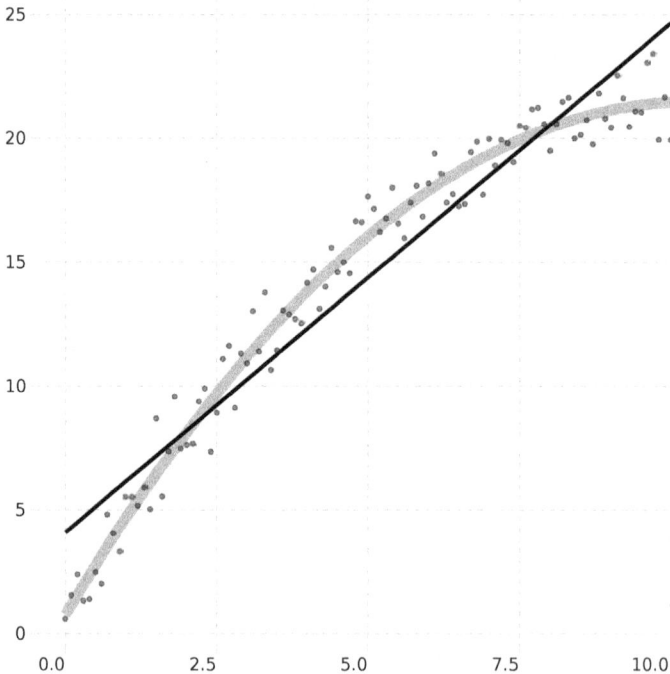

Figure 2.1. In-sample prediction.

Y on X and gets the following: $\hat{Y} = 4.42 + 1.99X$. These predictions—the $\hat{Y}$s—are represented by the black line. After plotting, the analyst realizes that the predictions generated by this regression model do not fit the data as well as they might. That is, she *misspecified* the regression model that she estimated. The gray curve represents the predictions generated by an OLS regression with a quadratic term (the correctly specified model). Such in-sample predictions are thus useful for the researcher to figure out the most appropriate model of the data when she doesn't know *ex ante* what it is.

Alternatively, our goal might be to predict the outcome of a case from some other dataset to which we may or may not have access. An obvious example would be predictions regarding future cases, about which we obviously do not have data. This is referred to as *out-of-sample prediction*. With out-of-sample prediction, the analyst is not interested in making predictions within their existing data except as a diagnostic for how well the model of the data will perform in other, similar datasets.[7] For example, suppose a researcher has access to Ninth Circuit data from 1995 to 2004, but not after 2004. They could build a model for predicting outcomes in the 1995–2004 time period and then use it to make a prediction about what happens from 2005 to 2010. One might expect the early data would provide decent (but not perfect) predictions for the latter period.

This is a somewhat subtle point, but we must emphasize that researchers using machine learning in contemporary research are almost always doing some form of out-of-sample prediction. We will discuss this in more depth below, but suffice to say, out-of-sample prediction is not only used to predict

---

[7]We use the terms "model" and "predictive model" loosely. As is common in machine learning applications, a "model" is an estimated relationship between variables that can be used to generate predictions in- or out-of-sample.

outcomes in data that are not available (such as forecasting the future). It turns out that the process of conducting out-of-sample predictions on subsamples of data not used in the initial machine learning analysis has important methodological benefits that improve the quality of predictions. These subsamples, which are randomly chosen and held out from the larger sample during the initial analysis, are referred to alternatively as *held-out samples, validation sets*, or *test sets*. As these names imply, these subsamples are untouched until they are used to evaluate the quality of the machine learning predictions that are generated using the portion of the sample that is not held out.

What Are "Good" Predictions?

We now ask what it means for predictions to be "good." The challenge is that an analyst can choose among several (and a potentially infinite number of) methods for generating predictions. Even restricting attention to regression, an analyst can choose between various types of regression—such as linear OLS or logistic—and also can choose which variables to include in the estimation.

One trivial solution for maximizing the accuracy of one's predictions is to estimate a *saturated model*, which returns a specific prediction for Y for every combination of the predictors. And more precisely, for a specific combination of the predictors, the prediction a saturated model returns is simply the actual value of Y in the dataset for that combination of predictors. In this case, the predictions would be the least biased estimates of the actual outcomes, given the available data.[8] This is because they yield highly tailored and specific predictions. However, despite the fact that those models perform very well in the sample from which they are estimated, they can be expected to perform very badly on other samples drawn from the same population. This is because, in any given dataset, there will be a very small number of observations, or no observations at all, for each combination of the predictors. This problem is often referred to as *overfitting*, and will lead to (out-of-sample) predictions with high variance. At the other extreme, an analyst could estimate a model that predicts Y by its mean. The predictions from this model will have very low variance but will be very biased. Analysts working with limited data always face a choice between reducing the bias and reducing the variance in their predictions. This is referred to as the *bias-variance trade-off*.

~ 33 ~

To find the optimal balance of bias and variance, one must evaluate a model's predictions outside of the sample used to generate the model. As a result, most supervised machine learning applications follow a specific recipe. First, the analyst randomly partitions the dataset into two subsets, a *training set* and a *test set*. The analyst uses the training set to estimate several predictive models of the outcomes in the training set. The analyst then takes each predictive model and assesses its performance in the test set in order to see whether it accurately predicts actual outcomes on data that were not used to build the model. One approach to doing this assessment is to calculate the *mean-squared error* (MSE), which, roughly speaking, measures the difference between the model's predicted outcomes and the actual outcomes. If the MSE is high, then the predictive

---

[8]Note that we use the term "bias" as in James et al. (2013): "Bias refers to the error that is introduced by approximating a real-life problem, which may be extremely complicated, by a much simpler model" (35).

model generates "bad" predictions because it cannot accurately predict data that were not used to build the model. The analyst chooses the model that performs best.

Validation, Validation, Validation

There is nothing inherently novel or difficult about splitting a dataset into training and test sets and performing model assessment via MSE. The advantage brought by machine learning techniques comes from the fact that computational power now allows us to quickly estimate complicated models and assess their quality repeatedly. This is advantageous in two ways. First, it allows analysts to specify a wide variety of candidate models, or even a combination of models (called an *ensemble learner*), and then let the machine choose which model (or combination of models) optimizes on the bias-variance trade-off. Second, for any given model, it allows the analyst to perform an especially robust form of evaluation known as *cross-validation*, which relies on the idea that an analyst should repeatedly (and independently) divide the data into training and test sets, estimate models, and assess model performance.

Cross-validation allows an analyst to use as much data as possible to both build and evaluate predictive models, thus reducing bias in estimates. It also insulates an analyst against concerns that the initial split of the data unintentionally generated training and test sets that are unrepresentative.[9] The process proceeds as follows. First, the analyst randomly partitions the data into K subsets, each called a "fold" of the original dataset. Then, for each subset k, the analyst estimates a model on all of the data *excluding* k. Then the analyst treats k

---

[9]For example, recall that at the 95% confidence level, randomization will fail to produce groups that are identical in expectation 5% of the time.

as the test set and calculates the MSE of the model generated without k. Once this is done for each of the K subsets, an analyst can calculate the average MSE across the K folds to either assess overall model performance or to assess a particular model against an alternative one. The number of folds, i.e., K, is determined by the analyst. If K is low, then the benefits of cross-validation are limited, and an analyst may tend to overstate the average MSE across the K-folds (i.e., it will be biased upward). However, if K is high, the procedure can become computationally difficult, which is a serious constraint in many applications. As a result, it is conventional for researchers to perform either five- or tenfold cross-validation, since "these values have been shown empirically to yield test error rate estimates that suffer neither from excessively high bias nor from very high variance" (James et al. 2013, 184).

As we have described in this section, machine learning offers a principled way to perform statistical analysis with the goal of generating accurate predictions. In later sections, we explore some of the ways this is useful for researchers. Before proceeding, however, we address its relationship to the credibility revolution.

Design Still Trumps a Machine Learning Analysis

The credibility revolution turned researchers' focus toward obtaining credible *causal* estimates. A causally credible estimate of the effect of some variable T (a *treatment*) on an outcome Y requires that it is only T, and no other confounding variable, that fully explains differences in outcomes across the subjects being studied. The most obvious way to establish this would be to examine an outcome for a particular subject with and without the treatment and compare those outcomes. Of course, the *fundamental problem of causal inference* rules out the possibility of observing an outcome when the subject was both exposed *and* not exposed (Holland 1986). In lieu of this, researchers compare average outcomes across "treated" and "untreated" (or "control") groups. In order for the effect on the outcome to be completely explained by the treatment, as required by our conception of causality, the two groups must be comparable and thus identical (on average) in all relevant characteristics.

With groups of research subjects formed "in the wild" it is often unknowable whether two groups are comparable, since that conclusion may depend on information unavailable to the researcher. But if the researcher controls the construction of the groups, she can randomly create the groups to ensure comparability on average—a *randomized experiment*. Or, if she knows something about the world that guarantees two groups were "as-if" randomly generated in the wild, then she can assume they are comparable—a *quasi experiment* or a *natural experiment*. Either way, a researcher's ability to make credible causal inferences depends on her outside knowledge of the comparability of her comparison groups. That is, the credibility of her measured effects is established by the *design* of her causal research study.

Estimates without such a design were largely discredited by social science researchers focused squarely on causal inference (see Angrist and Pischke 2010). The traditional approach of running an OLS regression, controlling for "other factors," and reading the coefficient on the variable of interest frequently failed to generate accurate causal estimates. When researchers attempt to obtain credible estimates by statistical modeling rather than design—an *analysis-based approach*—the researcher must, at minimum, rely on the assumption that, conditional on observed variables, treatment is independent of the outcome (i.e., "selection on observables") to identify causal effects. With the credibility revolution, researchers largely lost faith in the plausibility of that assumption. For example, Robert LaLonde famously compared the results of an analysis-based approach with the results of a randomized experiment to assess whether an employment program had a causal effect on wages (LaLonde 1986). He showed that the experimental results could not be replicated by a standard analysis-based approach.

But with machine learning and the large-n, variable-rich datasets that text processing makes possible, it is tempting to believe that the selection on observables assumption can be revived. Hopes should be tempered. Even with an immense number of variables, it is exceedingly difficult to know if we have measured every possible confounder, and all it takes is one missing variable to violate the independence assumption and thus the credibility of an estimate. Moreover, even if all relevant variables have been measured, machine learning algorithms may fail to recover the true form of the relationships between covariates and the dependent variable, and machine learning will not correct for a researcher's poorly theorized selection of variables (e.g., if the researcher includes post-treatment or collider variables in the dataset she feeds to an algorithm, there is no reason to expect machine learning to correct for those mistakes).

. . . But Machine Learning Improves Analysis

At the same time, there is still arguably a place for observational studies to supplement experimental and quasi-experimental research, particularly given recent concerns regarding the reliability of design-based studies (for example, see Open Science Collaboration 2015) and well-known limitations to the external validity of experimental studies. We do not wish to wade too deeply into the debate among researchers about the appropriate standards to apply for evaluating the credibility of empirical research—especially research that does not rest on strong designs (in the sense we have defined it). But, we do think one thing is clear: insofar as researchers are proceeding without strong designs, they should be using machine learning in their analysis.

Machine learning techniques can improve traditional econometric methods for estimating average treatment effects under the conditional independence assumption by generating better estimates for any prediction component of a larger estimation problem. An obvious example is estimating the propensity score: the "conditional probability of assignment to a particular treatment given a vector of observed covariates" (Rosenbaum and Rubin 1983). At its heart, the probability of treatment is a prediction question, and as such is well suited to a machine learning approach in situations where machine learning classification algorithms often outperform standard parametric modeling techniques (Westreich, Lessler, and Funk 2010). In much of the published literature, the propensity score is estimated using maximum likelihood logistic regression models. However, the correct model of treatment is generally unknown, and a misspecified propensity score model can increase bias, even if the conditional independence assumption is valid (Drake 1993). A popular package for conducting matching in the R statistical program, GenMatch is an example of a machine learning–based matching method that requires no manual programming and performs better than propensity scores generated from logistic regressions (Diamond and Sekhon 2013).

More generally, model misspecification is a core problem for researchers working with observational data. Doubly robust estimation techniques attempt to insulate an analyst from concerns about model misspecification by combining a propensity score model with a traditional model of the outcome, such as an OLS regression of Y on T and $\mathbf{X}$. The key advantage of doubly robust techniques is that they will generate consistent estimates of the average treatment effect *even if one of the models (but not both) is misspecified* (Bang and Robins 2005). Technical details aside, a doubly robust estimate is an average treatment effect derived from predictions generated by a propensity score model and predictions generated by an outcome model. It is unbiased if one is willing to believe there are no unmeasured confounding variables and at least one of the models is correctly specified. Since machine learning itself reduces the possibility of model misspecification, doubly robust estimates generated via machine learning will be *even more* insulated against misspecification. One example of a doubly robust approach that incorporates data-adaptive machine learning is Targeted Maximum Likelihood Estimation (TMLE, see Van der Laan and Rose 2011).[10]

Machine Learning Improves
Design-Based Causal Inference

Even design-committed researchers have something to gain by incorporating machine learning into their causal inference research. Research on the connection between machine learning and causal inference is rapidly expanding. For example, machine learning can be used to adjust for covariates to improve the precision of estimates of average treatment effects in randomized controlled trials, can increase power and reduce bias in instrumental variables research resulting from violations of the monotonicity assumption, and can help recover heterogeneous treatment effects. For extensive

---

[10]Augmented inverse-probability weighting (AIPW) is another popular doubly robust estimator introduced in the missing data literature by Robins, Rotnitzky, and Zhao (1994).

discussions, we refer readers to Athey and Imbens (2017) and
Mullainathan and Spiess (2017). Here, we focus on an issue
that is of particular interest to legal scholars: estimating the
heterogeneous effects of decision-makers on case outcomes.

A core concern for the legal system, as well as policymakers in other parts of government, is whether front-line
officials resolve cases *consistently*. The empirical literature
on inconsistency in adjudication is rapidly accumulating.
We now have studies of interjudge disparities in asylum
cases (Ramji-Nogales, Schoenholtz, and Schrag 2007), social
security disability awards (Nakosteen and Zimmer 2014),
criminal sentencing in the federal courts (Abrams, Bertrand,
and Mullainathan 2012), and patent examiner grants in the
US Patent and Trademark Office.[11] Some of the findings
have been unsettling. Evidence of inconsistency in criminal
sentencing, for example, facilitated the development of the
Federal Sentencing Guidelines (Stith and Cabranes 1998). And
large disparities in asylum grant rates have given rise to calls for
institutional reform (Ramji-Nogales, Schoenholtz, and Schrag
2007), as have the findings of inconsistency in the courts of
appeals (Tiller and Cross 1999).

Most empirical studies of inconsistency in decision-
making compare the decision rates of judges who have been
randomly or as-if randomly assigned cases. Although the
random assignment of cases allows for causal inference (e.g.,
the effect of Judge A on the reversal rate as compared to
Judge B),[12] the traditional approach to studying inconsistency

---

[11]See O'Neill, J., "Visualizing Outcome Inconsistency at the
USPTO" (2018), http://www.ipwatchdog.com/2018/10/31/
visualizing-outcome-inconsistency-uspto/.

[12]Recent studies have cast doubt on the randomization assumption invoked
by many legal and judicial politics scholars (for example, Chilton and Levy
2015). We note here that even if *judges* are not randomly or even as-if

can dramatically understate inconsistency. By comparing decision rates on a single, intuitively specified dimension— such as whether a case was decided in a liberal direction or whether a case was reversed—scholars are potentially missing much interjudge disagreement (Fischman 2014). A judge's decision-making may vary across different kinds of cases, and the nature of that variation might also depend on the judge she is being compared to. In essence, there are heterogeneous treatment effects. For example, two judges might have identical reversal rates overall but have very different reversal rates in subsets of cases: one judge may reverse more often when the plaintiff prevails, while the other may reverse more often when the defendant prevails. We could, as some researchers have done, subjectively code certain types of reversals differently from other types (e.g., in employment discrimination cases code a reversal as liberal/conservative if a defendant/plaintiff won at the trial level). And while we might fully capture the disagreement between Judge A and Judge B by changing the analyzed outcome to "liberal" rather than "reversed," that coding scheme might poorly describe the form of disagreement between Judge A and Judge C. Moreover, such a time-consuming effort may miss important distinctions or stress distinctions that are not actually relevant.

Machine learning can be used to aggressively search for interjudge disagreement, minimizing the amount of undetected disagreement. Copus and Hübert (2017) provide a full and technical explanation, but we recount the core intuitions here. At a fundamental level, estimating inconsistency involves estimating

---

randomly empanelled in federal circuit courts, this does not mean that *cases* are not randomly assigned to panels. In Copus and Hübert (2017) we rely on the assumption that only groups of cases are randomly assigned. Even so, out of an abundance of caution, we also perform adjustments to guard against possible threats to randomization.

the causal effect of assigning different judges on case outcomes. The contribution of our paper is to cast the decision about *which* case outcome to analyze as one of prediction. Indeed, when a researcher intends to measure inconsistency and chooses an intuitively coded outcome (such as whether judges affirm or reverse cases) she is implicitly *predicting* that outcome will best capture disagreement between judges. Our starting point is to ask, What if our intuition about the best outcome for the task is a bad one? Instead, we turn this predictive task over to an optimized machine learning algorithm, which uses all available data on cases to generate prediction about how likely each judge is to vote to reverse each case. In this step, our goal is *not* to determine what factors "caused" judges to make their decisions but rather to simply predict whether a given judge would affirm or reverse a given case. Our trick is to use these predictions to generate bespoke outcome measures for each pair of judges. That is: when we estimate the causal effect of assigning Judge A to a case as opposed to Judge B, our technique allows us to compare the rates that each judge makes "Judge A-ish" decisions (as opposed to intuitively coded outcomes, such as rates of "liberal" decisions).

To illustrate empirically why one cannot simply rely on intuitions about which outcomes best capture the causal effect of assigning different judges, consider figure 2.2. In each plot, we compare two pairs of judges: Judges Leavy and Reinhardt (top panel), and Judges Kleinfeld and Pregerson (bottom panel). On each axis, we plot a specific judge's predicted probability of reversing a case.[13] Each dot represents a single

---

[13] To generate these predicted probabilities, we use Super Learner, a machine learning ensemble method (Van der Laan, Polley, and Hubbard 2007). The Super Learner method generates estimates for a set of user-specified machine learning algorithms and then optimally weights these algorithms based on their performance in order to generate its final predictions. In each

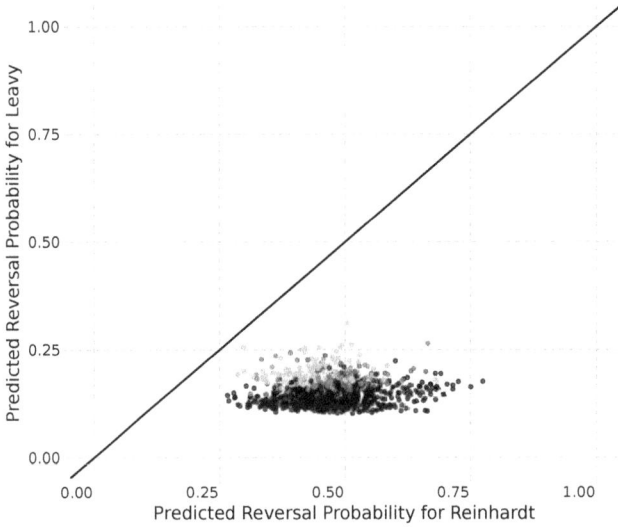

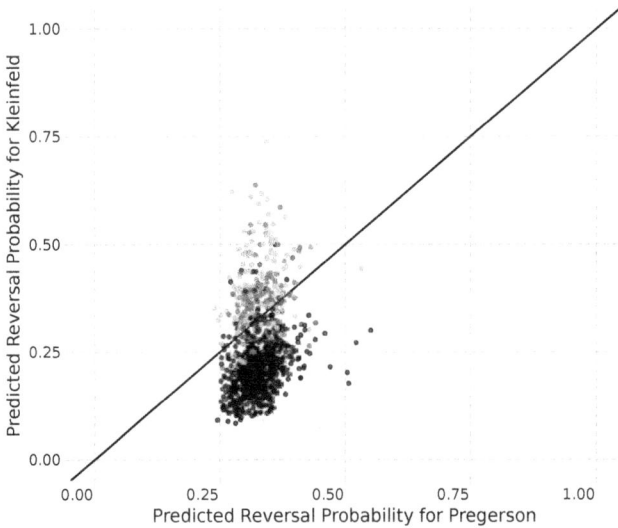

Figure 2.2. Predicting the votes of Ninth Circuit judges.

case, and the shading of the dot corresponds to whether the defendant or the plaintiff won in the district court. Darker dots indicate that the defendant won, whereas lighter dots indicate that the plaintiff won.[14] First, note that if two judges voted to reverse cases in similar ways, then all of the dots would be arranged on the 45-degree line. Judges Kleinfeld and Pregerson have roughly the same propensity to reverse (although Judge Kleinfeld's is noisier), while Judge Reinhardt is much more prone to reverse than Judge Leavy. However, while Judge Pregerson is equally prone to reverse cases won by the plaintiff or defendant, Judge Kleinfeld is much more likely to reverse cases won by the plaintiff. A similar, although much less pronounced pattern exists for Judge Leavy: he is somewhat less likely to reverse cases won by the defendant than those won by the plaintiff. To be clear, we had no *ex ante* theoretical expectation about the model that most accurately describes the way that case-level factors affected outcomes. Instead, we used machine learning to find the model that best predicted each judge's *actual* votes on cases, abstracting away from the precise reasons for those votes.

Using these models, we are able to say, for each pair of judges, whether the outcome of each specific case in our dataset is more consistent with Judge A's decision-making or more consistent with Judge B's decision-making. Consider a hypothetical case from our dataset won by the defendant at trial and reversed on appeal. The predictive models depicted in figure 2.2 would indicate that outcome is more "Judge Pregerson-ish" than "Judge Kleinfeld-ish" and more

---

model we include a set of appeal-specific and trial court–specific variables derived from the Ninth Circuit's docket sheets.

[14] For reasons we do not need to recount here, the outcome in the trial court is expressed as the probability the defendant won rather than a binary variable. That is why the dots are several shades of gray.

"Judge Reinhardt-ish" than "Judge Leavy-ish." Importantly, the former distinction is due to the fact that Kleinfeld has a pro-defendant tilt in his decision-making, whereas the latter distinction is due to the fact that Reinhardt is simply more prone to reverse overall. By definition, the difference between the rates at which Judge A and Judge B make Judge A-ish decisions will best capture the disagreement rate between those two judges. Then, if Judge Reinhardt makes "Judge Reinhardt-ish" decisions in 80% of cases, while Judge Leavy makes "Judge Reinhardt-ish" decisions in 50% of cases, then they disagree 30% of the time.

This exercise in statistical modeling may strike some as overly complicated. Suppose, instead, that a researcher simply compared the reversal rates between these judges. She would estimate a large degree of disagreement between Judges Leavy and Reinhardt but less so between Judges Kleinfeld and Pregerson. If, instead, she compared the proportion of pro-plaintiff decisions between these judges,[15] she would estimate a large degree of disagreement between Judges Kleinfeld and Pregerson but less so between Judges Leavy and Reinhardt. Our procedure shows how researchers can use the predictive powers of machine learning to guide their choice of outcomes instead of relying on intuitive, and possibly incorrect, guesses about which outcome coding scheme best captures disagreement among a heterogeneous set of judges. More specifically, our procedure generates predictions about how each decision-maker would decide each case, obviating the need to figure out which variables "explain" or "capture" disagreement. We emphasize that our technique embraces the spirit of machine learning by explicitly minimizing the importance of knowing

---

[15] For example, affirming a pro-plaintiff lower court decision or reversing a pro-defendant lower court decision.

which features explain disagreement between pairs of judges. Indeed, because estimates of the explanatory power of specific features are not causally identified, our procedure treats this information as fundamentally unknowable and shifts the analyst's focus to the accuracy of predictions.

The goal of Copus and Hübert (2017) is to *accurately measure* disagreement among judges, not to assess the reasons for that disagreement. Measuring disagreement among public officials—which is more feasible with machine learning—is important for those who are concerned about the quality and consistency of governance. Moreover, while advances in machine learning enabled us to make higher quality predictions, we were able to make these predictions only because we were *also* able to construct a high-quality dataset on Ninth Circuit decision-making. To do so, we used computational methods to turn large amounts of unstructured text in the court's docket sheets into a quantitative dataset. This underscores the promise of computational methods for helping resolve tough policy or management problems. But the lessons learned from predictions are only as good as the predictions themselves. In the following section, we describe how prediction alone can be useful for scholars, and we emphasize that prediction-based researchers, just like causal inference researchers, should aim to generate highly credible estimates.

Designing Good Predictions

Machine learning methods were developed and optimized for prediction, and, unlike causal inference questions, few assumptions are required for off-the-shelf machine learning prediction techniques to work. As machine learning techniques are moving from the domain of computer science to empirical legal studies and the social sciences, a nascent body of work has pointed to causal-adjacent prediction problems that have important policy applications (Kleinberg et al. 2015). In what follows, we describe some examples of promising applications of prediction problems relevant to empirical legal scholars. At the same time, we also emphasize that even if the problem is purely one of prediction, the researcher must still consider data and design issues that will impact the validity of the predictive model and the questions to which it is applied. As such, there are parallel lessons from the credibility revolution that should be heeded.

Prediction for Policy

One example relevant to the legal domain is the use of machine learning prediction to improve criminal justice decision-making (see e.g., Berk, Sorenson, and Barnes 2016). Risk prediction is centrally embedded in every aspect of the criminal justice system: police target areas where crime is most likely; judges assess defendants' risk of flight and risk to public safety when determining bail; prison administrators segregate inmates according to risk scores; and parole release determinations hinge on forecasts of inmates' future dangerous behavior. As such, better prediction via machine learning offers the potential to generate more efficient, effective, and equitable decisions and interventions. Kleinberg et al. (2018), for example, built a machine learning algorithm to predict criminal risk among defendants awaiting trial. They argue that the use of such an algorithm in bail decisions could reduce crime by up to 25% without any change in jailing rates, reduce the population jailed by 42% without any increases in crime, and all while reducing the percentage of African Americans and Hispanics jailed. Goel, Rao, and Shroff (2016) use machine learning methods to examine stop-and-frisk practices in New York City, arguing that if the police conducted only the 6% of stops that are statistically most likely to result in weapons seizure, they could recover the majority of weapons and mitigate racial disparities.

In addition to predicting external outcomes to help guide decisions, machine learning can also be used to predict decisions themselves—and those predictions can in turn be used to guide future decision-making. In many legal contexts, an outcome variable like reoffending is not available: there often is no better indicator for the right decision than the decision that a judge actually made. Federal and state judges, along with an army of front-line bureaucrats such as administrative law judges, food safety inspectors, and tax auditors, regularly interpret and apply centrally promulgated rules, but the application of those rules is often riddled with inconsistency. Some judges may make different types of decisions than other judges, and some judges may even be internally inconsistent, making different decisions based on their mood or cognitive biases like the gambler's fallacy (Chen, Moskowitz, and Shue 2016). Researchers can use prediction to distill decision signals and dispense with the noise, and such predictions can in turn be used to improve future decision-making.

Laqueur and Copus (2016) explain how predictive models of decisions can pool the judgment of many decision-makers and how that pooled judgment can be used to regulate and guide the decision-making of individual decision-makers, improving the consistency and overall quality of decisions. The key insight is that excluding factors that are statistically unrelated to the merits of cases (e.g., judicial identity and judicial mood) from a predictive model allows that model to smooth over and cancel out the influence of those arbitrary factors. The purified model of historical decision-making can then promote more consistent and better decisions in the future. The authors use text parsing methods to extract a robust set of variables from the transcripts of all parole hearings conducted by the California Board of Parole Hearings. They then show that a predictive model of California parole decisions could be implemented to target the most abnormal judicial decisions for secondary review.

Prediction is also useful outside of the adjudication context. Kleinberg et al. (2015) points to a number of other policy-relevant prediction problems. For example, in health policy, there are resource allocation questions such as which elderly patients should receive hip replacement. Often a doctor may want to know whether a specific patient will respond positively to a new treatment, and it is less of a priority to know *why* that patient responds positively. In government regulatory policy, building or hygiene inspection problems are questions about *where* to inspect as opposed to *why*. In the context of criminal law, a parole board may wish to know whether a convict is more or less likely to reoffend before granting parole but is less interested in what causes an inmate to reoffend.

Data and Design Considerations

Policy-oriented researchers can, should, and are focusing more on prediction problems. As compared to causal inference, prediction is easy. A turn toward prediction does not mean that researchers must abandon the policy issues they currently focus on, but they can alter their approach to leverage the clean power of prediction and avoid the messy complications of causal inference. With limited funding or time, a researcher primarily interested in a causal question might find it advantageous to convert their question to one of prediction. Consider, for example, a researcher interested in the relationship between probation services and violent reoffending. Without a source of randomization, any estimates of effects are likely to be unreliable. A researcher instead might put their efforts toward prediction: Which probationers are most likely to commit a violent offense? Those predictions could then be used to direct more resources toward the high-risk individuals.

At the same time, even when the task is mere prediction, machine learning cannot be applied blindly. There are data and design considerations that pose potential threats to model validity. We now turn to discuss these potential threats, using forecasts of criminal risk as an illustrative example.

GIGO: The Data

Computer scientists popularized the term GIGO—"garbage in, garbage out"—and this is a paramount concern in any prediction policy question. Risk assessment instruments in the criminal justice system aim to help judges make decisions in bail, parole, and even sentencing by predicting an offender's risk of return to crime. However, measuring whether a crime has occurred is not a straightforward matter. It requires relying on officially recorded criminal justice events, such as a crime report, an arrest, a conviction, or a return to prison, none of which may be consistent proxies for criminal behavior. Take, for example, Pennsylvania's recent efforts to develop a sentence risk assessment instrument. The initial instrument design included *any* rearrest or any reincarceration, including for a technical violation, as the measure of recidivism that the model aimed to predict. This broad definition of recidivism results in an overly broad model that is not actually forecasting the outcome judges are most concerned about—serious crime and violence. Moreover, using any arrest or any technical violation as a measure of recidivism can compound racial disparities by producing artificially high scores for individuals in heavily policed and supervised minority communities. Indeed, a recent review by the Pennsylvania Commission on Sentencing has pointed to racial bias in the state's risk assessment instrument in its current form.[16]

The Selective Labels Problem

The second, and perhaps the most crucial, concern is the problem of accurately evaluating algorithmic predictions in the presence of unobservables. Take, for example, a predictive model built to aid parole decisions. The model can only be built using data on individuals who are at risk of reoffending— those individuals whom a parole board has decided to release from prison. But the model aims to be applied to the full population of parole-eligible inmates. There is a potential mismatch between the dataset used to build a predictive model and the set of individuals to whom the model is applied. Information about paroled individuals may well not provide accurate forecasts for individuals that a judge would not have paroled. Judges do not, presumably, release observably similar individuals randomly, so there is reason to worry about the application of forecasts of paroled inmates to the entire population of parole-eligible inmates. The problem of unobservables invalidating predictive accuracy parallels the problem that plagues valid causal inference with observational data.

---

[16]See Pennsylvania Commission on Sentencing, "Risk Assessment Update: Arrest Scales," http://www.hominid. psu.edu/specialty_programs/pacs/publications-and-research/ research-and-evaluation-reports/risk-assessment/ risk-assessment-update-february-2018-arrest-as-predictive-factor.

The issue may be surmountable, but it requires careful attention and research design. For example, Lakkaraju et al. (2017) describe this "selective labels" concern and propose exploiting the heterogeneity of decision-making as a means to accurately evaluate the predictive performance of models in the presence of unobservables that influence the human decision and thus the observed outcome.

Campbell's Law

The psychologist Donald Campbell explained in 1979, "The more any quantitative social indicator is used for social decision-making, the more subject it will be to corruption pressures and the more apt it will be to distort and corrupt the social processes it is intended to monitor"(Campbell 1979, 85). The adage— sometimes referred to as "Campbell's law"—applies to prediction for policy. A publicly available algorithm may alter the behavior of individuals to whom it is applied, rendering the algorithm less accurate. For example, consider a risk assessment instrument used in bail decisions. The presence of a defendant's family at the bail decision hearing has been found to be predictive of the defendant successfully returning to their next court date and not being rearrested in the meantime (Kleinberg et al. 2018). The presence of the family at the hearing is likely associated with unobservable characteristics of the defendant that are not included in the predictive model. If inmates know they will have a lower risk score if their families attend the hearing, more defendants may ensure their families join them. Given that family presence is unlikely to be what *causes* the individual to reappear in court but, rather, is merely predictive and associated with unobserved factors that make them lower risk, this could artificially lower the risk scores for inmates who otherwise would be classified as higher risk.

Prediction for Description

The credibility revolution has raised the bar for causal inference research. One promising, but underused, response is for researchers to engage in more descriptive research—research that eschews causal inference altogether. Some of the most useful research is descriptive in nature (for a recent discussion of this, see Grimmer 2015). For example, the Martin–Quinn scores for Supreme Court justices (Martin and Quinn 2002) have been widely influential across several disciplines, as well as news reporting about the court. Machine learning is a valuable tool for building new measurements that can help researchers describe the world. In this section, we outline two particularly promising ways to use supervised machine learning techniques for measurement: classification and proxy variables.

~ 53 ~

CLASSIFICATION

Supervised classification involves sorting observations into known categories.[17] At its heart, classification is a kind of dimension reduction exercise that allows a researcher to generate aggregated variables based on fine-grained distinctions in the underlying dataset, which may be substantively useful. Classification can be used to solve more "technical" problems, such as identifying well-defined features from a large amount of unstructured text, or it can be used for more "interpretive" problems, such as applying an expert's judgment about conceptually complex issues to a large dataset. In either case, a researcher starts with a subsample of hand-coded observations and trains a machine learning model to generate classifications for the remainder of the dataset.[18]

---

[17] A more general version of this problem is *scoring*, where an analyst assigns observations to a continuous scale. The same ideas apply in those contexts.

[18] To be clear, researchers need not code their own training sets. Samples of previously hand-coded observations exist in many places, such as

Plaintiffs	CALIFORNIA MEDICAL ASSOCIATION
	CALIFORNIA DENTAL ASSOCIATION
	CALIFORNIA PHARMACISTS ASSOCIATION
	NATIONAL ASSOCIATION OF CHAIN DRUG STORES
	CALIFORNIA ASSOCIATION OF MEDICAL PRODUCT SUPPLIERS
	AIDS HEALTHCARE FOUNDATION
	AMERICAN MEDICAL RESPONSE WEST
	JENNIFER ARNOLD
Defendants	TOBY DOUGLAS, *Director*
	Department of Health Care Services of the State of California
	KATHLEEN SEBELIUS, *Secretary*
	United States Department of Health and Human Services

Figure 2.3. Parties listed in Ninth Circuit docket sheet (Case 12-55315).

In Copus and Hübert (2017), we draw data from around fifty-four thousand Ninth Circuit docket sheets. Despite the fact that most docket sheets follow a particular template, some of the data in docket sheets is unstructured text that is difficult to parse. For example, consider figure 2.3, which presents a list of the parties to a specific Ninth Circuit case from 2012. Notice that there are many parties to this case, each of which could belong to a specific category of interest, such as "business," "government," "advocacy organization," or "private person." However, the format of these entries is inconsistent and thus difficult to categorize using a deterministic rule. For example, while one of the plaintiffs, the California Medical Association, is an advocacy organization, another one of the plaintiffs, American Medical Response West, is a business. Moreover, notice that Jennifer Arnold is a private person, whereas Toby Douglas is a government official. Without significant resources,

LexisNexis's keywords. Moreover, researchers should be attuned to inconsistencies in the hand-coding of a test set and account for them as best as possible.

it is not feasible for researchers to categorize each party to each case by hand. Instead, a researcher could draw a random sample of the total number of parties—say, one thousand—and have a human coder classify each party into a specific category. Then the researcher could use this set of human-coded parties as a test set to derive a classification model that can be applied to the rest of the data. There are currently models that are already trained to identify named entities and are ready for out-of-the-box use. One of the most famous examples is the Stanford Named Entity Recognizer,[19] which is a Java-based tool that has a variety of interfaces to other programming languages, such as Python, Perl, and Ruby.

PROXIES

Another measurement-related use for machine learning is the creation of data-driven proxy variables. More specifically, a researcher may use high-quality predictions from machine learning to serve directly as a proxy variable for some other quantity of interest. In Copus and Hübert (2017), we use case-level data to generate a predicted probability of reversal for each case and each possible panel of judges that could have been assigned to that case. We then use these predicted probabilities to generate a case-level disagreement score by measuring the spread in the predicted probability of reversal across the panel types. By using the predicted probabilities generated from our machine learning method, we were able to create a new, and substantively useful, measure of how much disagreement particular cases elicit among judges. This measure could serve as a proxy for whether a case is an "easy case" or a "hard case" since it provides information about whether different judges would come to different decisions. This measure has many other

---

[19] Available at https://nlp.stanford.edu/software/CRF-NER.shtml.

potential applications. For example, a study of Supreme Court *certiorari* decisions might include this variable as a proxy for the salience and/or complexity of a case.

We wish to emphasize, however, that the primary benefit of using machine learning to generate new proxy variables is *not* its ability to help researchers determine which predictors are the most relevant. Machine learning methods seek to optimize predictions, not pin down which variables do most of the work. Making inferences about the most predictive variables using machine learning techniques will inevitably cause problems. Researchers are *almost always* constrained by their available data, and the fact that a variable is especially predictive in one particular machine learning application does not mean that, at a theoretical level, it is a suitable proxy. Rather than relying on intuition about what variables should be most predictive (even if guided by an estimated model), we suggest that researchers use the predictions directly.

Conclusion

In this chapter, we have introduced the basic idea of machine learning and argued that its core contribution is its ability to make high-quality predictions. We have also emphasized that while machine learning is not a solution to the fundamental problem of causal inference, it can be a powerful tool for aiding researchers with causally oriented research. That said, ultimately the greatest promise of machine learning for the legal research community (and more broadly) is likely in its ability to solve prediction-policy issues and aid descriptive research. Many of the interesting questions confronting legal scholars lend themselves to high-quality prediction tasks. How can we effectively classify a large set of documents, such as docket sheets or legal opinions, into useful categories?

What is the probability that a convict will reoffend if granted parole? How do judges sitting in the same court differ in their decision-making on individual cases? Indeed, as scholars explore the massive amount of new data made available through digitization of legal texts, they can and should better exploit the power of machine learning to answer questions like these. ⮌

')'

TEXT AS OBSERVATIONAL DATA

Marion Dumas, London School of Economics and Political Science
Jens Frankenreiter, Max Planck Institute for
Research on Collective Goods

Quantitative research has traditionally been focused on estimating the parameters by which different variables are related, with an emphasis on establishing causal relationships. It is well known that any study using observational data has to overcome fundamental challenges to its internal validity and precisely elucidate the reasoning and assumptions made about the counterfactual. In this chapter, we address this crucial concern in the context of the new wealth of textual observational data now available for quantitative approaches to legal studies. We also discuss at a high level the opportunities made possible by these new data, as well as the methodological considerations raised by the use of quantitative techniques in a domain heretofore dominated by qualitative approaches.

The Challenge of Causal Inference

Most traditional work in the field of quantitative empirical legal studies is concerned with estimating parameters that describe the relationship between different variables, with a particular focus on establishing a causal relationship. This is most evident for research attempting to measure the effects of certain policies (e.g., Levitt 1997) or interventions in the legal process (Greiner and Wolos Pattanayak 2012; Ho, Sherman, and Wyman 2018). Yet, much quantitative research in social science also falls in this category, even when the research is not explicitly framed in causal terms. For example, the research investigating differences in the behavior of individual decision-makers (e.g., Sunstein et al. 2006) amounts to estimating the effect of the involvement of individuals with a specific background in a legal case.

Causal estimation is difficult because it invariably involves a counterfactual element. In principle, a researcher interested in estimating a causal relationship wants to compare the state of the world after a certain intervention with the state of the world had there not been an intervention. As this is impossible, the main challenge in causal estimation is to find a way to extrapolate how those parts of the world affected by an intervention would look had they been unaffected by it, and the other way around.

As is widely discussed in the literature, the task of extrapolation faced by all causal research is especially hard for research using observational data (Angrist and Pischke 2008; Imbens and Rubin 2015). The main reason for this is the lack of control the researcher has over the process of administering the treatment to the units of observation. This has led some to conclude that for causal research to produce credible results, there needs to be evidence that the treatment variable is exogenously determined (Angrist and Pischke 2010). In the absence of exogenous variation, the researcher must make some assumptions about the counterfactual, and it is crucial to be explicit about this counterfactual reasoning to assess the validity of the causal claim (Morgan and Winship 2014).

The Role of Textual Data in Investigating the Relationship between Variables

How does the availability of large corpora of textual data affect our approach to parameter estimation and causal inference?

On a very basic level, this development implies a change in both the scale and the nature of the data available to researchers. To understand this change, consider how empirical legal research was traditionally conducted. Before, researchers used rather sparse datasets in their analysis. Legal texts had to be hand-coded according to certain predefined coding schemes in order for them to be used in quantitative analyses. Such hand-coding often captured only very limited information about the individual legal texts that formed the basis of such research.[1] Many computational tools, by contrast, allow for a representation of individual texts or large sets of texts as (potentially) high-dimensional vectors of numbers encoding many different characteristics. As computational tools are used to convert text into data, there are almost no limits on the scale of the data and the complexity of the representation.

~ 61 ~

This change brings with it a number of opportunities, but also challenges. As mentioned before, it allows researchers to capture the contents of a potentially larger number of documents in a much more fine-grained way than before. At the same time, traditional statistical methods are not well suited to work with such high-dimensional data. Machine learning models, while built specifically to work with high-dimensional data, are not designed to investigate causal relationships between parameters but to perform prediction and classification tasks (Mullainathan and Spiess 2017).

In recent years, researchers have increasingly been exploring ways to employ machine learning techniques in an attempt to use high-dimensional data (including textual data) in social

---

[1] For example, in research on decision standards of individual judges, decisions were often coded as zero or one, depending on whether they came out in favor of the party which had asked the court to rule in the more "conservative" or "liberal" direction.

science research. It is possible to differentiate between two different types of applications. The first type uses these tools to broaden the scope and/or improve the accuracy of traditional statistical tools. For example, it is possible to reduce the dimensionality of textual data in order to generate new variables for use in regression analysis. This can be done by fitting latent variable models, by using other classification techniques, or by training machine learning algorithms on training sets containing variables that are not available for all observations. This last technique can be particularly useful if the researcher's goal is to create a large dataset but he or she can only afford to hand-code a subset of cases. This strand of research, by and large, stays within the traditional parameter estimation framework, which also implies that it faces the same challenges regarding internal validity.[2]

A second type of application, to which we now turn, attempts to apply machine learning to textual data to generate results that are of direct interest to social scientists. Different from traditional regression techniques, machine learning tools are better able to process high-dimensional data. However, as mentioned before, these models serve to answer different questions than traditional quantitative work in social science, which has been focused on investigating causal relationships between different variables. The typical social science question is something along the lines of "Does policy intervention x cause a change in outcome y?" Machine learning tools, on the other hand, are primarily geared toward prediction and classification. In other words, these tools are best at answering questions such as, Given a set of variables x, what is the most

---

[2]Chapter 2 in this volume discusses ways in which prediction algorithms can directly enhance the credibility of causal inference, such as in the use of instrumental variables. See also Mullainathan and Spiess (2017).

likely outcome y? Machine learning algorithms are not built to solve causal inference questions, and they generally are not well equipped for this task.

The reasons why machine learning algorithms often fail at providing meaningful answers for causal inference questions in principle mirror those that address the difficulty in using traditional regression techniques to conduct causal research. In at least one important way, however, machine learning models perform worse than traditional regression techniques in investigating the relationship between variables. In regression analysis, researchers can specify exactly the functional form of the regression model. This allows them not only to control for observable variables, but also to implement assumptions about the impact of unobserved variables in the model. Because machine learning algorithms usually adjust the model and the variables included in it in an attempt to maximize the predictive power of the model, researchers generally have less influence over the functional form of the model. This arguably makes it harder to reason transparently about the counterfactual assumption made by the model and also often renders it impossible to implement, in any quantitative way, assumptions about a counterfactual.[3]

Against this background, one important challenge for researchers seeking to exploit the full potential of machine learning techniques is to identify questions that can be answered meaningfully by means of prediction and classification (see also Kleinberg et al. 2015). This constitutes a rather new epistemolog-

---

[3] In line with this, Mullainathan and Spiess (2017) show that the parameters estimated by machine learning algorithms are often nonrobust (although the prediction is robust), which precludes using them to measure the relationship between individual variables. Also, the exclusion of (relevant) variables through the regularization process inherent to all machine learning techniques can lead to biased results.

ical approach for social scientists, and research agendas based on predictive inference are just starting to emerge. Here, we want to warn against one potential trap for social scientists: On a semantic level, it is possible to rephrase almost any question about the existence of a causal relationship in terms of classification or prediction. For example, instead of asking, Does policy intervention x cause a change in outcome y?, we can ask, Are the predictions for y different if one includes policy intervention x? This does not change the fact that this question involves a counterfactual element, that is, that the nature of the inquiry is causal (see also Mullainathan and Spiess 2017). In such a case, it is important that the framing of the question as a task of prediction or classification does not keep the researchers from clearly spelling out the assumptions needed to draw inferences about the causal effect. For example, in Chapter 7, one of us investigates, *inter alia*, whether judges from countries that have a similar legal tradition write in a distinctive style compared to judges from other countries. Let us assume here that this analysis had shown that, with respect to style, judges from countries with a similar legal tradition tend to cluster together. In order to interpret this finding as evidence supporting the hypothesis, one has to assume that the use of function words, which serves as a proxy for the style of writing, is solely influenced by the involvement of a specific judge in the case. If, by contrast, certain case characteristics also influenced the use of function words, any observed effect could be due to the fact that judges from countries with similar legal traditions are assigned to author opinions in similar cases.

One example of work that brings prediction techniques to social science in a meaningful way is Kleinberg et al. (2018). This paper shows that algorithms are better than human judges at predicting the chances of a defendant committing another crime when released on bail. As part of their analysis, Kleinberg et al. (2018) also answer questions about the cases in which judges make bad assessments. Note that this paper does not attempt to analyze the impact of a change in one variable on another variable, which is the hallmark of causal research. Instead, it is mainly interested in assessing the risk of defendants committing another crime. This task is, in principle, purely predictive in nature and does not involve a counterfactual question.[4]

To sum up, machine learning algorithms are not based on clear counterfactual assumptions and are therefore in no way better equipped, and in fact are worse equipped, to tackle concerns about endogeneity than traditional regression analysis. Thus, when used on their own, prediction and classification tools are generally not well suited to answer questions about the relationships between different variables (causal relations, in particular), unless we can spell out clearly the counterfactual assumptions on which the analysis is based.

Use of Corpora to Describe Processes and Develop New Theory

So far, we have contrasted the use of text as quantitative data in prediction, classification, and causal inference tasks. All these forms of analysis have in common that they draw on variation in the value of some variables across units of observation to learn about the relationship between these variables. However, such analyses do not exhaust all forms of inquiry that are of interest to social scientists. In particular, investigating the relationship

---

[4]Note that the paper does involve a counterfactual element stemming from the fact that the risk of jailed defendants cannot be observed. The authors of this study solve this problem by exploiting different decision standards of individual judges and the fact that cases are assigned to judges in a quasi-random way. However, different from causal research, this counterfactual element does not form the core of the researcher's interest.

between such variables often reveals little about causal *processes* and *mechanisms.*[5] In other words, while causal inference is typically aimed at answering the question of whether a change in variable x leads to a change in variable y, and emphasizes the internal validity of this inference, it usually leaves unanswered the question of *how* and *why* such a change occurs and the range of contexts and populations in which we can expect these processes to occur and matter.

We stress here that there is no fundamental epistemological opposition between causal inference, the description of processes, and the discovery of mechanisms. In a full analysis, they would all be integrated. However, these forms of inquiry differ in their methodological approach and, due to the methodological trade-offs they imply, in the degree of emphasis they place on measuring causal effects versus understanding process. For example, consider mediation analysis, which seeks to identify variables that mediate a cause-effect relationship between x and y. There are many types of models for mediation analysis, and those that place more emphasis on the details of the mechanism (for example, by including dynamics and multiple mediating variables) offer less guarantee of the internal validity of causal inference (Imai, Keele, and Tingley 2010; Preacher and Hayes 2008).

Traditionally, the empirical description of processes and mechanisms in the social sciences has been the domain of qualitative research that focuses on in-depth within-case analysis and the tracing of processes (Bennett 2008; Brady and Collier 2010; Abbott 2001). There is no inherent reason

---

[5] By process, we mean the sequencing of events in time through a set of causal steps that influence outcomes (Abbott 2001). By mechanism we mean the set of objects and their interactions that together provide an explanation for an outcome (Craver and Tabery 2017).

why research on causal processes should be solely associated with qualitative research. For example, the field of stochastic processes shows the value of quantitative models of processes and their capacity to illuminate social phenomena (Helbing 2010). The reason we have historically associated empirical research on mechanisms with qualitative research is that uncovering mechanisms typically requires more fine-grained data than causal effect inference, including mediating variables and a sufficiently fine temporal resolution to *trace* sequences of events or sequences of effects. Additionally, some variables, such as beliefs, preferences, and perceptions, which are often important mediating variables in social mechanisms, have been difficult to measure at large scale.

~ 67 ~

However, the advent of "big data" is opening new opportunities to do quantitative empirical work on social processes. Textual data seem particularly relevant for understanding social processes because of the central role of communication in social life. Any theory that assumes or implies a certain flow of information, pattern of communication, or distribution of beliefs could, in principle, be tested using textual data. This idea is, of course, not new: historians and case study researchers make ample use of textual archives. The difference here is that the coupling of large-scale digitization projects with computational methods opens up the possibility of using large corpora of text that may encompass all written communication within a certain institution or field.

In principle, such corpora can be used to shed new light on social processes by revealing large-scale patterns of communication and the negotiation and encoding of shared meaning on a macro level. Importantly, it is possible to obtain meaningful insights merely from describing such patterns. For example, Klingenstein, Hitchcock, and DeDeo (2014) analyze over 150 years of the archives of the Old Bailey, showing that trials for acts of theft and trials for acts of violence were initially semantically indistinguishable. Yet, over the course of the nineteenth century, in ways that may reflect the establishment of the state's monopoly on violence, these two types of crimes began to be treated differently, with increasingly distinct semantic fields and concepts used to describe and judge them. This study uses a large corpus to describe a large-scale process of gradual differentiation of meanings and rules. Such a description of a process can provide meaningful insights without having to rely on counterfactual reasoning.

Descriptive inference is particularly useful in the early phases of using textual data in quantitative analyses, but as the field matures, researchers can and should do more than just describe the flow of information. Instead, they should aim to form theories about how specific causal mechanisms manifest themselves in textual data, which, in turn, will allow them to test hypotheses about which causal mechanisms are at work in a specific setting.

Early examples of work in this field include studies that mix elements of theory testing (deductive) and data discovery (inductive) to test the reach of a theory while learning from the data how the theory can be refined (McFarland, Lewis, and Goldberg 2016). In one of the first quantitative studies of consensus formation in science, Shwed and Bearman (2010) use the modularity of scientific citation networks to operationalize Bruno Latour's theory of scientific consensus. With this metricization, the authors are able to describe the temporal structure of consensus formation in several contentious scientific areas (such as climate change, smoking, etc.). Vilhena et al. (2014) build on a simple model of communication to determine from scientific texts the efficiency of communication across fields of science, revealing patterns of balkanization in some fields and integration in others. These examples have in common that they start from a broad, highly qualitative theoretical framework to constrain the analysis but let the data fill in areas where the theory is silent. The full potential of these types of analysis awaits better theories about how language mediates social life, and how different social mechanisms operate through and leave an imprint on language and text.[6]

Although theoretical developments may assist future analysis, researchers have already found creative ways to use textual data to make inferences about causal mechanisms. Jensen et al. (2012) build a panel dataset of partisan phrases from the Google Books Ngram Corpus and the Congressional Record and show that the use of polarized speech increases first among the cultural elite (authors of books in the Google corpus) before increasing in congressional speech, suggesting that elite discourse influences politics.[7] Rzhetsky et al. (2015) analyzed millions of biomedical publications over thirty years to infer typical research strategies of scientists. They show that these strategies are highly conservative: researchers favor experiments that expand existing research incrementally instead of venturing deep into uncharted territory. Using a computational model, they argue that the strategies probably slow scientific advance at the aggregate level. Note that these studies may not provide *conclusive* evidence of the causal effect they purport to investigate (the influence of elite discourse and the influence of conservative research strategies, respectively). Nevertheless, they still represent potentially important steps

---

[6]Early theoretical developments that could support this effort include Axelrod (2015) and Rubinstein (2000).

[7]Clearly, this finding is subject to the same internal validity challenges discussed earlier.

toward a better understanding of the phenomena under investigation. For example, given the findings of the first study, it seems hard to argue that the polarization of congressional speech caused the polarization of speech in and among the elites.

Despite these limitations, these examples suggest exciting new avenues to answer a number of important questions about the law that have, so far, largely gone unanswered. For example: What are the most influential doctrines at a given moment in time, and how do they spread through the legal system? What beliefs do judges, litigants, and lawyers bring to bear on a situation? How does law reflect the beliefs and values of society? These questions concern process. To use textual data to answer them quantitatively will almost certainly first require a good amount of process description. Eventually, however, such descriptive work has the potential to inform the development of a more micro-based theory of the use of language in law, which will in turn allow for the use of textual data to estimate structural models of language and communication. ໒

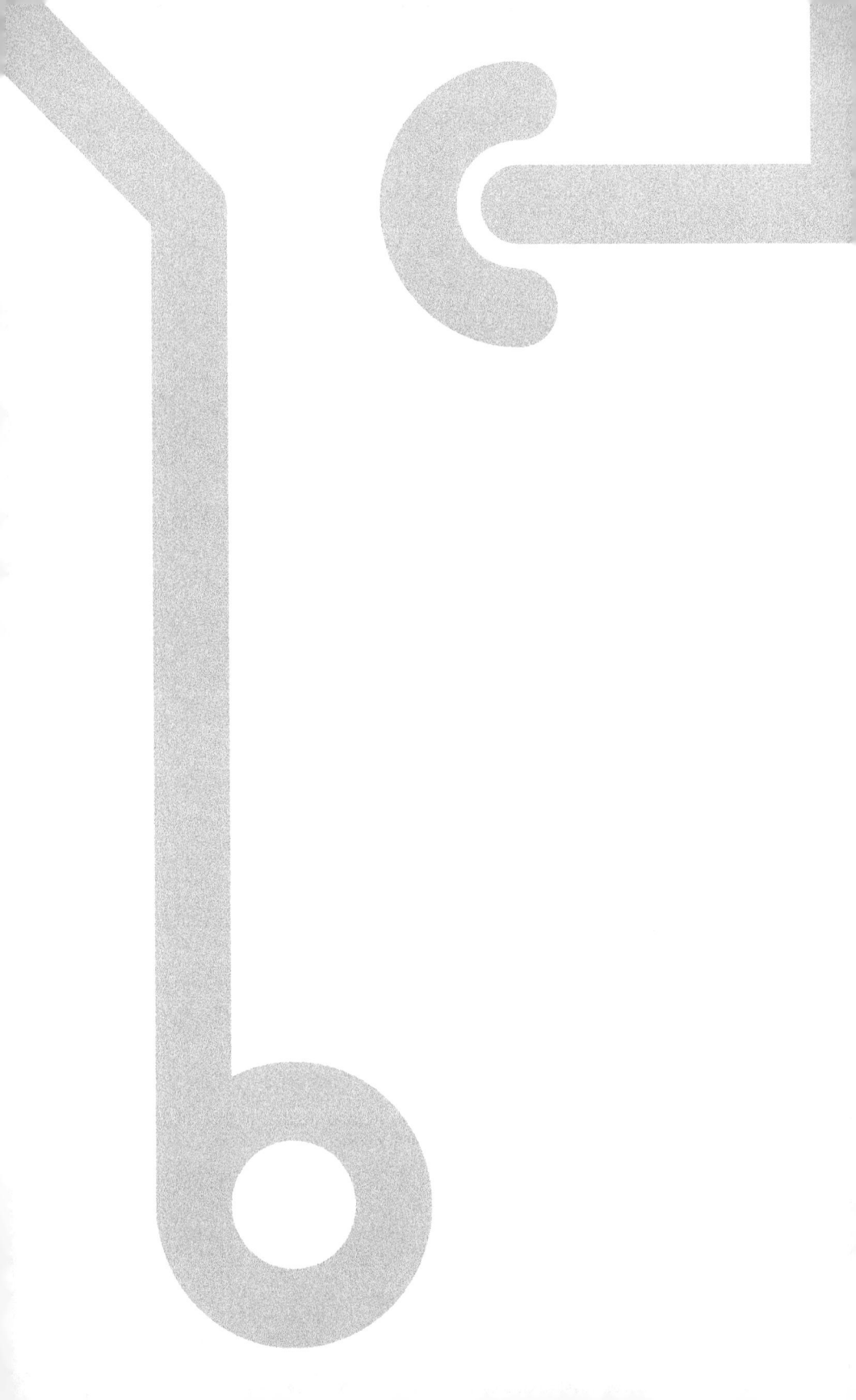

ƪ

PREDICTION BEFORE INFERENCE

Allen Riddell, Indiana University Bloomington

Competing probabilistic models of past events can always
be evaluated in terms of how well they predict ("retrodict")
events using a measure of out-of-sample predictive accuracy.
Are there settings where it is worthwhile to devote time and
energy to developing models designed, in particular, to identify
causal effects from observational data? Liters of ink have been
spilled debating this question. In the specific common case of
searching for credible probabilistic narratives of patterns in
observational data—what Gelman (2011) labels "reverse causal
inference"—insisting on formal models of causal inference is
unhelpful. (Those designing and conducting field experiments,
by contrast, *should* concern themselves with causal inference.)
Models which aim to describe associations or make predictions
without aiming explicitly at causal inference are often useful.
Even when they are not practically useful, evaluating competing
models of past events in terms of predictive performance is
frequently valuable work and, in the final accounting, essential.
The spectre of human and machine error as well as questionable
research practices (e.g., p-hacking, HARKing, and publication
bias) requires that all analyses, including those which claim to
have made causal inferences, be evaluated using measures of
usefulness and reliability which look beyond immediate formal
proprieties of models.

The comments presented here arise from an interest
in learning from observations of past events that can be

described quantitatively and used in probabilistic models. Because credible narratives of past events often complement each other—someone else's findings may help members in another intellectual community—everyone benefits when researchers in the social sciences and humanities use data-intensive methods that yield inferences that stand up to scrutiny. Narratives which prove, in retrospect, to be unreliable risk diverting time and resources away from more productive pursuits. Recent experience has shown that it is not a foregone conclusion that one learns much from an arbitrary piece of published research (Ioannidis 2005; Angrist and Pischke 2010; Camerer et al. 2016; Munafò et al. 2017; Ioannidis, Stanley, and Doucouliagos 2017). In fact, the false discovery rate in social science research featured in prestige venues such as *Nature* and *Science* appears to be higher than 30% (Camerer et al. 2018).

In this context the question of what sorts of methods should be allowed, encouraged, or discouraged is one of general interest. So, too, is the question of whether or not the use of certain classes of models yields reliable characterizations of past events.[1]

The comments here are organized as follows: after noting arguments against fetishizing models of past events designed to infer causal effects, this chapter argues for appreciating the value of work which attends primarily to the comparison of models using measures of out-of-sample predictive accuracy.

Consistency's Costs

Insisting on the use of models aimed at inferring causal effects from nonexperimental data is obviously unhelpful if, in practice,

---

[1] Reliable by the standards of the relevant intellectual community. One way of identifying reliable research in the case of research involving experiments involves independent replication of the experiment. In the case of nonexperimental research, independent (re)analysis of data can be used to gauge reliability (Silberzahn et al. 2018).

the findings which emerge from using these models tend to yield false discoveries at a higher rate than research using other methods. Models can yield invalid inferences for many reasons, including formal mis-specification, researcher error, and programming error. Thanks to growing interest in meta-analysis, we will eventually have a better sense of the association between specific statistical practices in certain disciplines and false discovery rates.

Some examples of this kind of work are already available. Young (2017) looked at thirty-four instrumental variable (IV) models in journals of the American Economic Association. Young (2017) found that the labor invested in careful construction of IV models was likely counterproductive: simpler models would have produced better estimates of quantities of interest and yielded fewer false discoveries. Combined with concerns about the general quality of published research, this result gives us reason to question demands for formal purity—and, in particular, demands for the use of IV models—without accompanying concern for substantive validation.

The demand that researchers use models that aim for causal inferences using observed data—such as the potential outcome framework associated with Splawa-Neyman (1990) and Rubin (1974)—also risks limiting the types of questions that tend to be asked. Encouraging researchers to devote resources to developing models in this framework risks discouraging investigation of hypotheses that cannot easily be formulated using the framework. Many observed patterns merit investigation, but not all are easy to express using the potential outcomes framework. Exploring these patterns often involves asking questions that *do not* take the form, "What is the effect of X on Y?" Many questions, including most questions in the political economy, cannot be expressed in this form. Here are some examples: Why did real GDP in the United States decline between December 2007 and June 2009? Why is the return on capital greater than the return on labor? Why are so many US prison inmates African American men? Why are so few women professors promoted in economics departments at US universities?[2] Researchers should be not discouraged from investigating these sorts of questions merely because the models they would naturally use do not belong to a favored class.

The Value of Prediction

Researchers can and do solve problems and address matters of concern by focusing on prediction, setting aside, for the most part, questions of causation. Predictively accurate models of local weather, traffic conditions, flu incidence, loan defaults, and economic recessions have obvious utility. So, too, do models which reliably "predict" how an expert human translator will translate an English sentence into French. More germane to interests here, predictively accurate models of individuals', legislators', and judges' votes—including those models that use textual data—are useful to many people working in politics and the law. Provided that their predictions are reliable, the models' utility is not diminished much if they fail to identify causal factors. That models used in these and similar settings do, in fact, tend to ignore concerns about causal inference is obvious to anyone with passing familiarity with the application of probabilistic modeling in the contemporary technology industry. One illustration of the disproportionate concern for prediction could be found at the 2017 meeting of the most important academic and industry conference in machine learning, Neural Information Processing Systems (NIPS). At a conference with over 7,800 registered attendees, Judea Pearl, a Turing Award

---

[2]See "Barriers to Entry" in the May 10, 2018, issue of *The Economist*: https://www.economist.com/finance-and-economics/2018/05/10/barriers-to-entry.

winner and prominent figure in the development of formal models for causal inference, delivered a talk which was not particularly well attended.[3] (This is all to say that, had the conference been addressed to a different audience, the lecture hall likely would have been packed.) In the world of machine learning, formal models designed for causal inference are rarely mentioned. Models used to guide self-driving cars or recognize human speech are invariably and typically exclusively compared in terms of out-of-sample predictive accuracy.

Research concerning itself primarily with prediction—that is, on the association between variables where one is a predicted quantity—should also be of interest to those who are ultimately interested in studying the causal effects of specific interventions. This is because the study of effects invariably begins with the observation of some association. Documenting a correlation between two variables is often not trivial. It is especially difficult if the number of observations is small or the association is weak. If there is, in fact, no association between two variables, then investigating the causal effect of one variable on the other is certain to waste time and resources. Establishing that one can, in fact, predict the value of Y after conditioning on the value of X at a rate better than chance is therefore valuable preliminary work. For example, demonstrating that there is a reliable pattern in the voting behavior of senior federal judges appointed by Republican presidents is likely to face a number of challenges, including controlling for regional variation (e.g., differences between federal districts) and accounting for small sample sizes. Similar obstacles would confront someone trying to establish that the appearance of a pattern of argument or speech in oral arguments can be used to predict how a judge votes better than a baseline model.

---

[3] See V. Dignum, "Food for thought . . ." on Twitter (2017), https://twitter.com/vdignum/status/939440384010280960.

Researchers who dedicate labor to this sort of work facilitate future studies of phenomena of interest, including studies that use the potential outcome framework.

In the common case of when a researcher is evaluating two or more competing probabilistic models of past events, predictive performance can almost always be used to assess whether or not one of the models better accounts for observations than another (Press 2009, 266).[4] In certain fields, in particular machine learning and fields where Bayesian methods are popular, this procedure is ubiquitous (Gelman and Imbens 2013). The parameters of each model are inferred using a fraction of the data (say, 90%), and then the model is asked to predict observations in the remaining held-out fraction. Each model's held-out performance (typically measured in terms of accuracy or the likelihood it assigns the unseen data) can be used to compare the candidate models (Vehtari, Gelman, and Gabry 2017). So, if a researcher can formulate at least two narratives of observed events, model comparison is almost always an option.

Discouraging research focused on investigating associations between variables using out-of-sample predictive accuracy rather than formal models for causal inference risks delaying valuable lines of inquiry. For example, doing so will discourage researchers from exploring reliable associations between textual data and events. Although individual words or phrases may be reliably tied to practices and durable dispositions, it is exceedingly difficult to incorporate their occurrence or nonoccurrence into models aimed at causal inference. Deprecating the use of models other than those which aim at causal inference runs other risks as well. For example, doing so may encourage the use of research methods

---

[4]"Almost always," because there are rare cases where calculating the likelihood of held-out observations is nontrivial. For an example involving topic models, see Buntine (2009).

with a questionable record of application. In light of these risks, it is prudent to permit the use of a range of methods in the study of past events, including the comparison of competing explanations using out-of-sample predictive performance. ❧

INVESTIGATION

STYLE AND SUBSTANCE ON THE US SUPREME COURT

Keith Carlson, Dartmouth College
Daniel N. Rockmore, Dartmouth College
Allen Riddell, Indiana University Bloomington
Jon Ashley, University of Virginia
Michael A. Livermore, University of Virginia

The United States Supreme Court is a singular institution within the American judiciary. It has many unique institutional features, such as the ability to select the cases that it will decide, and it plays a unique role in American political and social life. However, while as an institution the Supreme Court is distinctive, it remains recognizably a court of law. The Supreme Court shares certain rituals with other US judicial institutions, such as the black robe and gavel. It also shares many procedures with other courts, including adversarial hearings and restrictions on *ex-parte* contacts. Perhaps most important is that its mode of decision-making is through case-by-case adjudication, typically in the course of hearing an appeal from a lower court decision. As such, when the court creates, amends, or clarifies legal obligations, it does so not through directly stated rules (as in a statute or regulation) but through the justificatory documents that accompany a disposition in a particular case, largely embodied in the written opinions of the court and its justices.

The judicial opinions of the US Supreme Court serve as among the most important pieces of "data" in understanding the evolution of legal thought and sociopolitical dynamics in the United States. The judicial opinions issued by the Supreme Court have provided legal scholars (and others) with fodder for analysis since the dawn of the American legal academy. Recent advances in natural language processing and computational text analysis provide new ways to examine and understand the work of the court. In this chapter, we will discuss our research using these kinds of tools on two sets of questions concerning the court that are particularly well suited to such analysis. First, we examine trends in writing style on the court through sentiment analysis and a form of stylometry based on the frequency of function words in a text. Second, we examine whether the Supreme Court has begun to carve out a unique judicial genre by examining how the content of Supreme Court opinions differs from the federal appellate courts that it supervises.

Why Study Style?

Judges, lawyers, legal academics, and law students have frequently turned their attention to noncontent, stylistic features of legal writing. For example, legal writing courses at American law schools evidence a desire to teach students appropriate writing style, in addition to facilitating a mastery of legal content (Romantz 2003). Practicing lawyers are often called on to persuade through the written word, and stylistic features of a text can contribute to (or detract from) its persuasive force. Guides on legal writing, geared toward law students and practicing attorneys, often pay substantial attention to noncontent textual characteristics (Garner 2013). A host of stylistic conventions distinguish legal writing from standard written English, and a lawyer's competence is judged, in part, by the degree to which his or her individual stylistic voice conforms to this particular "professional discourse community" (McArdle 2006, 501).

Judge Richard Posner has defined writing style as "the range of options for encoding the paraphrasable content of a writing" (Posner 1995). Essentially, under Posner's definition, writing style amounts to the individual imprint that judges leave on their writings, holding the legal content constant. Perhaps in part because judges are individually responsible for drafting their opinions (although there is often a good measure of group editing), there is substantial stylistic variation within judicial writings.

Writing style in judicial opinions is important for a variety of reasons. Style may serve as an indicator of judicial temperament or disposition. Stylistic norms may constrain judicial writing in ways that ultimately affect judicial reasoning, and in turn, legal outcomes. The evolution of writing style may indicate broader substantive trends on the court. Style can affect the comprehensibility and usability of the law. Finally, style may be deserving of study simply as an empirical feature of an important cultural artifact. In fact, judicial writing style has long been the object of qualitative analysis, with commentators frequently examining (and criticizing) opinions both for basic clarity as well as (sometimes) their literary quality (Ferguson 1990).

There is a nascent movement among legal scholars to bring quantitative tools to bear on the analysis of writing style. As it turns out, an early important application of computational stylistic analysis had something of a legal (or, more properly, constitutional) context, as Mosteller and Wallace (1963) brought statistical methods to the problem of identifying the authors of individual Federalist Papers. As for quantitative work directed at opinions, Little (1998) uses a coding procedure to identify "linguistic devices that obscure" meaning and analyzes Supreme Court cases on federal jurisdiction; Black and Spriggs II (2008) examine opinion length over the entire period of the court's existence. More recently, Owens and Wedeking (2011) examine "cognitive clarity" in recent Supreme Court cases using the "linguistic inquiry and word court" (LIWC) software package. Long and Christensen (2013) examine the use of "intensifiers" and readability scoring to test their theory that justices broadcast weak legal position through use of language. Johnson (2014) examines readability over time in the court, comparing Flesch–Kincaid scores in 1931–1933 and 2009–2011 terms. Black et al. (2016b) use computational tools to examine how Supreme Court writings alter language usage according to context and audience.

In addition to scholarly investigations, computational analysis of judicial writing style has even found its way into pop culture. For example, one analysis used token analysis—a measure of sophistication in language use—to compare the vocabulary of several justices to famous rappers and Shakespeare.[1] The authors find that Jay-Z and most of the justices have similar vocabulary use, while rapper Aesop Rock and Justice Holmes have exceptionally large vocabulary use, and DMX and Justice Kennedy are on the low end.

In this chapter, we report two stylistic analyses that contribute to the growing judicial stylometry literature. We make use of an original dataset of US Supreme Court opinions: Human researchers conducted a series of "by-year" searches

---

[1] Adam Chilton, Kevin Jian, and Eric Posner, "Rappers v. Scotus: Who Uses a Bigger Vocabulary, Jay Z or Scalia?" Slate.com (June 12, 2014), https://slate.com/news-and-politics/2014/06/supreme-court-and-rappers-who-uses-a-bigger-vocabulary-jay-z-or-scalia.html.

on a commercial database to download digitized versions
of all Supreme Court cases. All proprietary information
was stripped out. A series of iterative human and Python-
based analyses were then carried out to separate majority,
dissenting, and concurring opinions and to assign an authoring
justice and year to each opinion. *Per curiam* decisions were
removed from the dataset, as were opinions with a file size
smaller than one kilobyte. Jonathan Ashley, research librarian
at the University of Virginia, was primarily responsible for
identifying resources, collecting cases, and providing the
markup needed for analysis.

The resulting data cover all opinions for the years
1792–2008.[2] Our data include 25,407 decisions. We exclude
footnotes from our analysis. There are roughly eight thousand
dissents and 4,600 concurrences. We have data for 110 justices:
Justices Sotomayor and Kagan were appointed after the end of
our study period. We have partial data for justices who began
their terms prior to 2008 but either retired after our study
period or remain on the court. Our analysis was conducted
when Justice Scalia was an active member of the court.

We first conduct a basic sentiment analysis of US Supreme
Court opinions and arrive at the interesting conclusion that
the court's language has become decidedly more "grumpy" over
the course of the past two centuries. Our second analysis of
the court's writing style is more detailed and attempts to gain
purchase on longstanding questions concerning the role of
clerks in influencing the work of the court. For that analysis,

---

[2] We define a "decision" as the set of opinions that relates to a case,
identifiable through a citation in the United States Reporter, for example,
"347 US 483 (1954)." A decision can include multiple opinions, including
a majority opinion, plurality opinions, and one or more dissents or
concurrences. In our data, we do not distinguish majority from plurality
opinions.

data concerning the number of clerks employed in chambers were provided by the Supreme Court Library.

Judicial "Friendliness"

Sentiment analysis is a form of natural language processing, which is a broader field within computer science and computational linguistics focused on human–computer interactions through language. At the heart of sentiment analysis is the concept of sentiment, which is a relation between a person and a target. Simply and intuitively, the sentiment of A toward target X is whether A likes or dislikes X. Although, in theory, sentiment analysis could distinguish between nuanced emotional flavors, the general tendency in the field (to date at least) has been to reduce sentiment to a single dimension that ranges between positive and negative poles.

A leading researcher recently defined sentiment analysis as "the field of study that analyzes people's opinions, sentiments, appraisals, attitudes, and emotions toward entities and their attributes expressed in written text" (Liu 2015). Prior work has focused primarily on text-relevant commercial "entities"— goods and services like movies, or products for sale on Amazon. The law, broadly understood, is also full of entities of various stripes capable of generating the opinions, sentiments, appraisals, attitudes, and emotions that are amenable to sentiment analysis.

Sentiment can matter a great deal in the law when it is part of an actual human evaluation, attitude, or affect. A recent example helps illustrate the point. *Bowers v. Hardwick* was the 1986 decision of the US Supreme Court that upheld a Georgia antisodomy statute. In his concurring opinion, Chief Justice Burger wrote separately to "underscore" his views on the matter. To establish the pedigree of antisodomy statutes,

Burger's opinion quotes Blackstone's treatise on the English common law describing an "'infamous crime against nature' [...] an offense of 'deeper malignity' than rape, a heinous act 'the very mention of which is a disgrace to human nature,' and 'a crime not fit to be named.'" This language conveys an extraordinary degree of negative sentiment and was used to justify the rejection of any constitutional protection for the private consensual sexual conduct of same-sex couples. In *Lawrence v. Texas*, the US Supreme Court revisited this decision, and the shift in sentiment is striking, with Justice Kennedy writing:

> [A]dults may choose to enter upon this relationship in the confines of their homes and their own private lives and still retain their dignity as free persons. When sexuality finds overt expression in intimate conduct with another person, the conduct can be but one element in a personal bond that is more enduring.

Further along that line of cases, *Obergefell v. Hodges* found that the constitution guaranteed access to marriage for same-sex couples. In justifying this decision, Justice Kennedy, again writing for the majority, stated that:

> The nature of marriage is that, through its enduring bond, two persons together can find other freedoms, such as expression, intimacy, and spirituality. This is true for all persons, whatever their sexual orientation. There is dignity in the bond between two men or two women who seek to marry and in their autonomy to make such profound choices.

It is hard to imagine a more significant shift in sentiment from the court's characterization of same-sex relationships in Bowers and Obergefell. The changing sentiment in the texts, which presumably mirrors shifting attitudes of the court's majority, matters both in terms of the change in constitutional status for same-sex couples that it accompanied, but also in the expressive function it serves when communicated on behalf of a major institutional voice in American politics.

The difficulties of the sentiment analysis problem and the appropriate technique to be deployed are related to the degree to which a researcher attempts to zero in on targets within a single document. The simplest approach considers sentiment at the document level, which reduces all of the text within the document to a single sentiment score. Because there is no attempt to determine whether different targets are referenced, this type of analysis might be seen as that which best captures "sentiment" as the overall mood of the author. Take, for example, this sentence:

> Lousy Day: "I had a lousy day because my commute was blocked up by a terrible accident, and I had to wait around all morning for a boring meeting with my boss."

There are different flavors of sentiment (terrible, boring) as well as several targets of negative sentiment (day, commute, accident, meeting) that could be extracted, but it is clear at the document level that the overall mood is pretty sour.

Sophisticated forms of sentiment analysis can attempt to accomplish a more finely grained analysis by either developing more specific and focused measures of sentiment or extracting the targets of sentiments within a document, or both. Imagine that the Lousy Day sentence above appeared on

Twitter with geotagged information. If a traffic monitoring and predictive service wanted to use real-time social media information to improve its performance, it would be important to extract sentiment concerning some of the targets in the text (commute, accident) while ignoring other irrelevant sentiments.

An important component for sentiment analysis at any level is a "sentiment lexicon" that categorizes words according to the sentiment that they convey. In a sentiment lexicon, some words (e.g., wonderful, intelligent, great) will be classified as positive while others (e.g., terrible, stupid, bad) will be classified as negative. There are two general ways to construct a sentiment lexicon. The first is the thesaurus approach. For this, a researcher starts with some positive and negative seed words that have an obvious valence and then identify associated words. The second is the natural corpus approach. Here, the researcher starts with a set of documents produced in the real world, like Amazon reviews or (for our purposes) judicial opinions. If there is already metadata related to sentiment— such as how many stars are in the review—this can be used to determine words associated with that metadata. Without this kind of metadata, seed words with known sentiment can be used to identify a larger set of related words based on whether they often co-occur in the corpus with the seed words. As sentiment analysis has become more common, there are also now off-the-shelf lexicons available. These have the obvious benefit of saving time, enabling comparability and replicability, as well as reducing concerns that the lexicon is overfit to the data.

For our analysis, we use an off-the-shelf lexicon made publicly available by Liu and Hu.[3] In this lexicon, there are roughly seven thousand English words that are characterized as either positive or negative. Some examples of negative words are "admonish" and "problematic"; positive words include "adventurous" and "preeminent." A Python script was programmed to determine for each justice the total number of negative words and the total number of positive words in opinions he or she authored. The numbers of negative and positive words were then each expressed as percentages of the total number of words authored by a justice. The percentage of negative words was subtracted from the percentage of positive words to generate what we call a "friendliness score."

This analysis—while based on measures of sentiment that have been used in a variety of other contexts—should be approached with a healthy dose of skepticism. Comparing texts over a long time horizon may be problematic for a variety of reasons, including that a text that reads as relatively friendly in one time period may read as downright nasty in another (or vice versa).

With these caveats in place, figure 5.1 contains a plot of the "friendliness score" of each justice across time, with justices located at their median year of service. We constructed the score by subtracting the percentage of positive words from the percentage of negative words. By this measure, it is clear that over time the Supreme Court has gotten grumpier.

The obvious time trend is striking and raises a number of interesting questions. Would a bespoke lexicon generate the same results? Do lower courts exhibit the same trend toward

---

[3] Bing Liu and Minqing Hu, *Opinion Mining, Sentiment Analysis, and Opinion Spam Detection*, UIC, http://www.cs.uic.edu/~liub/FBS/sentiment-analysis.html.

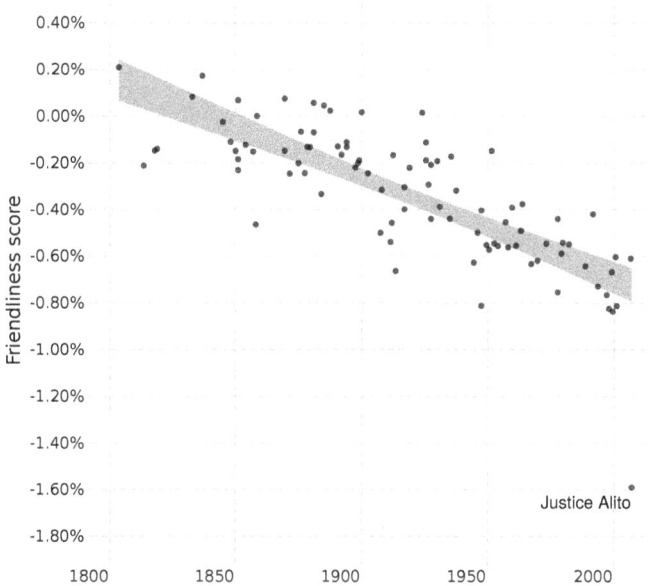

Figure 5.1. Sentiment score by authoring justice.

negativity? Is the trend driven by the growth of dissents? Is there a larger cultural trend toward grumpiness? These and other questions are worthy subjects for future inquiry.

The Role of Clerks

As judicial clerks have become an enduring feature of the operation of the federal courts, the role of these recent law graduates has been the subject of both scholarly and public debate (Peppers 2006). An important empirical predicate to this debate is the belief that clerks play a substantial role in authoring opinions. At least for the Supreme Court, there is a long history of anecdotal evidence supporting the claim that law clerks exert some influence over judicial decision-making. There is also a nascent literature that uses quantitative techniques to address the question of clerk influence over both substance and style. Techniques that have been used include the use of plagiarism software (Sulam 2014), analysis based on the party affiliation of clerks (Peppers and Zorn 2008), compression software (Choi and Gulati 2005), and identified stylistic features of opinions, such as the type–token ratio (Wahlbeck, Spriggs II, and Sigelman 2002). In Rosenthal and Yoon (2011) a method is used similar to the one we describe below based on variability of writing style but on a smaller set of data, and with a different strategy to identify clerk influence.

Our stylistic analysis relies on the use of function words in a document to serve as a broad proxy for a range of stylistic characteristics. Function words, such as "the," "a," and "at," do not directly carry content but instead serve syntactic and grammatical functions. For purposes of the following analysis, the most important characteristic of function words is that they have been found useful as the basis for a stylistic "fingerprint" that can be used for author attribution, and we therefore use it as a proxy for writing style more generally (see, e.g., Hughes et al. (2012)). Our study relies on 307 standard function words (also referred to as "content-free words" or CFWs). The individual occurrences of each CFW can be aggregated to construct feature vectors based on some object of interest. For example, a feature vector can be constructed for each justice or each year.

For our analysis, we rely on an intuitive model of the process of drafting and editing judicial opinions on the Supreme Court. One of the peculiar features of the contemporary clerkship is that it is so short, typically lasting a mere year. We take advantage of this clerk turnover as a source of variability. We construct two measures: one a measure of intrajustice interyear variability; the other a measure of writing style consistency for the court as an institution. We then compare our measure of consistency to a set of time periods based on a historical analysis of the role of clerks from Peppers (2006).[4]

The first measure of variability that we introduce is centroid distance. We construct feature vectors for each text, as well as for the year, and calculate the distance between the text vectors and the year vector.[5] This provides a measure of how tightly clustered the court's style is in a given year: the greater the centroid distance, the bigger the stylistic "spread." There is a clear time trend in our data, with intrayear consistency on the court increasing over time.[6] To examine whether the overall trend toward greater consistency differed as the institution of the modern clerk developed, we conducted a structural break test on the data. The point of a structural break analysis is to determine whether there has been an underlying shift in the data-generating mechanisms, such that the distribution of data from the period after the "break" is systematically different from the distribution prior to the break. We first ran a Chow structural break test, which is a standard tool to determine whether there are changes in the relationships between time and another variable over different time periods (Hansen 2001; Chow 1960). The Chow test rejects the null hypothesis that there are no structural breaks in the centroid distance data at the Peppers group's dates. For an additional text, we do not hypothesize any dates and rather use statistical tools to test

---

[4] These same groupings were used in Black and Spriggs II (2008).

[5] Our distance measure is cosine similarity, which is a representation of distance in a multidimensional vector space.

[6] For this analysis, we use a simple ordinary least-squares regression. For time as a predictor of centroid distance, the p-value is less than 0.01% and the R-squared value is 0.5. We do not report coefficients as the distance measure itself is somewhat difficult to interpret.

whether there is a structure break and, if so, the estimated break date.[7] The estimated break date that was returned was 1926—very close to the year that clerks took on a greater substantive role, as indicated by the Peppers group transition from "stenographers" to "assistants."

We also examine interyear, intrajustice variability in writing style. For purposes of our analysis, a chamber in a given year can be thought of as a "team" made up of a justice and several clerks. A team coproduces the opinions in a given year. When clerks turn over, it changes the composition of the team. In chambers with a larger number of clerks that turn over more frequently, there will be a higher percentage of team turnover from year to year.

Although some interyear stylistic variability can be expected even with a single author, we hypothesize that clerk turnover will decrease interyear consistency. The dependent variable in our analysis is an interyear consistency score. To construct the consistency score, we rely on the feature vectors based on the texts authored by a justice in each year of his or her tenure. To calculate the consistency scores, we calculate the Kullback–Leibler divergence (relative entropy) between each year's vector (interpreted normalized to a probability distribution) and a feature vector based on the remainder of the justice's writings. We examined the relationship between consistency score and the number of clerks that served in a justice's chambers over the course of his or her tenure, controlling for time through a quadratic function as well as each justice's total production in words (under the theory that justices who produce more words may be more consistent, and there will be less statistical noise between years). For this analysis, we examine the period after

---

[7]For this analysis, we used the Supremum Wald test in Stata. For additional background on this test, see Perron (2006).

1885, with the introduction of clerks as "stenographers" under the Peppers grouping.

Based on this model, we find the number of clerks is (statistically) significantly negatively related to the consistency score, with the interaction term indicating that clerks have had decreasing influence over time. This temporal effect may be associated with declining marginal influence of an additional clerk, as the court has institutionalized the practice of each justice having between four and five clerks.

It is worth remembering the difficulty of fully distinguishing the effects of unobserved time-related variables from the effects of clerks. Nevertheless, over the course of the twentieth century, the intrayear stylistic consistency of the court as an institution has increased, while the interyear consistency of writing style for individual Supreme Court justices has declined. Over the same period of time, law clerks have become ever more integrated into the substantive work of the court. Because the institution of the modern law clerk in the US Supreme Court evolved gradually over time, it is hard to know the degree to which clerks have contributed to changes in writing style independent from some other set of time-related variables. But we have some evocative information that provides some evidence, at least, that the institution of judicial clerks appears to reduce intrajustice writing style continuity that might otherwise exist while at the same time reducing the apparent stylistic differences between justices.

The Judicial Genre

Our second set of analyses moves from writing style to the content of US Supreme Court opinions.

One of the classic paradoxes of the American political system is that, in a country that purports to hold democratic values dear, final decision-making authority on some of the most hotly contested political issues is vested in a body that is almost entirely free from formal democratic accountability. Political scientists and sociologists have offered a variety of theories to explain the counterintuitive fact that the court enjoys a high level of support by the public even though, from time to time, it reverses the policy choices of democratically elected branches (Gibson and Caldeira 2009). This support can be both "diffuse"—i.e., a "reservoir of favorable attitudes or good will" toward the institution, as well as "specific," which is based on happiness with individual decisions (Easton 1965). Although there is some disagreement on this point, there is evidence that the court enjoys at least some diffuse support that is resistant to change, even in the face of disagreeable outcomes in individual cases (Caldeira and Gibson 1992).

There are a variety of theories about why the court might enjoy diffuse support. These include the "myth-of-legality" hypothesis, which holds that popular support for the court is grounded in a widespread misperception that all legal questions can be resolved impartially and dispositively based on the neutral application of relevant law (Scheb and Lyons 2000). A more nuanced version of this hypothesis is that the public accepts that there is a degree of discretion involved in judging but believes that the justices exercise their discretion in a principled, public-regarding fashion rather than strategically to benefit themselves (Gibson and Nelson 2014). A related theory is that the judicial symbols, such as the robe and gavel, help activate a positive frame that predisposes audiences in favor of the court. In Gibson, Caldeira, and Spence (2003) this effect is referred to as "positivity bias," and they argue that judicial symbols function in this way by signaling the difference between courts and other less favorably perceived official decision-makers, such as Congress or agency bureaucrats.

Relatively little empirical work has been done to assess the importance of judicial symbols in affecting perceptions of the court. One recent study (Gibson, Lodge, and Woodson 2014) examines how exposure to judicial symbols affects the level of support given to the court and willingness to challenge the court's rulings. In the study, one participant group was exposed to judicial symbols—a gavel, the Supreme Court courthouse, and the justices in their robes—while the other was not, and a survey elicited information about their responses to various judicial rulings. In general, the authors found that exposure to these symbols enhanced levels of support, especially for those with relatively less prior awareness of the court.

At the margins, it may be difficult to distinguish between functional and symbolic characteristics of courts. To the extent that there can be a purely symbolic feature of the judicial role, robe-wearing and gavel-wielding seem like strong candidates. But it is possible that other characteristics of the court that are more functional in nature could similarly trigger a positivity bias. For example, perhaps the ritual of oral argument serves a similar, positivity-bias-triggering function while at the same time (at least potentially) affecting substantive outcomes.

One of the most obvious distinguishing characteristics of courts is the form of the textual outputs through which their power is exercised and expressed. Judicial opinions are quite different from other textual manifestations of lawmaking, such as the statutes adopted by legislatures or the regulations promulgated by administrative agencies. Statutes and regulations take the form of more or less clearly stated rules, whereas opinions consist of narrative explanations for a decision in a particular case. The practice of issuing judicial opinions is among the most recognizable defining features of courts, especially appellate courts. Symbolically, judicial opinions may serve a role similar to robes or gavels by signaling courts' separation from the political branches.

Whether the practice of issuing recognizably judicial opinions actually reinforces the legitimacy of the court and if so, what characteristics of opinions are responsible for that effect, are empirical questions that we do not address here. But if issuing judicial opinions helps trigger support, in part, by demarcating the court as a judicial (as opposed to political) institution, the ability to do so would be bound up with how well the court's opinions conform to public expectations of the form more generally. It is possible to think of the judicial opinion, then, as a legitimating genre. By conforming to the norms and conventions of that genre, the court marks itself as a nonpolitical institution and, in doing so, triggers positive associations in the relevant public that reinforce feelings of support, even for that portion of the public that might disagree with a specific decision. But, if the court's opinions fail to conform to the judicial genre, then their value in marking the court as a nonpolitical (and therefore more legitimate) institution may be compromised. In the following sections, we discuss our quantitative exploration of whether the court's opinions do or do not conform to the judicial genre and whether the degree of the court's genre conformity has changed over time.

ESTABLISHING A BASELINE

To investigate how well the court conforms to the judicial genre, we need to establish some baseline for comparison. Rather than attempt to generate an *a priori* account of the genre (which would doubtless be controversial), we take as a starting place a less controversial judgment about the members of the class of judicial opinions. We identify federal appellate court opinions as a baseline. Starting with this body of documents, we then examine whether the court's opinions are distinguishable based on their semantic content, without imposing a theory about what is or is not a relevant characteristic.

Our approach can be illustrated through a simple thought experiment. Imagine a hypothetical law student walking the corridors of a law library. This law student notices on the floor a few pages torn out of the previous year's Federal Reporter. The document lacks information identifying the authoring court. The student tries to guess whether the opinion was written by the Supreme Court or an appellate court. Our hypothetical law student's ability to guess correctly will be related to the distinctiveness of Supreme Court opinions.

At one extreme, if Supreme Court opinions were written in Latin while appellate court opinions were written in Greek, the classification task would be trivial. At the other, if the Supreme Court's docket were selected at random from all appellate court cases and the justices employed similar reasoning and writing styles to appellate court judges, it would be extremely difficult to improve on the prior probability estimate based purely on background frequency. If it is relatively easy to distinguish Supreme Court opinions, then by our measure they depart from the more general genre of judicial opinions, which is defined according to the baseline corpus of appellate opinions.

Considered statically, it would be very difficult to interpret findings of distinctiveness or ease of classification other than to note departure from pure chance. But a dynamic understanding allows for comparison between time periods. If our hypothetical student is better able to classify cases from 2004 than cases from 1954, it is fair to infer that the Supreme Court has grown more distinctive over time. This conclusion does not imply that the appellate courts have remained steady while the Supreme Court has veered off in uncharted territory. But we can say that the Supreme Court has become more distinctive relative to the appellate courts.

There are two basic mechanisms through which the opinions of the court may come to be systematically different from those of the appellate courts that it supervises: the *certiorari* process and the process of opinion drafting. The hierarchal structure of the judiciary generates a vast winnowing of cases and issues before they reach the pages of the US Reports. Each year, roughly one million cases are filed in federal courts. In reporting year 2012, there were 35,302 federal appeals terminated on the merits, disposing of a number of cases roughly equivalent to 10% of the nonbankruptcy filings in the federal court (Administrative Office of the United States Courts 2013). The vast majority of these appellate dispositions were not accompanied by a published opinion. From this pool, several thousands of petitions for *certiorari* were submitted, with the court granting just over one hundred. The court's control over its docket allows it substantial ability to influence its own agenda.

Given the consequences of the court's *certiorari* jurisdiction, it is not surprising that it has long been a subject of study by social scientists and academic lawyers (Tanenhaus et al. 1963; Caldeira, Wright, and Zorn 1999). Based on this prior work, Yates, Cann, and Boyea (2013, 852) conclude that "[a] wealth of judicial politics literature suggests that justices have an interest in taking on cases that are salient, resolve important legal conflicts, and, in fact, do map well onto justices' distinct ideological preferences." Because the court's docket differs in systematic ways from the general pool of appellate cases, we should expect that the opinions the court issues will be distinguishable based on the unusual characteristics of the underlying cases. If the court uses its *certiorari* jurisdiction to focus its attention on certain legal questions (such as constitutional claims or statutory interpretation) while avoiding others (such as family law issues), then its opinions will naturally reflect that emphasis in its docket. Purely through the operation of the *certiorari* process, the body of Supreme Court opinions will reflect the issues that most capture the court's attention.

The second mechanism that could lead to differences between the court's opinions and those of the appellate courts is the opinion drafting process. Once *certiorari* has been granted, a case typically proceeds through merits briefing and oral argument, followed by drafting and editing. During the drafting phase, versions of the majority opinion, and any concurrences of dissent, are circulated within the court, spurring additional deliberations, occasional vote-shifting, and redrafting and editing. All of these internal operations are governed by both formal rules and entrenched conventions. Most distinctly from the lower appellate courts, the court always sits as a whole rather than in panels, so that opinions serve as part of a running conversation among the group that has the potential to create a unique culture, especially during a period when the court's membership is relatively stable.

There are many ways that the drafting process could lead to the court producing opinions that are distinct from the appellate courts, even holding the underlying cases constant. The court has considerable leeway to decide which of the legal questions presented in those cases to explore or emphasize. When the court grants *certiorari*, it frequently limits review to specific questions.

Furthermore, justices have considerable discretion when drafting majority opinions and even more when authoring dissents or concurrences (when they are less constrained by each other). During this process the justices face different incentives than lower court judges because their decisions cannot be appealed and will serve as the final word on the legal questions that they decide. Justices may, accordingly, be freer in their language or view themselves as addressing a broader public or posterity rather than a reviewing court. The justices may also make different choices in the language that they use, perhaps deploying certain rhetorical moves, such as personal anecdotes, metaphor, humor, or colloquialisms, that are less common in the lower courts. Given the unique processes employed by the court, the distinctive nature of the court's role and the audience that it addresses, and the peculiar nature of the justices engaging in bargaining, drafting, and editing, it would not be surprising if the types of reasoning or the language that is used in the court's opinions differ from those that are used in appellate court opinions, even when the set of legal issues is the same.

A TOPIC MODEL APPROACH

It is theoretically possible to carry out the thought experiment discussed above in real life by presenting students with randomly generated snippets of text and asking them to classify the documents as issuing from either an appellate court or the Supreme Court. But such an exercise would pose substantial technical and logistical challenges. To avoid these problems, we employ a principled application of statistical computational textual content analysis called topic model analysis. Given a textual corpus, a topic model produces topics that in the technical topic modeling sense are probability distributions over a vocabulary, where each word in the vocabulary is assigned a non-negative weight (such that all weights sum to one). Each document is in turn summarized as a probability distribution over the topics, creating a compact but descriptive representation of the semantic content of documents. The data generated by the topic model substantially reduce the number of dimensions needed to characterize the content of documents, allowing us to engage in useful statistical analysis.

The highest-weighted words within a topic provide a sense of the subject matter that the distribution represents. For example, in topic 3 generated by our model, the words "election," "political," "party," and "candidates" are weighted highly, which led us to hand label that topic as "elections." (Topics are generally hand labeled.) Thus the representation of a given document as a distribution over topics summarizes the document as weighted mixtures of intuitively understood themes. These distributions—both of the topics and the words they comprise—are produced as the best fit to an underlying generative probabilistic model for the observed simple word frequencies.

The canonical topic model is a latent Dirichlet allocation (LDA) mixed-membership model. The LDA model posits some number of topics (distributions over the vocabulary) that account for all words observed in a corpus according to the following generative story: For each document in the corpus, a set of topic proportions (or "shares") is drawn from a global probability distribution; then, each word in the document is drawn from a topic distribution in which the topic distribution in question is selected according to the previously mentioned document-specific set of proportions. Topic models are often fit using an iterative algorithm (known as a variational approximation) or by using a Markov chain Monte Carlo approach. In the case of the topic model, the parameters of interest are typically restricted to the topic-word distributions describing the association between topics and words, and the document-topic distributions that describe, for each document, the probability of finding words associated with each topic. See Blei (2012) for a general overview and formal description of an LDA topic model.

While numerous incremental improvements to LDA topic modeling have emerged in the intervening years, the essence of the original model remains, and the LDA topic model persists as a general industry standard for text analysis, serving, with minor variations, as a building block in more elaborate models of text data.[8] More than a decade after the model's introduction, researchers using topic models and closely related models may be found in almost every field where machine-readable text data are abundant. Topic models are now a familiar part of the methodological landscape in the human and social sciences, from political science to German studies (Quinn et al. 2010; Riddell 2014).

The data for our analysis are drawn from the private not-for-profit corporation Public.Resource.Org, which has created a digital version of the Supreme Court and federal appellate court corpus based on the noncopyrightable information within the Westlaw database. CourtListener, an effort within the Free Law Project, has augmented the information contained in the bulk resource data and created a user-friendly interface that is accessible to the public. We relied on CourtListener as the source for all the texts for the Supreme Court and appellate court decisions. The set of Supreme Court documents used in this study includes the opinions associated with all formally

---

[8]For our analysis, we use a nonparametric topic model based on the Pitman-Yor process in place of the traditional Dirichlet distributions. The "hca" software we use is authored by Wray L. Buntine and is open source. See "hca 0.61," Machine Learning Open Source Software, http://mloss.org/software/view/527/ (accessed 20 September 2015).

decided full opinion cases. There are 7,528 documents in this set. The set of appellate court documents used consists of all published opinions issued between 1951 and 2007, for a total of 289,550 documents. To reduce the computational burden of fitting the topic model, we randomly selected twenty-five thousand documents from within the appellate court set. In addition to the twenty-five thousand randomly selected appellate court opinion documents, 4,180 appellate court documents that are associated with cases selected for review by the Supreme Court decisions are also included. In total there are 29,180 appellate court opinion documents.

To identify the set of cases that were selected for review by the court, we gathered information from Lexis/Nexis, which provides "prior history" and "disposition" fields for Supreme Court decisions. The vocabulary associated with the corpus comprises those words occurring at least twenty times in the entire Supreme Court corpus. There are 21,695 total words in the vocabulary. For purposes of the current analysis, the number of topics is not central to our inquiry so we select one hundred topics, which is large enough to capture a great deal of the semantic variability of the corpus and small enough to make fitting the topic model computationally straightforward.

To generate the topics, all documents (i.e., both appellate court opinions and Supreme Court opinions) are treated as a single corpus and subjected to the topic model. The top words for the first ten topics generated (the order in which topics appear is not meaningful) are presented in table 5.1 (the labeling was done by the authors).

Our analysis then breaks out three sets of documents: all opinions published in the Federal Reporters during the study period; the subset of those opinions associated with cases that were selected by the Supreme Court for review; and the

Table 5.1.

Labels	Top words
Labor	union board labor employees employer NLRB company bargaining relations national local act unfair
Family	ms mrs did told husband time testified asked sexual home stated fact received mother daughter
Elections	election political party candidates candidate campaign parties primary elections contributions ballot
Narcotics	united drug cocaine government cir defendant conspiracy evidence drugs marijuana possession
Immunity	immunity officers officer official police law county qualified officials city conduct rights liability
Prisons	prison inmates inmate prisoner prisoners officials confinement conditions security jail amendment
Procedure	motion district judgment appeal rule order filed summary party appeals judge final fed notice rules
Medical	dr medical hospital mental treatment health care patient drug expert patients physician condition
Criminal	trial defendant plea guilty indictment united jeopardy criminal double prosecution government
Insurance	insurance policy company insured coverage insurer life ins policies liability loss judgment mutual

subset of those opinions that were published in the US Report (i.e., Supreme Court opinions). We build on the motivating thought experiment introduced above through a machine learning algorithm that mirrors the prediction task given to the hypothetical law student. Using only the information contained in the topic distributions, the goal of the algorithm is to predict whether a randomly selected opinion has been drafted by the court. We ask both whether prediction is possible and whether the corpora of Supreme Court opinions has been growing more distinctive over time. We generate a single metric of distinctiveness based on the predictive accuracy of a logistic classifier algorithm.

This estimate of predictive accuracy has an intuitive interpretation as a measure of distinctiveness. To generate this metric, we begin by limiting the corpus of appellate court and Supreme Court opinions to a single year. Because the number of decisions in each year varies considerably—there are far more appellate court opinions in 2000 than in 1960— we randomly sample year-specific corpora of equal sizes. We then hold out 50% of the appellate court and Supreme Court opinions and train a basic logistic regression classification model using the remaining opinions. The only information that the classification model uses is the topic proportions in the opinions. Once the classification model has been fit, we evaluate it for accuracy on the held-out 50%. This task is repeated many times, each time randomly sampling the 50% of cases that are held out. From this procedure, we construct a distribution of predictive accuracy for the classifier for that year. We repeat these same steps for each year in our sample, providing a means of evaluating whether the distributions change over time.

There is some risk that a naïve classifier will become quite good at prediction based on relatively insignificant differences, for example, in the usage of a few characteristic words, such as the justices' names or the courthouse address. A high degree of predictive accuracy, if based on these small differences, would not necessarily imply substantial and meaningful distinctiveness. Use of only the topic model proportions as the basis of prediction reduces this risk. There is a substantial level of aggregation involved in moving from all words to one hundred topics, and this aggregation lowers the risk that trivial differences will substantially affect the success of the classifier.

The loss of information associated with topic modeling helps reduce the risk of accentuating minor differences. If we find that the model has an easier time predicting whether opinions are authored by the court based on topic model proportions alone, we can say with a reasonable degree of confidence that the two corpora are growing more distinct from each other in a meaningful way.

The Idiosyncratic Court

Figure 5.2 displays the results of our analysis. We confirm that the logistic regression classifier performs reasonably well in predicting the difference between appellate court and Supreme Court opinions, and we find that prediction is improving considerably over time.

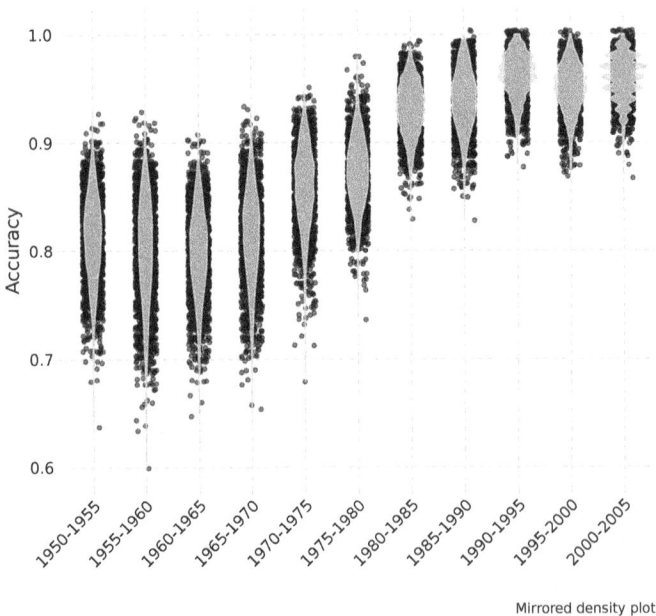

Figure 5.2. Prediction of Supreme Court opinions (mirrored density plot).

The center of the distribution of the accuracy of held-out prediction starts at roughly 80% in the 1950s but over time increases to over 95% in the 2000s. This is a highly significant result.[9] By the end of the study period, a quite simple classifier, using only topic proportions, achieved nearly perfect prediction. These results quite clearly indicate that Supreme Court opinions are growing more distinctive compared to those in the appellate courts. From this analysis, we know that the mix of topics present in each corpora in each year provides increasing information about the identity of the authoring court in the sense that the classifier improves over time.

As discussed above, both case selection and opinion drafting could result in Supreme Court opinions that are distinct from the general-pool appellate court opinions. From prior work we know that there is an important winnowing effect during the *certiorari* process, and the cases that come before the court are far from randomly drawn. At least part of the reason that the court's opinions are distinctive is that they are based on a nonrepresentative set of cases. It is worth considering whether the contribution of the *certiorari* process to the distinctiveness of the court's opinions is growing, declining, or remaining relatively flat. The converse question is whether the opinion-drafting process has changed over time such that, holding the underlying cases constant, the court's discussion of those cases has become increasingly distinctive over time.

To investigate these two questions, we carry out the same logistic regression classifier analysis on three different corpora: the set of all appellate court opinions, the set of appellate court opinions associated with cases selected for review, and the court's opinions. We then carry out two sets of analyses using the cases selected for review set as an intermediary corpus.

---

[9] Pearson product–moment correlation between year and accuracy is 0.79.

We first examine whether the court is using its *certiorari* power more aggressively than in the past in the sense of selecting cases that are more distinct from the pool of all appellate court cases. If so, we should find that the performance of the classifier would increase over time. We then examine the opinion-drafting process by analyzing whether the court's opinions are growing more distinctive *vis-á-vis* the intermediary corpus of appellate opinions associated with cases selected for review. In essence, this analysis holds the underlying legal issues constant to determine whether the court is discussing those issues in a more distinctive fashion. The results of these two analyses are reported in figure 5.3.

We do not find any evidence that there is any change over time in the representativeness of the group of cases being selected for review. Although this analysis cannot rule out the possibility that a more sensitive textual analysis would identify some temporal change, we fail to find any such effect using the same model that identifies an overall growth in the distinctiveness of the court's opinions. We can therefore say with confidence that the increasing distinctiveness that we identify is not caused by a change in the level of representativeness in the cases selected for review.

Since the court's opinions are growing more distinctive, yet the underlying cases selected for review are not, the natural inference is that the court's opinions must be becoming more distinct from the appellate court cases selected for review. We confirm this conclusion, finding that when comparing Supreme Court opinions and the intermediary corpus of appellate opinions associated with cases selected for review, the performance of the classifier improves over time. Starting with accuracy centered at roughly 75% in the 1950s, performance increased to well over 90% by the 2000s. The lesson from this analysis is that although the

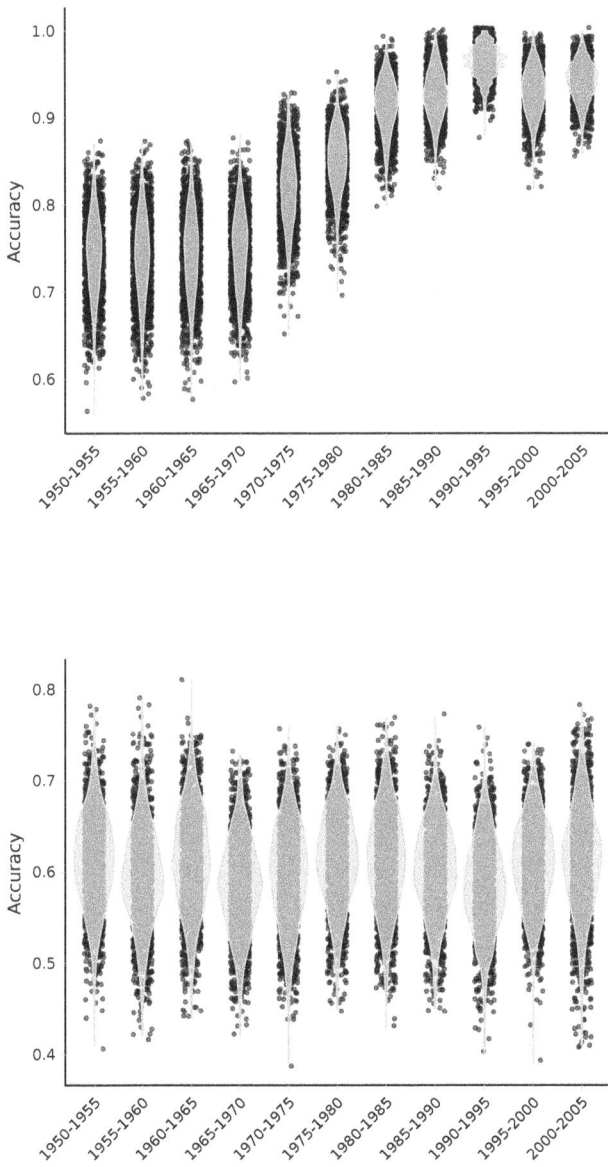

Figure 5.3. Prediction of Supreme Court and appellate court opinions (mirrored density plots).

cases selected for review in recent years are no more distinct from the pool of all appellate court cases than in the past, the way the Supreme Court analyzes and discusses the legal issues presented in those cases has grown increasingly idiosyncratic over time.

This finding is quite striking and indicates that, at least according to the measure developed and discussed above, the court's opinions conform less well to the genre of judicial opinions than in the past, and this change is due to the opinion-drafting process in the court. Opinions written by the Supreme Court are more characteristic and easily identifiable than in the past; they are, on their face, less-obviously associated with the opinions drafted by the appellate courts.

Concluding Thoughts

In this chapter we explore how several computational text analysis tools can be deployed to better understand the US Supreme Court. Both the stylistic analysis based on sentiment analysis and function words and the substantive analysis based on topic models hold substantial potential to contribute to further empirical study of the law. In the past, quantitative analysis of law has traditionally been hampered by the lack of attractive mechanisms for estimating case characteristics or the legal features of opinions. Style analysis and topic modeling provide promising avenues to estimate difficult-to-capture variables related to the legal content of opinions and judicial temperament. These tools also avoid some of the pitfalls of human readers, including error, bias, and, most important, time and attentional limits. By naïvely characterizing the relevant features of judicial opinions, topic models and stylistic analyses provide a quantitative and computationally tractable method to represent the text of the law. The corpus of the law—the published case law in the state and federal reporters, and other legal texts as well—is an enormous and rich dataset, and computational tools provide an effective means of capturing important characteristics of that data that can be subjected to analysis. With researchers continually introducing new tools and refining existing approaches, there is an ever-expanding frontier in empirical legal scholarship that has substantial potential to improve understanding of the law. ❧

Acknowledgments

This chapter draws from Carlson, Livermore, and Rockmore (2016) and Livermore, Riddell, and Rockmore (2017). Our thanks to the editors at the *Washington University Law Review* and *Arizona Law Review* for their excellent editorial support on those articles. We also thank Quinn Curtis, Michael Gilbert, Andrew Hayashi, Richard Hynes, John Setear, Jed Stiglitz, David Zaring, and participants of the 2016 Conference for Empirical Legal Studies in Europe for helpful comments.

PREDICTING LEGISLATIVE FLOOR ACTION

Vlad Eidelman, FiscalNote
Anastassia Kornilova, FiscalNote
Daniel Argyle, FiscalNote

Federal institutions in the United States, such as Congress and the Supreme Court, play a significant role in lawmaking and, in many observable ways, define our legal system. Thus, legal scholarship has been largely focused on understanding these entities and the role they play in our society. As federal legislative and regulatory data have become more readily available, political scientists and legal scholars have become increasingly quantitative, adopting objective data-driven methods for characterizing political and legal behavior and outcomes. Computationally driven analysis has extended into all areas of law, including analyzing the behavior of Supreme Court justices (Katz, Bommarito, and Blackman 2017; Lauderdale and Clark 2014), congressional legislators (Poole and Rosenthal 2007; Slapin and Proksch 2008), and administrative agencies (Livermore, Eidelman, and Grom 2018; Kirilenko, Mankad, and Michailidis 2014). The aim of most of this research is to move away from purely subjective analysis that is limited in its ability to quantitatively measure and empirically explain observable legal phenomena.

Although many issue areas are regulated primarily at the federal level, state governments also wield significant power, and an increasing number of issues are now being decided at the state or local levels, including emerging industries and technologies such as the gig economy and autonomous vehicles (Hedge 1998). In fact, the total quantity of state legislative activity dwarfs that of Congress. There are 535 members of Congress who introduce over ten thousand pieces of legislation a session,[1] of which less than 5% is enacted. In contrast, in the aggregate there are over seven thousand state legislators introducing over one hundred thousand pieces of legislation, with over 30% being enacted.

All US state legislatures work according to a committee system in which, for a bill to be enacted, it must pass through one or more legislative committees and then be considered on the chamber floor (which we refer to as "floor action"). The final step is pivotal (Rosenthal 1974; Hamm 1980; Francis 1989; Rakoff and Sarner 1975), and reaching it is not a given: on average only 41% of bills receive floor action, with most legislation languishing in committees.[2]

Legislative policymaking decisions are extremely complex and are influenced by a myriad of factors. These include the content of the legislation; legislators' personal characteristics, such as profession, religion, and party and ideological affiliations; constituent demographics; governor agendas; interest group activities; and even world events (Canfield-Davis et al. 2010; Hicks and Smith 2009; Talbert and Potoski 2002). While there has been substantial scholarship to understand the possible influences on legislator behavior (Canfield-Davis et al. 2010), it would likely be impossible to obtain data on or even create an exhaustive list of all possible influences. But with the advent of large-scale digitization of much of the public record, including proposed legislation, as well as the development of increasingly sophisticated computational tools (some of which

---

[1] A session is the period of time a legislative body is actively enacting legislation, usually one to two years.

[2] By comparison, 13.3% of bills receive floor action at the congressional level.

are specific to text), it is nonetheless possible to construct accurate predictive models, even if the true causal factors at play are unknown or unobserved.

In this chapter we explore how machine learning and natural language processing tools can be used to better understand state lawmaking dynamics and the state legislative process. We apply these tools to the problem of predicting the likelihood that legislation will reach the floor in each state across all fifty states and the District of Columbia. As there are many dimensions underlying the content of the legislation, such as the policy area and ideology of the sponsor (Linder et al. 2018), that may affect the likelihood of floor action, we focus on two sets of features: (1) contextual legislature and legislator-derived features, and (2) legislative text. Specifically, we follow previous literature and derive several established contextual features. These features describe the legislative environment or identity of the legislator, such as committee membership, party affiliation, or controlling party, and can be easily extracted from publicly available data. We take a computational text analysis approach to legislative texts, and instead of relying on in-depth manual analysis, process the text automatically using natural language processing technologies to identify salient content. Using several machine learning algorithms, we build predictive models in each state from different subsets of the features, quantitatively modeling the floor action process across all fifty states. We find that despite the complexity and diversity of legislative processes at the state level, we can fairly accurately predict legislative success and achieve large improvements in accuracy over naïve baselines.

Congressional Related Work

Much of the research on analyzing the federal legislature is aimed at understanding legislator preferences through the use of voting patterns. One of the most popular techniques in political science is the application of spatial, or ideal point, models built from voting records (Poole and Rosenthal 1985, 2007). These are often used to represent unidimensional or multidimensional spaces that represent the ideological stances of the relevant actors (Clinton, Jackman, and Rivers 2004).

As most of this literature is aimed at building descriptive models with explanatory rather than predictive capacity, it presents a few shortcomings. The first shortcoming is that most of these models have been limited to in-sample analysis. In other words, the model is only applicable to the data that were used to construct it. For example, the ideal point models mentioned above can infer ideology from past votes and predict a legislator's vote for an in-sample bill. But they are incapable of making out-of-sample predictions for new bills because the information used to predict the vote of any given legislator is the votes of his or her peers—information that does not exist for a newly arrived piece of legislation.

The second shortcoming (or missed opportunity) is that previous work mostly ignores a fundamental aspect of lawmaking, namely, the text of the laws themselves. So while these models can assess legislator preference, there is no indication of what that preference is based on. Unstructured textual artifacts, such as legislation, floor debates, and committee transcripts, are a much richer representation of the lawmaking process and the law than are structured artifacts such as observable votes. By including these data in our models we hope to achieve a better quantitative understanding of the broader dynamics of legislatures.

In recent years a wide variety of primary and secondary legal data, both structured (e.g., votes) and unstructured (e.g., text), has become increasingly available. Coupled with advances in natural language processing and machine learning, these data have enabled the construction of richer statistical models for multidimensional ideal point estimation. For example, researchers in the computer science community have turned their attention to congressional roll call prediction and created various novel models that account for both the text of the legislation and the voting records to predict out-of-sample votes: Gerrish and Blei (2011) use topic models to construct multidimensional ideal point models, Nguyen et al. (2015) use both the legislative text and floor debates to construct a hierarchical topic model, and Kornilova, Argyle, and Eidelman (2018) use a neural network to create continuous embedding vector representations of both bills and legislators.

While the focus of the above is still on roll call prediction through improvements upon multidimensional ideal point estimation, other work has started to emerge with different, but related, aims. Yano, Smith, and Wilkerson (2012) introduced the problem of congressional bill survival, where the task is to predict whether a congressional bill will be reported out of the committee to which it was assigned. They utilize a logistic regression model with several different feature sets, combining basic contextual binary features—sponsor and committee— with an *n*-gram representation of the initial bill text version, showing that the combination leads to the best performing predictive model. Nay (2017) focused on the overall passage problem, predicting whether a congressional bill would be enacted into law. They take a different approach to modeling the text, using Word2Vec (Mikolov, Sutskever, et al. 2013) to create word embeddings, which are combined to construct a continuous vector space representation of the bills. This representation is used for one predictive model and stacked in an ensemble with gradient boosting and random forest classifiers trained on a set of contextual features. They also show that the best model combines text and contextual legislature and legislator-derived features.

Moving beyond primary legislative texts, floor-debate transcripts have been used in a number of applications. Thomas, Pang, and Lee (2006) use transcripts to predict voting. They frame the task as sentiment analysis, employing an SVM classifier trained using the text of the transcripts. In addition to the text, they compute agreement links between different legislator's transcripts, showing that both contribute to vote prediction. Iyyer et al. (2014) use transcripts to build a recursive neural network to detect ideology bias.

There is also established literature examining broader legislative dynamics, such as measuring legislative effectiveness (Harbridge 2016), evaluating the impact of legislation on stock prices using legislator's constituents (Cohen, Diether, and Malloy 2012), creating cosponsorship networks (Fowler 2006), and examining the role of lobbying (Bertrand et al. 2018; Hill et al. 2013).

State-Related Work

As noted above, while Congress has received much of the attention, many important issues are decided at the state or local level. There are many areas where, traditionally, the state enacts regulation, including land use, insurance, housing, and many others. In addition, as Congress has grown more polarized and less able to act (Hedge 1998), states have seen increased opportunities to take more of a leading role in regulating emerging industries and technologies, for example, the gig economy and autonomous vehicles.

While there is an increasing amount of state legislative research, state legislation continues to receive significantly less scholarly attention than congressional legislation (Hamm, Hedlund, and Miller 2014). One major reason for this is that quantitative methods require data, and the availability of data for Congress far exceeds that of state legislatures. Yano, Smith, and Wilkerson (2012) describe the "challenge" of moving beyond the relatively well-studied and data-laden US Congress to "a larger goal of understanding legislative behavior across many legislative bodies (e.g., states in the US, other nations, or international bodies)" as severely hampered by the difficulty "of creating and maintaining such reliable, clean, and complete databases."

A second reason is that modeling and comparative analysis across fifty separate localities poses unique challenges. For instance, spatial models require individuals to have expressed a preference on the same item and thus are not applicable across instances where different sets of actors are expressing preferences on different items (i.e., they would require all state legislators to vote on the same bills). Thus, while there has been scholarship quantifying voting, the role of committees, and other legislative processes, it has been limited in scope to a few sessions or states or is reliant on survey data (Francis 1989; Rakoff and Sarner 1975; Rosenthal 1974; Hamm 1980). For example, Hamm (1980) constructed a multivariate model using intra- and extra-legislative factors for determining the important variables affecting committee survival in Wisconsin and Texas over three sessions. Rakoff and Sarner (1975) performed a similar analysis for three sessions in New York, while Francis (1989) uses survey responses from over two thousand legislators across fifty states to compare committee performance.[3]

More recently, as different kinds of state data have become more accessible, it has enabled larger comparison studies. Squire (2007) studies the effects of professionalism on legislatures, Gray and Lowery (1995) examine the influence of interest groups on legislative activity, Shor, Berry, and McCarty (2010) and Shor and McCarty (2011) apply spatial models to state legislators, and Linder et al. (2018) compare the textual similarity of legislative language. This chapter builds on this prior work to study state legislative dynamics by evaluating the predictability of state lawmaking and the factors that most enhance the performance of predictive models. We create a novel task—predicting the likelihood of legislation to receive floor action—and utilize a corpus of over one million bills to build computational models of all fifty states and the District of Columbia. We present several baseline models utilizing various features and show that combining the legislative and legislator contextual information with the text of bills consistently provides the best predictions. In addition, our analysis generates across-state comparisons of the predictive models and finds that, although there are some consistent patterns, there are many variations and differences in what best predicts the likelihood of floor action in each state.

Data

There is state-to-state variation in the legislative procedure of how a bill becomes law, but the path is largely similar. Legislation is introduced by one or more members of the legislature in their

---

[3]Similarly, Shor and McCarty (2011) rely on surveys to derive a scaling for all legislators for ideal point modeling.

respective chambers[4] and assigned to one or more standing subject committees. Committees are made up of a subset of members of their respective chambers and are chaired by the majority party. Once in committee, legislation is subject to debate and amendment only by the committee members, with the successful outcome being a favorable referral, or a recommendation, to be considered by the full chamber on the floor. Depending on the state, other groups, including legislative committees, legislative delegations, the governor, or non-elected individuals, can introduce legislation. For the purpose of this work we focus on legislator-sponsored legislation.

The primary data we use to model floor action were scraped directly from each state legislature's website. For each state, we downloaded legislation, committee, and legislator pages for all sessions that were publicly accessible. Legislation pages were automatically parsed to determine legislative contextual metadata, which includes text versions of the bill, sponsors, committee assignments, and the timeline of actions. Legislator pages were parsed to obtain sponsor contextual metadata, which includes party affiliation, committee assignments, and committee roles.

States demarcate legislative status in the timeline of actions differently, so we automatically map and normalize all textual descriptions of legislative actions to a finite set of statuses.[5] These statuses are used to determine whether a piece of legislation survived committee and received consideration on the floor. All bills with a status of either having passed in their introductory chamber or having had a recorded floor vote are treated as positive examples, while any status prior to floor action is considered failed,

---

[4]All legislatures are bicameral, with either a House or Assembly as the lower chamber and the Senate as the upper chamber, except the District of Columbia and Nebraska, which are unicameral.

[5]The normalized statuses include "introduced," "assigned to committee," "reported from committee," and "passed."

including legislation that was reported out of committee but not considered on the floor.

Finally, since each state follows their own conventions with regard to classifying the type of legislation, we normalize all legislation across states to two types, resolutions and bills, using the state-assigned classification to group legislation. Legislation described by its state as an "appointment," "resolution," "joint resolution," "concurrent resolution," "joint memorial," "memorial," "proclamation," or "nomination" is mapped to resolution, while legislation classified as a "bill," "amendment," "urgency," "tax levy," "appropriation," or "constitutional amendment" is mapped to bill.

Figure 6.1 shows the total number of bills introduced and receiving floor action for each state. In total, our dataset consists of 1.3 million pieces of state legislation, broken into one million bills, with 360,000 receiving floor action at an average rate across states of 41%, and 275,000 resolutions with 210,000 receiving floor action. On average, we have ten legislative sessions of data per state. Bills represent substantive legislation with a much lower floor action rate, while resolutions are much more likely to receive floor action; thus, for the rest of this chapter we focus on bills only and refer to bills and legislation interchangeably. We include fifteen sessions of US federal legislation in our data for comparative purposes, with 23,000 of 172,000 bills receiving floor action.

Figure 6.2 presents the percent of bills receiving floor action. It is interesting to note the difference in difficulty for legislation to receive floor action in different states. For example, in New Jersey and Massachusetts, fewer than 15% of bills reach the floor, whereas 75% reach the floor in Colorado and Arkansas.[6]

---

[6]Our average across states, chambers, and sessions is in line with previous single state and session findings; in examining five states Rosenthal (1974) found between 34% and 73% of legislation did not survive committee.

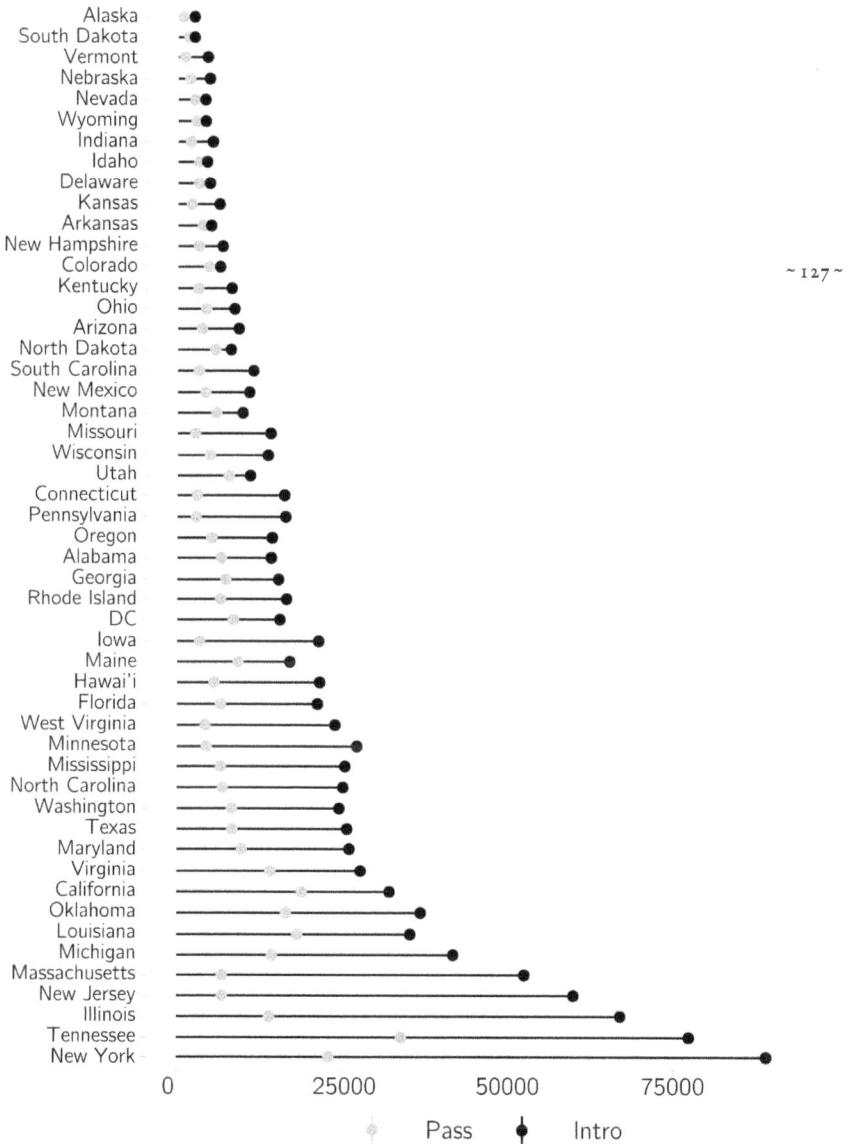

Figure 6.1. Number of bills introduced and receiving floor action for each state.

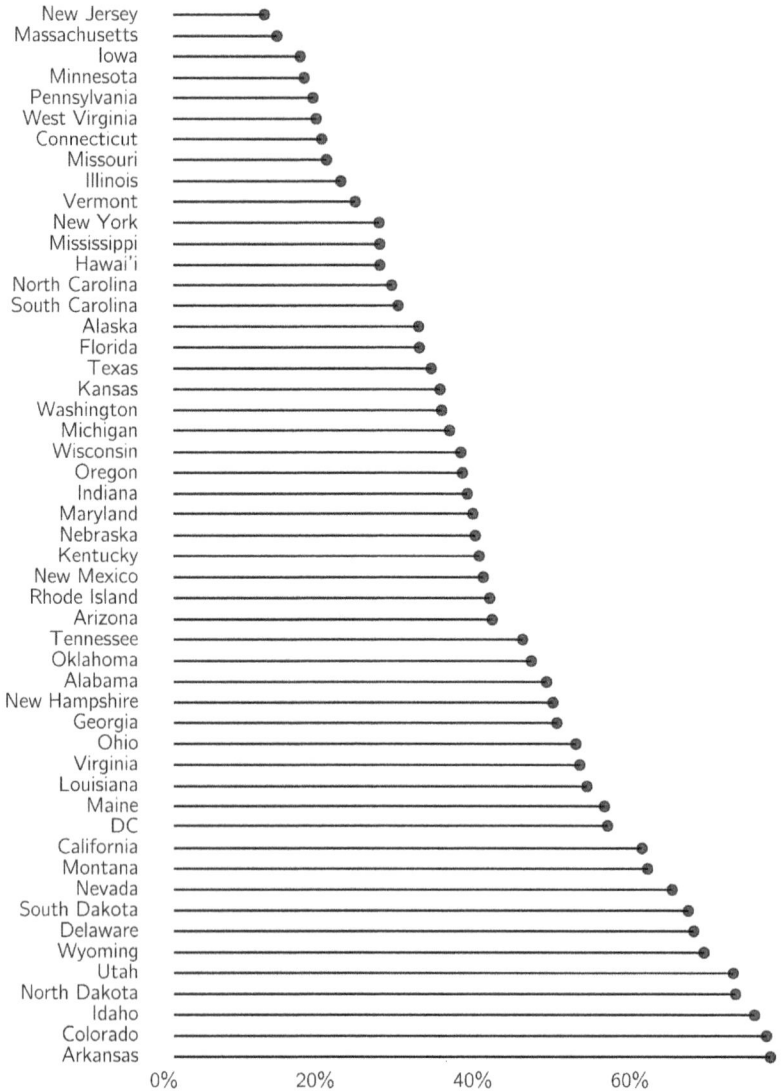

Figure 6.2. Percent of bills reaching the floor per state.

Methods

In order to be able not only to predict but also examine the importance of features to our prediction, we chose three relatively interpretable models for our modeling framework. Formally, let our training data (X, Y) consist of n pairs $(x_i, y_i)_{i=1}^n$, where each x_i is a bill and each y_i a binary indicator of whether x_i received floor action. Let $F(x_i)$ be a feature vector representation of x_i and w the parameter vector indicating the weight of each feature learned by the model. The first two models are linear classifiers—a regularized log-linear model and NBSVM (Wang and Manning 2012)—and the third is nonlinear— a tree-based gradient boosted machine (Friedman 2000). We use the `scikit-learn` (Pedregosa et al. 2011) implementation for the log-linear and gradient-boosted models and implemented NBSVM based on the interpolated version in Wang and Manning (2012).

Hyperparameters in the machine learning algorithms, such as learning rate and regularization strength, have a significant impact on model performance. We use a Bayesian hyperparameter optimization (Bergstra et al. 2011) to select the optimal hyperparameters for each model on a held-out development set. We used the tree-structured Parzen estimator (TPE) algorithm implemented in `hyperopt` for our sequential model-based optimization (Bergstra, Yamins, and Cox 2013). After individually optimizing hyperparameters and training each of the three base models, we use their outputs to train a meta-ensemble model—a regularized conditional log-linear model—forming a linear combination over the three base model predictions (Breiman 1996).

As the lawmaking process in each state, and even within each chamber, is different, we divide the problem space by state and chamber, building separate models for each subset.

Specifically, we consider each of these as separate problems: upper chamber bills and lower chamber bills. We have two predictions per state, and each prediction is comprised of four model outputs, three from the base models and one from the meta-ensemble, resulting in four hundred models.[7]

FEATURES

As there are many dimensions underlying bills that may affect the likelihood of floor action, we compute and utilize several established contextual legislature and legislator-derived features. Previous literature has proposed various factors that may affect legislation, including the content of bills,[8] number of and identity of sponsors, extra-legislative forms of support, timing of introduction, leadership's position, seniority, identity of chairperson of the committee, identity of one's own party, and membership of the dominant faction (Hamm 1980; Rakoff and Sarner 1975; Harbridge 2016; Yano, Smith, and Wilkerson 2012).

In order to quantitatively evaluate these factors and establish a strong baseline from which to measure the predictive value of text, we include the contextual features shown in table 6.1. These indicator features derived from the sponsors, committees, and bills are meant to capture many of the major factors that are proposed in the literature.[9]

---

[7] There is only upper chamber legislation in the District of Columbia and Nebraska, so we have 49 states x 2 predictions + 2 states x 1 prediction = 400 models.

[8] In most previous literature the content was determined via a manual analysis of each bill to establish the scope of impact, the complexity, or the incremental nature.

[9] Each count-based feature, such as number of sponsors, also spawns a number of discretized features, including ranks, percentiles, and deviations from the mean thereof. We automatically compute companion bills using a cosine-based lexical similarity.

Table 6.1. Contextual feature types and descriptions.

Feature Type	Description
Sponsor	primary and cosponsor(s) identity, primary and cosponsor(s) party affiliation, number of primary and sponsors, number of Republicans, number of Democrats, sponsors bicameral, sponsors bipartisan, sponsor in majority/minority, majority party Republican or Democrat
Committee	identity of assigned committee(s), number of committee assignments, number of sponsors members of the committee, sponsor same party as committee chairman, sponsor role on the committee, referral rate of committee(s)
Bill	chamber, bill type, session, introductory date, companion bill(s) existence, companion(s) current status

To strengthen the representation of legislators in our model beyond the basic features described above, we compute several measures of legislator "effectiveness." The effectiveness score is calculated from the sponsoring and cosponsoring activity of each legislator and is meant to represent where they stand in relation to other legislators in successfully passing legislation. This notion of legislator effectiveness is useful for our model but is quite limited. Legislators engage in a variety of tasks, and the overall efficacy of legislators is judged based on the idiosyncratic utility functions of the legislators themselves as well as their constituents. Stopping legislation, engaging in constituent services, performing oversight functions, or making symbolic statements on issues of public concern are all activities that might be judged as contributing to legislator effectiveness in a general sense but are not included in our model. However, given the goal of our model (predicting the survival of legislation to floor action), these other activities are either less important or unobserved, and so our admittedly limited notion of legislator effectiveness is sufficient for our purposes.

Similar to Harbridge (2016), the score we compute for each legislator is a combination of several partial scores, computed for each important stage of the legislative process. Each legislator gets a score for how many bills they sponsored and the performance of those bills as measured by their survival through committee, floor action, passing their chamber, passing the legislature, and being enacted. The score for each stage is further broken down by how many of those pieces of legislation were substantive, i.e., bills attempting a meaningful legal change, versus nonsubstantive (i.e., resolutions). This results in twelve factors for each individual. To compute a score for each legislator's relative performance as compared to the other members in the chamber, we create a weighted combination of that legislator's bills and resolutions, where bills receive more weight, and compute the ratio based on the weighted contribution of the other members in the chamber. All the stage scores are then combined into a second weighted combination, where each successive stage in the process receives more weight, to generate the final score. Finally, the scores are normalized to the range zero to ten. In addition to using the effectiveness scores directly as features, we further compute and discretize several statistics derived from them, including ranks, percentiles, and deviations from the mean thereof.

To further enrich the bill representation beyond contextual information, we utilize the textual content of the bills. The predominant source of text examined by the NLP community comes from relatively short and topically coherent documents, such as news articles, social media posts, and other online resources. Oftentimes the level of analysis will focus on individual sentences or paragraphs as opposed to documents, limiting the amount of text that needs to be processed.

Legal text differs from this in a number of ways that make it more challenging. Primarily, the legislation in our collection is comprised of long documents, with an average of eleven thousand words, often containing significant amounts of procedural language and pieces of extant statutes. As the length of the document can create challenges for a computational approach to be able to identify the salient points, typical NLP approaches reduce the size of the text being analyzed. For this work we chose to focus on a condensed amount of text, specifically, the state-provided title and description, that averaged seventeen and eighteen words, respectively. Although this is a coarse approximation of the bill content, removing most of the nuances of the scope and impact of the law that a qualitative analysis would focus on, it nevertheless captures some of the high-level substantive aspects of the bill. Both title and description are preprocessed by lowercasing and stemming. We treat each field as a bag-of-words and compute the tf-idf weighting (Jurafsky and Martin 2000) for n-grams of size $n=1,2,3$ on the training data for each prediction task, and select the top ten thousand n-grams from the title and description separately.

While we would like to study the predictability of reaching the floor upon first introduction, bills often change after introduction and are updated with additional information. Thus, we limit our features to those available at the time of first introduction.

Results

In order to clarify the impact that each set of features has on predictive performance, we create five different subsets of features described in table 6.2 and train models on each one of them separately.

The first condition, `combined`, contains all the contextual and text content features. The second condition, `no_txt`, removes text content, allowing us to study the importance of all contextual

Table 6.2. The five feature settings with contextual and lexical features.

Condition	Feature Set
combined	sponsor, committee, bill, text
no_txt	sponsor, committee, bill
no_txt_spon	committee, bill
just_txt	text
just_spon	sponsor

features. By comparing combined to no_txt we can evaluate whether the text has any information that is complementary to the contextual features. The third condition, no_txt_spon, further removes sponsor features, essentially allowing us to study the importance of committee information. By comparing no_txt to no_txt_spon we can evaluate what sponsors contribute. The fourth and fifth conditions use only sponsor and only text features, respectively, to study the importance of each individually.

All models for a given condition are built from the same training data and feature space. We measure and report several performance metrics of our models using tenfold cross validation. The baseline model is based on the majority class: for states in which the passage rate is greater than 50%, the model predicts that all bills will receive floor action; for other states, the model predicts that all bills will fail to receive floor action.

Although accuracy is informative with respect to how many correct binary decisions the model made, as noted in Bradley (1997), for imbalanced problems such as this where one class dominates, the baseline accuracy can be very high. As a supplement, it is useful to measure a probabilistic loss, where there is a cost associated with how aligned the model probability was with the correct class. Thus, we move beyond pure predictive performance and consider the actual probability distributions created by our models under different conditions. The log-linear and gradient-boosted models are probabilistic, while NBSVM is not; thus, we train a probability transformation on top of NBSVM using Platts scaling (Niculescu-Mizil and Caruana 2005) to obtain probability estimates. In addition to accuracy, we measure model performance on log-loss and area under the receiver operating characteristic curve (AUROC) (Bradley 1997).[10]

~135~

By considering the true positive (TP) and false positive (FP) rates at different values, we can construct a distribution, known as the receiver operating characteristic (ROC) curve, to visualize the model's performance at various TP and FP thresholds. The area under that curve (the AUROC) can be interpreted as the probability that the model will rank a uniformly selected positive instance (floor action) higher than a uniformly selected negative instance (failure). A random model will have an AUROC of 0.5 and a 45-degree diagonal curve, while a perfect model will have an AUROC of 1 and be vertical, then horizontal.

Table 6.3 shows the average accuracy, log-loss, and AUROC with standard deviations for each of the five conditions on bills. The average baseline accuracy is 68%. The just_txt model achieves an accuracy of 73%, outperforming the baseline by 5%, and notably, shows that there is a predictive signal even within the limited amount of text available in the title and descriptions.

To examine where text content is most and least predictive on its own, we disentangle the average performance of the just_txt model in figure 6.3, showing the per state and

---

[10]For further details see the appendix.

Table **6.3.** Average and standard deviation across states on accuracy, log-loss, and AUROC for bills on each feature set.

Feature Set	Accuracy		Log-Loss		AUROC	
	Avg.	Std. Dev.	Avg.	Std. Dev.	Avg.	Std. Dev.
baseline	0.68	0.1	0.6	0.09	0.5	0
just_txt	0.732	0.09	0.53	0.14	0.7	0.14
just_spon	0.759	0.102	0.48	0.16	0.74	0.15
no_txt_spon	0.81	0.113	0.39	0.18	0.8	0.18
no_txt	0.846	0.098	0.32	0.18	0.82	0.21
combined	**0.859**	0.093	**0.31**	0.17	**0.85**	0.21

chamber pair change from baseline. The states that improve the most over baseline, with 15% improvement or more using only textual features, are Oregon, Oklahoma, Tennessee, District of Columbia, South Carolina, Louisiana (lower), Georgia (lower), and Alabama (lower). On the other hand, text is least predictive in Connecticut, Wyoming, Idaho, New Jersey, Utah (upper), New Hampshire (upper), and North Dakota (upper), all underperforming the baseline.

The relatively small improvement over baseline of just_txt provides insight into the lawmaking process, raising the possibility that other contextual factors, outside the subject matter of the legislation, such as identity of the sponsors and the committee, may be more important than the subject of the legislation.

The just_spon model achieves an average accuracy of 76%, slightly outperforming just_txt with an improvement over baseline of 8%. This further indicates that knowing sponsor-related information, without reference to the subject of the legislation, is itself highly predictive. In fact, figure 6.4 shows that, except for New Hampshire (upper), almost all states achieve gains using sponsor-only information, with Oklahoma, Texas, and Ohio achieving gains of 30% or more. The committee information in

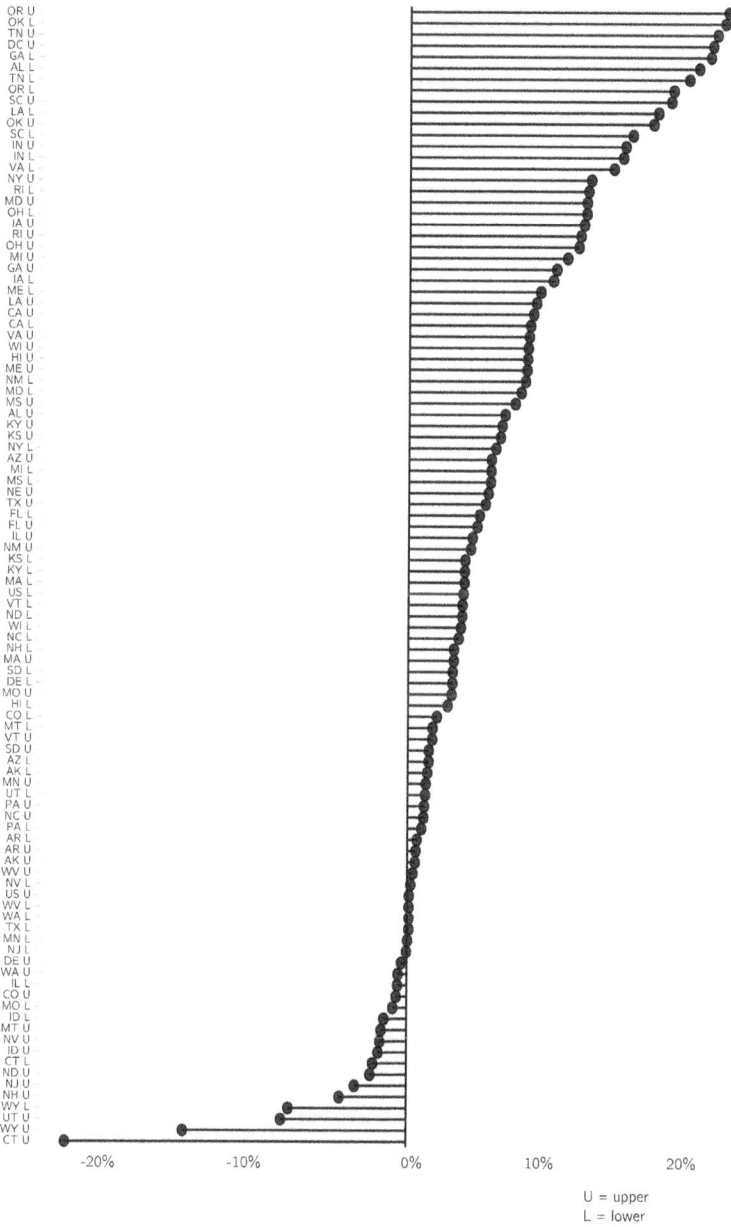

Figure 6.3. Change from baseline with text-only features.

no_txt_spon, which includes the sponsor committee positions, is even more predictive than sponsor and text only, and the addition of sponsors in no_txt improves performance by 3.5%.

Including text in the combined model further improves performance by 1.3% over no_txt and 18% over the majority class baseline, showing the complementary effects of contextual and lexical information, as this model consistently outperforms all others. Figure 6.5 shows the per state and chamber pair baseline and combined model performance. The AUROC performance follows a very similar trajectory.

On log-loss, the model performance follows a similar path, with all models showing improvement in probability estimates from the baseline. Log-loss almost doubles from the combined model's 0.31 to 0.6 on baseline. This reinforces that the combined model makes very confident correct predictions. Including text in the combined model improves performance slightly over no_txt, while having just sponsors or just text decreases the log-loss to around 0.5.

Table 6.4. Average accuracy, log-loss, and AUROC for bills using legislative events postintroduction.

Feature Set	Accuracy		Log-Loss		AUROC	
	Avg.	Std. Dev.	Avg.	Std. Dev.	Avg.	Std. Dev.
combined	0.859	0.093	0.31	0.17	0.85	0.21
combined+act	0.94	0.059	0.16	0.12	0.97	0.04

Analysis

All contextual and lexical features considered above are available upon the introduction of a bill, or shortly thereafter,[11] and thus the evaluation above indicates how well floor action can be predicted from the day of introduction. However, after the

---

[11] Some states do not indicate committee assignment immediately. For those we include the first assignment after introduction.

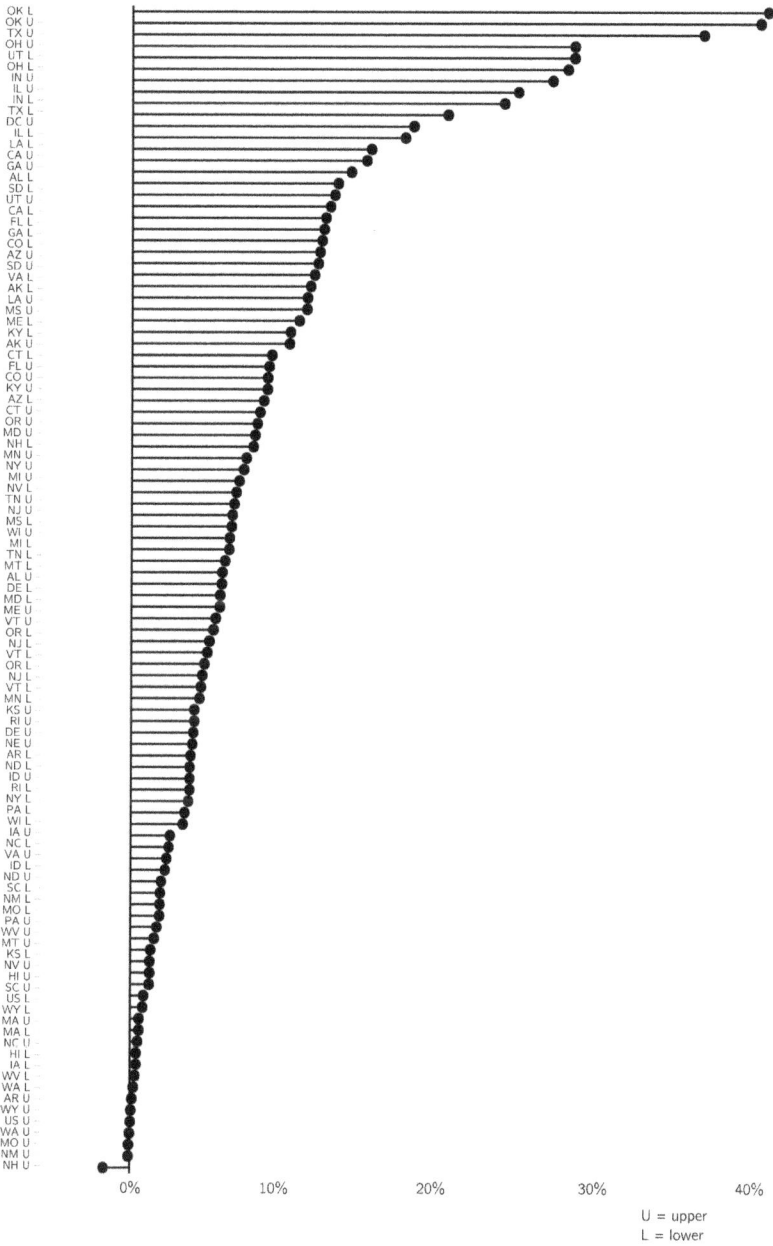

Figure 6.4. Change from baseline with sponsor-only features.

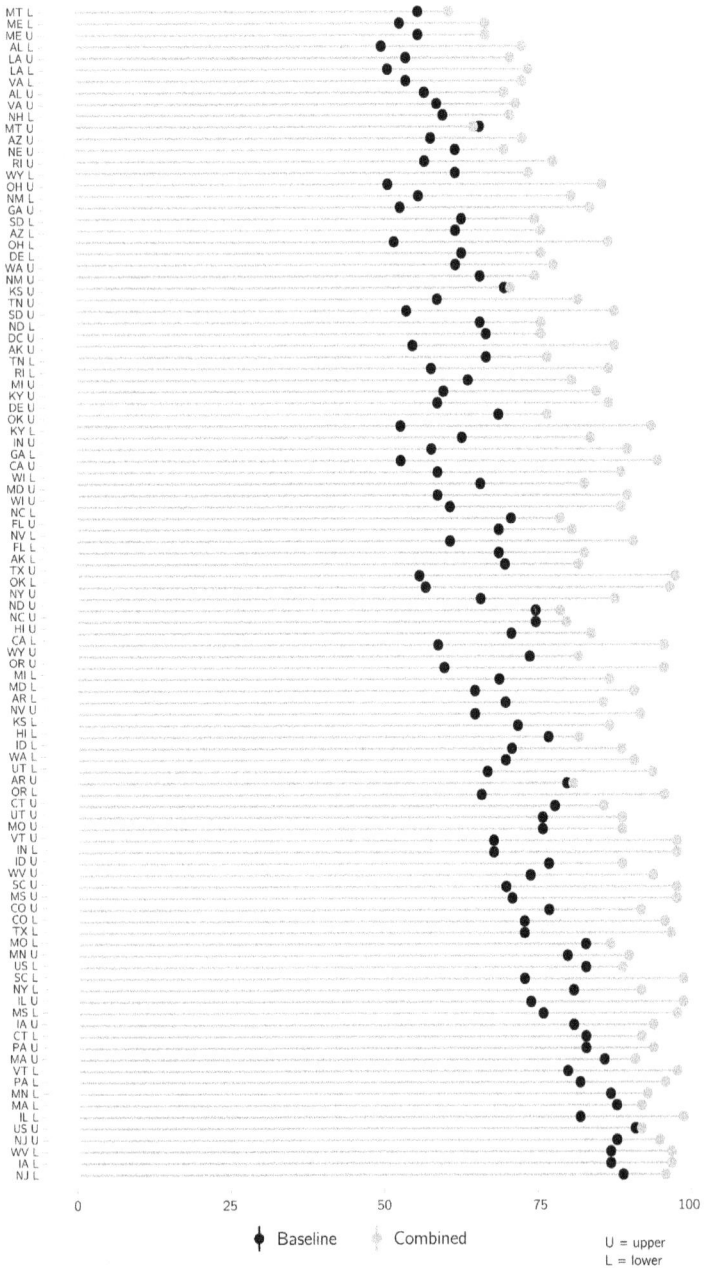

Figure 6.5. Prediction accuracy on bills with combined model.

bill is introduced, subsequent legislative actions provide further contextual information about the legislative process. As these actions may carry relevant predictive information, we further examine subsequent events in the legislative process in the `combined+act` feature set. We include a binary feature for the occurrence of amendment introduction and outcomes, votes, committee referral outcomes, and readings up to the point of floor action. By comparing `combined+act` to `combined` we can examine the information value of different events in the legislative process for predicting floor action (at least in our models).

Table 6.4 shows the results of the `combined+act` feature set. Accuracy improves to 0.94, while log-loss drops by half to 0.16, confirming that legislative events that occur up to the point of floor action carry significant complementary information to other contextual factors and are highly indicative of floor action. While `combined+act` confirms the predictive power of procedural factors outside the legislative text, sponsor, and committee assignment, the `combined` model is arguably the most important result, as it indicates how well we can predict various aspects of features that are available upon introduction.

Beyond the predictions, we are interested in identifying the different features that contribute to legislative success across the states. As there are a large number of models and features, in order to understand the relative predictive importance of contextual and legislature-specific dynamics, we choose several expert-informed factors deemed to be important for floor action and compare the rank and weight they received in each model.

We first examine the median rank by model weight given to the following features in the `just_spon` condition across all states: bipartisan, sponsor in minority, sponsor in majority, and the number of sponsors. While many of these contextual features are highly ranked, there are many variations and differences across states. The top half of table 6.5 shows the top states for which each feature was ranked among the top twenty. For example, the bipartisan feature is ranked in the top five in Missouri, Virginia, Maine, and Mississippi, accounting for up to 6% of the explanatory power. As a comparison, in South Dakota, Hawaii, Minnesota, Wisconsin, and Pennsylvania, bipartisanship ranked lower than two hundred. Whether the sponsor is in the minority is important in the US Congress, where it is ranked sixth, along with Delaware, Tennessee, and West Virginia. Being in the majority accounts for 10% in Kentucky and 7% in Wisconsin. This aligns with previous literature, as Wisconsin is known to have a strong party system (Hamm 1980), and indeed, we find sponsor in majority and minority features to be ranked first and ninth, respectively, while in Texas, which has a much weaker party system, those features are ranked among the lowest of all states.

Similar rankings are presented for committee features in the bottom half of table 6.5 in the no_txt_spon condition. The committee features play a similarly predictive role, with the sponsors holding membership positions on the committee accounting for over 10% of explanatory power in Delaware, Connecticut, Maine, and South Dakota.

To examine the difference in probability assigned by the models under different conditions, we compare the probability of floor action assigned by our model for legislation that received floor action and legislation that did not. We expect the median of the probabilities on legislation that received floor action to be higher than the median of the probabilities on legislation that failed, which is indeed the case. When comparing feature sets, we can compare how much the probability estimates are affected by different features. Taking a representative example where neither contextual nor lexical features dominate—Pennsylvania's lower chamber—the combined model's median and mean predictions

Table 6.5. Feature importance in sponsor-only and committee model. "Median" reports median ranking by weight assigned across states; "top" indicates how many states have that feature ranked within the top twenty weighted features; "top states" lists the states where each feature was ranked the highest and was one of the first twenty features; "bottom states" lists the ten states where each feature was ranked the worst.

	Feature	Median	Top	Top States	Bottom States
Sponsor-only model	Bipartisan	64	11	MO, VA, ME, MS, NC, SC, AK, DE, WA, US, NV	SD, MN, WI, PA, UT, NE, ID, FL, DC, AR
	in Minority	24	20	DE, US, WV, TN, IA, WI, AL, ND, MD, MI, RI, NC, MT, OH, PA, MO, SC, CO, WA, NY	CA, IL, HI, TX, NJ, UT, NE, FL, DC, AR
	in Majority	23	22	WI, MN, TN, KY, NH, NC, CO, IL, AL, OH, US, WV, GA, MI, MT, IA, ND, SC, DC, MO, PA, NY	AK, TX, MA, VA, NE, NJ, ME, UT, FL, AR
	Sponsor	28	20	CO, UT, IL, VT, IN, IA, OR, SD, OH, US, ND, HI, CA, LA, NJ, MT, KS, MA, TN, NE	PA, AZ, WA, WY, NM, NV, VA, MS, MN, AR
Committee model	Ranking Mbr	24	15	NE, VT, AR, KY, US, GA, OK, ME, NY, OR, SD, IL, MN, DC, WV	KS, MO, MT, OH, PA, RI, TN, UT, VA, WY
	No Cmte Mbr	17	23	AR, ME, NE, IL, SD, NC, NV, NM, DE, KY, WV, CO, OK, CT, NY, TX, MD, RI, WI, VT, NJ, GA, MA	HI, ID, IA, KS, MT, OH, PA, TN, UT, WY
	Members	6	33	DE, CT, ME, SD, NC, NV, OK, NY, GA, IL, KY, NJ, IN, WI, MD, WA, DC, MI, AR, OR, RI, TX, US, NE, CA, MN, NH, SC, MA, ND, WV, AL, NM	HI, ID, IA, KS, MT, OH, PA, TN, UT, WY

of success on bills receiving action are above 90%. That model also has the largest difference between the probability of success assigned to passing bills and the probability of success assigned to failing bills (close to 0%). The no_txt model has a similar mean, but the probabilities are more evenly distributed across both pass and fail. Removing sponsors significantly affects the distribution and shifts the mean lower to 70% while just_spon and just_text both drop the mean to around 40%.

Finally, we examine language ranked most and least predictive on the just_txt condition for New Mexico, Pennsylvania, and

Table 6.6. Top- and bottom-ranked phrases.

State	Top Phrases	Bottom Phrases
New Mexico (upper)	day, campus, recognit, month, defin, alcohol, date, recipi, procur, cours, registr plate, revis	tax credit chang, enmu, residenti, lobbi, statewid, or, abort, safeti, date for, test for, primari care, analysi
New Mexico (lower)	day, studi, length, citi, of nm, fingerprint, geotherm, fund project, dog, definit, loan for, month	of game fish, peac, senior citizen, math scienc, transfer of, state fair, self, bachelor, develop tax credit, nmhu, wolf, equip tax
Pennsylvania (upper)	provid for alloc, creation of board, manufactur or, an appropri to, of applic and, medic examin, fiscal offic, for request for, corpor power, within the general, for the	an act amend, as the tax, known the, act provid, known the tax, wage, act prohibit, citizen, of pennsylvania further, tax, youth, requir the depart
Pennsylvania (lower)	or the, contract further, and for special, memori highwai, within the general, in game, first class township, whistleblow, emerg telephon, offens of sexual, for	act amend titl, an act amend, act provid, known, act prohibit, amend the, pennsylvania, an act provid, code of, act establish, an act relat, the constitut
New York (upper)	fiscal year relat, memori highway, year relat to, implement the health, for retroact real, portion of state, the public protect, implement the public, inc to appli, budget author, program in relat,	languag assist, direct the superintend, the develop of, author shall, subsidi, automobil insur, such elect, limit profit, disabl act, polici base, polici to provid
New York (lower)	care insur, applic for real, physic educ, fire district elect, establish cfredit, to file an, abolit or, hous program, the suspens of, are necessari to, the membership of, relat to hous	appropri, fuel and, numer, school ground, vehicular, incom, tax for, prohibit public, tag, senat and assembl, on school, on school, class feloni

New York in table 6.6.[12] Previous literature has proposed several theories on how the policy content of legislation—such as whether it is perceived to have substantial distributional consequences or very broad scope—affects the likelihood of passage (Rakoff and Sarner 1975; Hamm 1980). While each state is likely to have idiosyncratic factors that affect the set of issues that are likely to be taken to the floor or left in committee, there is also evident overlap. In the top phrases, several states contain budgetary

[12]Additional states are presented in table 6.8 in the appendix.

issues, expressed with fiscal and appropriation language; the fact that states have to pass budgetary measures may account for this phenomenon. We also see commendation and procedural language, which may be less partisan or contentious. In the bottom phrases, several states have tax-related language and several education-related topics, which may indicate that these issues are particularly controversial. In future work we hope to explore the language differences identified by the model to examine questions related to differences in policymaking processes or political environments between states. Future work can also expand the textual base beyond the limited amount of text used in this analysis.

Conclusion

Although legislative policymaking decisions are extremely complex, the increasing amount of machine-readable data available capturing its salient aspects, from the legislative process itself to public reaction in social media, as well as the development of increasingly sophisticated computational tools, makes it possible to construct models that quantitatively measure and empirically explain observable legal phenomena.

In this chapter we explored how machine learning and natural language processing tools can be used to better understand state law-making dynamics and the state legislative process. Specifically, we introduced the task of predicting which legislation would be considered on the chamber floor across all fifty states and the District of Columbia. We presented several models capturing different contextual aspects, such as the identity of the legislator, committee membership, party affiliation, controlling party, and legislator effectiveness, and showed that combining contextual information about legislators and legislatures along with bill text consistently provides the best predictions, achieving an accuracy of 86% when predicting which legislation will reach the floor upon first introduction. We further analyzed these various factors, such as the identity of the sponsors and the committee and their respective importance in the predictive models across the states, gaining a broader understanding of state legislative dynamics. While the factors that influence legislative floor action success are diverse and understandably inconsistent among states, with some states largely driven by party and others by legislative policy content, by examining them quantitatively we can shed empirical light on the similarities and differences in state policymaking processes. ✒

Appendix

The first two models are linear classifiers, where the prediction of floor action, $\hat{y}_i$, is given by $\text{sign}(w^\top F(x))$. The first is a regularized conditional log-linear model $p_w(y|x)$:

$$p_w(y|x) = \frac{\exp\{w^\top F(x)\}}{Z(x)} \tag{6.1}$$

where $Z(x)$ is the partition function given by $\sum_y \exp\{w^\top F(x)\}$. The model optimizes w according to

$$\min_w \sum_i^n -\log p_w(y_i|x_i) + \lambda||w|| \tag{6.2}$$

The second model is NBSVM (Wang and Manning 2012), an interpolation between multinomial naive Bayes and a support vector machine, which optimizes w according to

$$\min_w C \sum_i^n \max(0, 1 - y_i(w^\top(F(x_i) \circ r)))^2 + ||w||^2 \tag{6.3}$$

where r is the log-count ratio of features occurring in positive and negative examples. The third model is nonlinear, in the form of a tree-based gradient boosted machine (Friedman 2000), which optimizes w according to

$$\min_w \sum_i^n l(y_i, \hat{y}_i) + \sum_{k=1}^K \Omega(T_k) \tag{6.4}$$

where K is the number of trees, l is the loss function (typically binomial deviance), and $\hat{y}_i$ is given by $\sum_{k=1}^K T_k(x_i)$, where T_k is a tree.

Log-loss, LL, is defined as

$$LL = -\frac{1}{n} \sum_{i=1}^n 1(y_i = \hat{y}_i) \log(p_i) + (1 - 1(y_i = \hat{y}_i)) \log(1 - p_i) \tag{6.5}$$

where $1(y_i = \hat{y}_i)$ is a binary indicator function equaling 1 if the model prediction $\hat{y}_i$ was correct, and 0 otherwise. LL equals

Table 6.7. Data statistics for number of bills introduced and receiving floor action for each state.

State	Bills Introduced			Resolutions			Sessions
	Floor Action	Introduced	Rate	Floor Action	Introduced	Rate	
AL	6697	14327	0.467	898	1201	0.748	16
AK	781	2527	0.309	372	609	0.611	4
AZ	3719	9308	0.4	367	868	0.423	24
AR	3809	5076	0.75	156	330	0.473	8
CA	18978	32143	0.59	2141	2554	0.838	17
CO	4808	6428	0.748	876	1014	0.864	11
CT	3044	16236	0.187	1258	1816	0.693	8
DE	3185	4858	0.656	1262	1386	0.911	6
DC	851	15593	0.546	11016	18422	0.598	9
FL	6592	21298	0.31	377	1177	0.32	15
GA	7416	15379	0.482	20913	23283	0.898	13
HI	5630	21615	0.26	1256	4196	0.299	5
ID	3259	4446	0.733	379	537	0.706	8
IL	14106	66926	0.211	19559	22884	0.855	10
IN	1958	5291	0.37	-	-	-	4
IA	3434	21457	0.16	874	1569	0.557	7
KS	2123	6324	0.336	889	1039	0.856	6
KY	3149	8185	0.385	4923	6032	0.816	20
LA	18346	35277	0.52	13231	14584	0.907	32
ME	9268	17095	0.542	-	-	-	8
MD	9857	26125	0.377	44	170	0.259	14
MA	6862	52467	0.131	-	-	-	7
MI	14520	41730	0.348	4434	10235	0.433	11
MN	4494	27240	0.165	169	1083	0.156	10
MS	6621	25450	0.26	4577	6324	0.734	21
MO	2736	14143	0.193	224	1244	0.18	8
MT	5910	9905	0.597	761	942	0.808	8
NE	1837	4829	0.38	1358	2422	0.561	6
NV	2614	4163	0.628	276	337	0.819	7
NH	3243	6793	0.477	47	202	0.233	6
NJ	6900	59861	0.115	1248	7059	0.177	8
NM	4253	10909	0.39	13	50	0.26	8
NY	23071	89072	0.259	30216	31346	0.964	4
NC	6922	25152	0.275	813	1309	0.621	10
ND	5735	8089	0.709	628	886	0.709	9
OH	4356	8605	0.506	5080	8706	0.584	9
OK	16579	36827	0.45	2601	4004	0.65	10
OR	5240	14404	0.364	475	1009	0.471	12
PA	2887	16414	0.176	5049	6001	0.841	5
RI	6596	16584	0.398	3400	3859	0.881	5
SC	3269	11532	0.283	9575	10800	0.887	6
SD	1647	2539	0.649	13	530	0.025	9
TN	33936	77331	0.439	28008	29732	0.942	12
TX	8371	25771	0.325	19321	20607	0.938	9
UT	7816	11072	0.706	436	620	0.703	23
VT	1035	4520	0.229	2247	2353	0.955	4
VA	14215	27813	0.511	3678	12562	0.293	24
WA	8317	24578	0.338	1501	2038	0.737	6
WV	4308	23917	0.18	2496	4297	0.581	12
WI	4982	13761	0.362	-	-	-	14
WY	2825	4223	0.669	102	185	0.551	11
US	22973	172921	0.133	6440	31067	0.207	15

Table 6.8. Top and bottom ranked phrases for New Jersey, Maryland, California, and Florida.

State	Top Phrases	Bottom Phrases
New Jersey (upper)	for farmland preserv, preserv trust, green acr fund, acquisit and, mmvv million from, vehicl from, budget for, fund for state, in feder fund, unemploy, for state acquisit, infrastructur trust	retir benefit for, of educ for, school board member, clarifi law, contract and, tax reimburs program, appropri mmvv for, to develop and, tax rate, credit under corpor, certain vehicl
New Jersey (lower)	environment infrastructur, to dissemin, farm to, dmva to, concern certain, and dhs, unsolicit, atm, contract law, link to, manufactur rebat, limit liabil	polit, import, all school, for water, facil to be, respons for, grant program for, relat crime, state administ, from tax, chair, to all
Maryland (upper)	financ the construct, festiv licens, issu the licens, grante provid and, to effect, advisori commiss, an evalu of, that provis of, financ statement, board licens, to borrow, defer	not to, phase, be use as, use as, facil locat in, to own, law petit, or expenditur, trust establish, expend match, and expend match
Maryland (lower)	counti alcohol, improv or, to financ the, termin provis relat, counti sale, sanction, alter, counti special tax, montgomeri counti alcohol, report requir repeal, length, licens mc	Grant to the, creation state debt, educ fund, state debt baltimor, establish the amount, elimin, disclos to, propos amend, incom tax rate, purpos relat, crimin gang, deced die after
California (upper)	ab, revolv fund, household, intent that, these provis until, restitut, counsel, employe, onli if ab, properti, if ab, would incorpor addit	legisl, cost of, veterinari, enact legisl, to the, law, regul econom, governor, incom tax deduct, hour, motor vehicl recreat, decis
California (lower)	add articl, to amend repeal, bill would incorpor, to add and, budget act of, urgenc statut, make nonsubstant, and make, as bill provid, relat the budget, and of the, ab	would make nonsubstant, enact legisl, make technic nonsubstant, would make technic, unspecifi, code to add, baccalaur degre, salari, fraud prevent, flexibl, of the state, would
Florida (upper)	ogsr, abrog provis relat, grant trust fund, govern act, person inform, to supplement, employ contribut to, legisl audit committe, jac, maintain by the, insur regul, financi inform	Senat relat to, senat relat, to, ssb, elder, school, municip that, and legislatur by, that law enforc, provid minimum, admiss to, local law enforc
Florida (lower)	etc, certain propos, re creat, repeal under, to qualifi, boundari, program revis requir, environment permit, counti hospit district, alcohol beverag licens, except under, ranch	hous relat, day, renew energi, provid for alloc, make recommend, for employ of, of damag, from particip, week, catastroph, dhsmv to develop, employ from
Delaware (upper)	uniform, would increas the, amend chapter volum, person convict, relat the delawar, dealer, child support, bureau, violenc, associ, charter chang, for fiscal year	rent, state languag, the content, act regul, certain licens, give local, assembl from, delawar code establish, for citizen, reimburs, propos constitut amend, salari
Delaware (lower)	of the th, tax refund, thi act also, amend of the, this section of, the titl, the act to, of member of, electron transmiss, for in the, and the date, parent quardian	predatori, hour per, relat state employe, unfair practic, communic, open meet, equal the, to the construct, the construct, medicaid, state agenc, relat to prevail

zero for a perfect classifier and increases with worse probability estimates. Specifically, *LL* penalizes models more, the more confident they are in an incorrect classification.

AUROC allows us to measure the relationship between a model's true positive (TP), that is, how many floor action bills were correctly predicted as floor action, and false positive rate (FP), that is, how many failed bills were predicted as floor action. It is defined by

$$AUROC = \sum_{i=1}^{N} p(TP)\Delta p(FP) + \frac{1}{2}(\Delta p(TP)\Delta p(FP))$$

(6.6)

N

WRITING STYLE AND
LEGAL TRADITIONS

*Jens Frankenreiter, Max Planck Institute for
Research on Collective Goods*

The European Court of Justice (ECJ)[1] in many ways is a
unique court, and it has undergone a remarkable transition in its
roughly fifty-five years of existence. While it originally consisted
of just seven judges, six of whom hailed from a country with
a French law tradition, it now consists of twenty-eight judges
from all major European legal families. Originally designed as
an international court with jurisdiction to adjudicate disputes
about the interpretation of an international treaty with a rather
narrow scope, it has developed into what might be the single
most powerful court in Europe. Despite the ECJ's importance,
the drivers of its decisions are not well understood. Commentators
take different views on whether its decisions are mainly motivated
by legalistic considerations (Sankari 2013), whether the ECJ is
best understood as a political actor with strong preferences toward
European integration (Rasmussen 1986), whether the political
background of individual actors plays a role (Frankenreiter 2017,
2018), and to what degree the French legal culture, which
dominated the ECJ at least during its early years of existence, still
influences the output of EU courts today (Komárek 2009; Zhang,

---

[1] In this chapter, the term ECJ denotes the highest branch of the European
Union (EU) judiciary, the Court of Justice of the European Union (CJEU).
The term EU is used to denote both the EU and its predecessor
organizations, the European Coal and Steel Community and the European
Community / European Economic Community.

Liu, and Garoupa 2018). The study presented in this chapter focuses on one specific aspect of the ECJ's output, namely, the writing style of its opinions, and uses a computational approach to explore whether the French influence has become less dominant over the years.

The style of opinion writing at the ECJ is a contentious issue. Those critical of the court accuse it of using an inscrutable style which does not reveal the true drivers of legal decisions (Brown and Kennedy 2000, 55), voicing concerns that this style might undermine the authority of the court (Arnull 2006, 13). Joseph Weiler once demanded that the ECJ "abandon the cryptic, Cartesian style which still characterises many of its decisions and move to the more discursive, analytic, and conversational style associated with the common law world" (Weiler 2001, 225). Its defenders admit that the style of the ECJ can seem apodictic but maintain that this style is a necessity given the collegial nature of decision-making at the court (Bobek 2015b, 171).

It appears that this dispute is related to different assumptions about the role of the French legal tradition in the evolution of the writing style of the ECJ. Different jurisdictions have produced different ways of opinion writing (Goutal 1976; Wetter 1960). For example, the French legal tradition is commonly associated with the short and seemingly apodictic form of reasoning which the ECJ is often criticized for. Common law judges, by contrast, are customarily described as being more transparent about the "real" motivations behind decisions. Those critical of the writing style of the ECJ essentially accuse it of not having adopted sufficient elements of the writing style of common law judges, or in other words, of writing in a style that is too heavily influenced by the French legal tradition. This reasoning would seem particularly plausible if judges from countries with a French legal tradition have had an outsize influence on the development of the ECJ's writing style. By contrast, it seems much more plausible to argue that the writing style today is a reasonable response to the institutional constraints the ECJ faces if its writing style is the result of a process in which the court was able to develop its own style, free from the dominant influence of any specific tradition.

Given this apparent disagreement about the role of the French legal tradition in the evolution of the writing style of the ECJ, it is worthwhile to explore the use of quantitative techniques for textual analysis to shed light on this question. Such an investigation also showcases the potential of such techniques to serve as an important complement to qualitative research. Only by using such tools alongside a digitized corpus does a researcher have a realistic chance to include more than a small sample of cases in such an analysis. Only by devising quantitative tests can the question of whether the French legal tradition has had an outsize influence on the development of the writing style of the ECJ be answered in a way that is independent from the subjective judgment of individual researchers. Arguably, computational tools also offer the benefit of being able to detect subtle trends that might be missed altogether by qualitative research. This chapter thus contributes to the growing body of literature that brings to bear the tools of "stylometry" to help better understand legal and judicial writings.

French Legal Culture and the Style of the ECJ's Opinions

The ECJ today is one of the most important courts in Europe, and its decisions have greatly influenced European law in many different areas. French legal culture has had an outsize influence on the ECJ, at least during the first decades of its existence (Mancini and Keeling 1995; Perju 2009). For one, the ECJ was designed to resemble the French Conseil d'État (Komárek 2009, 818). Second, its working language has always been French (McAuliffe 2012, 203), potentially putting native French speakers at an advantage over their peers from other countries (Mancini and Keeling 1995, 398).

Most importantly, however, a large majority of the ECJ's members came from countries with a French law tradition during the ECJ's early years.[2] Since its establishment, the ECJ has, in principle, always consisted of one judge per member state (Art. 19 Treaty on European Union [TEU]). Judges are appointed by the member state governments for periods of six years and can be reappointed an unlimited number of times (Art. 253(1) Treaty on the Functioning of the European Union [TFEU]). Although the law envisions the appointment of judges "by common accord of the governments of the Member States" (Art. 253(1) TEU), in practice, one slot is allotted to each member state government.[3] As a result, initially, six judges came from countries influenced by the French law tradition,[4] while only one judge came from a country with a German legal tradition.[5]

Commentators by and large agree that the prevalence of the French also affected the style of its early opinions, which not only used the same opening phrase, but were also, similar to

---

[2] For the classification of countries, I follow the classification in La Porta, Lopez-de-Silanes, and Shleifer (2008), with the exception that I treat countries formerly under Soviet influence as a distinct group of countries.

[3] Traditionally, member state governments would not question candidates proposed by other governments. The Treaty of Lisbon established a panel that evaluates the suitability of candidates before they are appointed to the ECJ (Art. 255 TFEU), and reportedly some governments have withdrawn candidates after they received a negative review by the panel (de Waele 2015, 44 et seq). For a more detailed account of the institutional setting, see Frankenreiter (2017) and Frankenreiter (2018).

[4] Belgium, France, Italy, Luxembourg, and the Netherlands (which was given two slots in order to ensure that the number of judges at the ECJ was uneven).

[5] Germany.

their French counterparts, relatively short and "assembled in a succinct, syllogistic structure, with a dry tone and abstract style" (Bobek 2015b, 169).

However, there are various reasons to believe that the French legal culture today is less influential at the ECJ than during its early years. First, a number of institutional reforms (most importantly, the introduction of a chamber system) that were implemented to allow the ECJ to deal with a dramatically increasing caseload (Gabel et al. 2003) changed the original institutional design. Second, and more crucially, multiple rounds of EU enlargements brought judges from other legal traditions to the ECJ, which today consists of twenty-eight judges from all major European legal families. Figure 7.1 contains a map of all countries in the EU, displaying both the year of the accession to the EU and the prevalent legal tradition in the country.

~157~

Commentators differ in their assessment of whether these changes have had a significant impact on the style of opinion writing at the court. Some argue that the increased diversity has changed the ECJ's style of reasoning and weakened the influence of the French legal tradition (Bobek 2015a; Mancini and Keeling 1995, 400–2; Perju 2009, 363), with one commentator even calling it "rather an exaggeration" to stress the French influence still today (Komárek 2009, 818). Others, mostly those critical of its writing style, contend that it is still heavily influenced by the French legal tradition and by the choice of French as the working language of the court. Arnull (2006, 12–13) even described the ECJ's writing style as being stuck in the 1950s and 1960s.

Who is right? Does the writing style of the ECJ today blend influences from all the legal traditions from which the ECJ members hail in a more or less egalitarian way, or does the

Figure 7.1. Map of the European Union. This figure shows a map of all countries in the EU, together with information on the respective year of accession. Colors and color patterns indicate a country's prevalent legal tradition. Dark gray: French law. Black: German law. Light gray: common law. Vertical lines: Nordic law. Horizontal lines: countries formerly dominated by the Soviet Union. White: nonmember countries.

adoption of some stylistic elements commonly associated with other legal traditions amount to nothing more than window dressing? It seems reasonable to assume that both parties in this debate have some valid arguments on their side. On the one hand, the ECJ seems to have adopted some of the features which are commonly associated with common law and the German legal traditions, such as including in its opinions references to prior decisions and embracing the so-called principle of proportionality (Mancini and Keeling 1995). On the other hand, the ECJ still writes in a strong version of what Posner (1995) calls the "formal style" of judicial reasoning. At the same time, it seems impossible to answer this question on the basis of qualitative research alone. How should one determine whether changes in writing style are sufficient to conclude that the newly created "internal diversity . . . had profound effects on the court's argumentative practices, greater than the court's alleged desire to remain faithful to its French origins" (Komárek 2009, 818)? Answering this question requires precision in the measurement and comparison of the different effects, a task for which qualitative tools are ill suited.

Using Computational Tools to Study Writing Style

Against this background, the study presented in this chapter explores the use of quantitative and, in particular, computational tools to improve our understanding of the development of the ECJ's writing style. In doing so, it also attempts to showcase the potential as well as the limitations of computational methods to enrich our understanding of the development of legal institutions.

This study makes two contributions to the discussion on the development of writing style at the ECJ. It attempts to track the development of the writing style of the ECJ over time to see whether the qualitative assessments reported above are reflected in the data. It also tests two different hypotheses regarding the development of the ECJ's writing style. First, in an attempt to test the impact of EU enlargements, it asks if the writing style of the ECJ as a whole has become more similar over time to the writing of judges from new EU member states during their first few years at the court. Second, in an attempt to estimate the continuing influence of the French legal tradition, it asks if the writing style of the ECJ has, on average, become more similar over time to the writing of newly appointed judges from countries with a French legal tradition, as compared to their peers from other countries.

In order to test these hypotheses, one first has to find a way to attribute opinions to individual judges. In the context of the ECJ, this is not trivial. Judges at the ECJ sit in chambers of three, five, or fifteen judges; cases are decided by way of majority voting; and all opinions are issued *per curiam*. Nevertheless, the judge responsible for writing the opinion is known in every case. Early in the proceedings, the president of the court[6] assigns one judge to act as *judge rapporteur*. This judge remains responsible for drafting the opinion irrespective of whether he or she supports the outcome favored by the majority of judges.

Also, other judges are reported to influence the drafting process (Edward 1995). This might call into question whether the *judge rapporteur* can really be seen as the author of an opinion or whether the influence of other judges is so strong that the identity of the *judge rapporteur* hardly matters. Another potential source of disturbance is the fact that every judge is assisted by a number of clerks, the so-called *référendaires* (Bobek 2015b, 168). Despite these potential

---

[6] The president of the court is elected by the judges for periods of three years (Art. 253(3) TFEU).

confounds, this study assumes that the *judge rapporteur* retains some amount of influence over the contents and style of the opinion so that the opinion can be attributed to him or her. Some of the results obtained below in fact do seem to confirm that the person acting as *judge rapporteur* has at least some amount of influence over the style of an opinion.

There are, of course, numerous ways to analyze and quantitatively compare the style of legal opinions. Examples of measures used in previous studies include simple metrics, such as the length of opinions (Goutal 1976), and slightly more complex measures, such as the average number of words in paragraphs or sentences, the average length of words, the diversity of words or sentence length, the number of different words in opinions, and the ratio of words that appear exactly once in an opinion (Wahlbeck, Spriggs II, and Sigelman 2002).

This study examines similar measures, but its focus lies mostly on another measure of writing style. In line with a number of recent contributions from various fields (Carlson, Livermore, and Rockmore 2016; Hughes et al. 2012; Rosenthal and Yoon 2011), it uses the frequency distribution of so-called function words as a stylistic fingerprint (Carlson, Livermore, and Rockmore 2016) of different authors or groups of authors. Function words are words that do not in themselves convey any meaning; rather, they establish the relationship between those parts of speech that do. Function word frequencies have a relatively long history of use in stylometry and have been used to relate, distinguish, and uncover authorship in other literary contexts (Hughes et al. 2012; Binongo 2003). Because of the possibility to express the same message in various ways, authors might differ in their use of function words even when they are arguing the same point.

At first glance, one might think that this measure is not well suited for an investigation of how different legal traditions influence the writing style of a court because it is admittedly hard to relate this measure to common notions about what constitutes a "typical" opinion written in a jurisdiction belonging to one of the various legal traditions. At the same time, it allows for the development of a rather objective measure for how different the styles of two different opinions or groups of opinions are. One can then use this measure of difference to document the development of a writing style over time and to link changes to other events such as judges from new member states taking office.

It seems worth pointing out that the use of function words in analyses of judicial writing style is particularly helpful if and insofar as the use of function words is independent from opinion content. The reason for this is that, like in many other courts (see Bowie, Songer, and Szmer (2014) and Maltzman and Wahlbeck (2004)), opinion assignments at the ECJ are not random. If one were to use features related to the content of a case as a measure of judicial style, one would risk conflating the effects of nonrandom opinion assignment with the effects of a judge's writing style.[7] To the extent that function words are unrelated to the content of a text, the same problem does not occur. However, it seems possible that certain function words appear more often in the context of certain topics than in others, which opens the way for opinion assignments to influence the occurrence of function words. This might

---

[7] To understand what this means, consider an example: Assume that one were to treat the example of the frequency of use of the word "liberté" as a measure of writing style. A judge who acts as *judge rapporteur* in many cases involving the four freedoms of the common market might frequently use this word merely to describe what the case is about; it would be a mistake to conclude that it is this judge's style to use this word more often than others.

be particularly true at the ECJ, where chunks of text are frequently copied from previous decisions.

This study uses a new dataset consisting of the text of all opinions issued by the ECJ between its establishment in 1952 and February 2015. As the working language of the ECJ is French, the dataset contains the French versions of the opinions.[8] All opinion texts were downloaded from the official databases of EU law using a Python script.[9] Of the 10,226 judgments issued during this time period, French opinion texts are available online for 10,131.[10] The date of the decision, *judge rapporteur*, and other information about the cases were obtained from metadata provided by the court databases.

Using this dataset, the study first documents that the writing style of the ECJ has changed over time. It tracks trends in different measures for style and uses the distribution of function words to show that, just like at the US Supreme Court (Carlson, Livermore, and Rockmore 2016) and in literature in general (Hughes et al. 2012), there seems to exist a "style of the time" in the decision-making of the court that is different from how judges write in other periods.

The following analyses examine two different possible causes for the observed changes in style. The first pathway

---

[8] Judgments by the ECJ are regularly translated into various languages after being issued.

[9] See https://curia.europa.eu/ (accessed December 1, 2017) and http://eur-lex.europa.eu/ (accessed December 1, 2017). Note that, for older opinions, accents are not displayed in the text provided by the websites. Therefore, in order to guarantee uniformity of the dataset, I have omitted the accents in all texts. This increases the noisiness of the data because it is impossible to differentiate between some words, for example, *a/à* and *ou/où*.

[10] Judgments whose French opinion texts are not available online cluster around 1990, with a second (smaller) cluster at around 1997.

operates through the various rounds of EU enlargements. In particular, judges from new member states may influence the writing style of all judges. In fact, there is evidence that two rounds of enlargements (1973 and 1995) were followed by substantial changes to the writing style of the ECJ. However, there is no evidence suggesting that sitting judges adopted features of the writing style of judges from new member states. Second, new judges from countries with a French law tradition may, on average, be more likely to influence the development of the writing style of the entire court in subsequent years. As will be discussed in more detail below, there is some evidence in support of this hypothesis. This can be interpreted as suggesting that the French legal culture has had an outsize influence in shaping the decision-making of the ECJ, even as the court became more diverse.

Evidence from Simple Measures of Style and Clarity

This first part of the empirical investigation explores how common notions about the writing style of the ECJ and its development are reflected in a number of quantitative measures of style. Figure 7.2 displays the development of four key stylistic features of ECJ opinions over time: opinion length, average sentence length, average number of syllables per word, and type–token ratio.[11] A brief inspection of the figure seems to confirm the contention by those commentators

---

[11] The *NLTK* and *Pyphen* packages in Python were used to obtain these measures. The number of tokens was determined by counting all numbers and words in the opinion's reasoning. The sent_tokenize function in NLTK was used to divide the reasoning into individual sentences in order for them to be counted. The average number of syllables was obtained by hyphenating all words in the reasoning using Pyphen, counting the resulting syllables, and dividing the syllable count by the total number of words. The type–token ratio was determined by counting the number of different words in the opinion's reasoning and dividing the result by the total number of words.

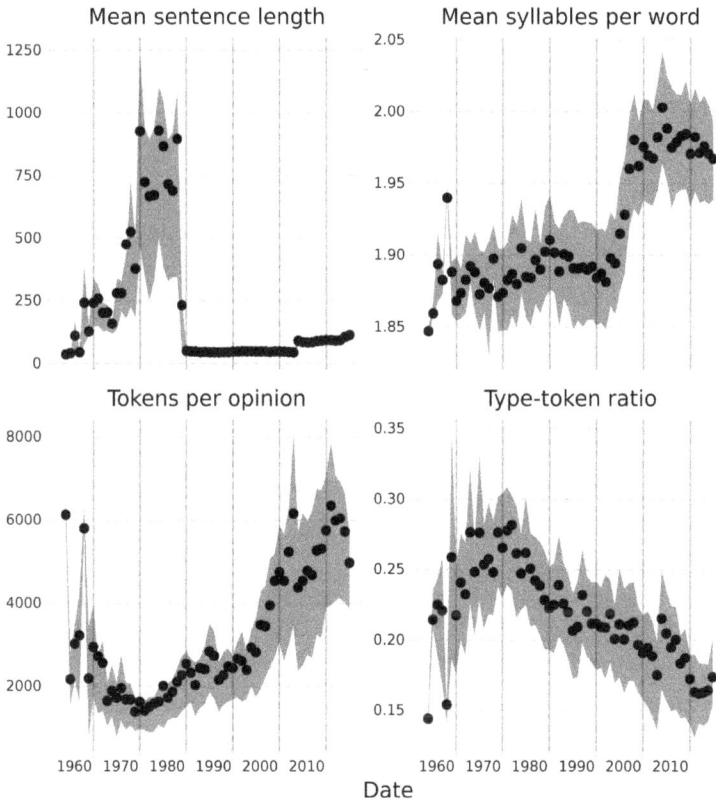

Figure 7.2. Development over time of four measures of style. Standard error smoothed using local regression; showing only mean points.

who argue that the style of the ECJ's writing has undergone a significant evolution over time.

Opinion length. The bottom left panel displays the number of tokens (words and numbers) per opinion. It is readily apparent that this number has grown steadily over time. During the first two decades of the ECJ's existence, opinions on average had below two thousand tokens; only a small number of opinions went beyond eight thousand tokens. Similar to other courts (Black and Spriggs II 2008), the length of opinions has grown considerably. By the mid-2000s, the average length of opinions had grown by more than 150%. Individual opinions have grown to the length of short books, with the longest opinion counting more than fifty thousand tokens.[12] On average, the length of opinions grows by around 105 words per year.[13] The growth in opinion length certainly is in line with the claim that the ECJ has, over time, distanced itself from the French model of opinion writing, which is known for its comparably short opinion texts (Goutal 1976). However, it can potentially also be explained by other factors, such as an increase in cases that raise complex factual and technical questions. Such cases likely became more common with the proliferation of secondary EU law, a considerable share of which deals with rather technical matters, and the increasing importance of the European Commission as an administrative agency.

Mean sentence length. The top left panel displays the average number of tokens per sentence. It is strikingly evident from visual inspection that while the average sentence length seems to increase during most periods in time, it took a sharp turn downward in the late 1970s. In fact, until early May 1979, many opinions had an average sentence length of several hundred tokens; the number of tokens per sentence in opinions published between June 1979 and May 1980 averaged "just" around fifty-two.[14] To understand what happened, it makes sense to have a closer look at decisions issued around

---

[12] Judgment of 15 October 2002, *Limburgse Vinyl Maatschappij NV and Others v. Commission,* joined cases C-238/99 P, C-244/99 P, C-245/99 P, C-247/99 P, C-250/99 P to C-252/99 P and C-254/99 P, ECLI:EU:C:2002:582.

[13] Linear regression indicates that this effect is highly statistically significant.

[14] A t-test comparing the average length of sentences in the time periods 1970–1978 and 1980–2000 indicates that this change is highly statistically significant.

that time. Before the change, many opinions were written in at most a few sentences, with different reasons given in the form of subordinate clauses. (Of all opinions written before 1979, more than 80% consisted of less than ten sentences and more than 60% consisted of less than ten sentences.) The ECJ would often begin its reasoning with the phrase *attendu que* and then lay out its reasons in what Michal Bobek has called the "succinct, syllogistic structure" typical of the French drafting style. The ECJ apparently abandoned this way of structuring its reasoning completely and started using many more main clauses.[15] This shift was followed by a period of time in which the ECJ used rather short sentences (the average sentence length between 1980 and 2000 was around forty-eight words per sentence). Around the time of the 2004 EU enlargement, the ECJ seems to have partly changed back to using longer sentences: the average sentence length has grown to more than ninety in all opinions issued since 2005.

Average number of syllables per word. A different development can be observed in the panel reporting the average number of syllables per word. The length of words appears to have been relatively constant until the late 1990s. Then the measure

---

[15]As an example, consider two decisions from May 1979, *Henningsen Food* (case 137/78, ECLI:EU:C:1979:117), and *Galster v. Hauptzollamt Hamburg-Jonas* (case 183/78, ECLI:EU:C:1979:143). Both decisions decided cases concerning the interpretation of the Common Custom Tariff; both were decided by the first chamber, featuring the same sample of three judges; and the Italian judge, Giacinto Bosco, acted as *judge rapporteur* in both decisions. In the French version of the judgment in *Henningsen Food*, a decision dated May 2, 1979, the legal reasoning commences with the phrase *attendu que*, followed by a list of reasons in the form of subordinate clauses separated by semicolons. This sentence does not span the entire judgment but still covers several pages. In *Galster v. Hauptzollamt Hamburg-Jonas*, by contrast, the court still writes in rather long sentences, but a sentence usually does not cover more than a single paragraph or argument.

ticks upward.[16] This change could indicate that the language of the ECJ's opinions has become more technical at this point in time, but it is beyond the scope of this chapter to investigate further how this shift manifested itself in the writing of the ECJ.

Type–token ratio. The last panel depicts one aspect of the variability of language use, namely, the so-called type–token ratio (obtained by dividing the number of different words in a text by the total number of words). The measure slopes downward slightly but appears to stay relatively constant over time, with a rather high amount of variation between different judgments issued at the same point in time.

What can we learn from an inspection of these measures? First, as noted before, the data seem to confirm that, on various dimensions, the writing style of the ECJ has changed since its early days. Apparently, the most far-reaching shift occurred in 1979, when the ECJ, in a rather abrupt change, stopped delivering its arguments mostly in the form of subordinate clauses.

Second, the data also confirm that the writing of the ECJ is relatively inaccessible. This can be demonstrated by calculating Kandel/Moles scores from the data, which are a version of the Flesch Reading Ease statistics,[17] adapted to the specifics of the French language (François and Fairon 2012).[18] Scores are commonly ranked between zero and one hundred, with zero

---

[16] A t-test confirms that the increase between the time periods 1990–1994 and 1998–2002 is highly statistically significant.

[17] The Flesch Reading Ease statistics has been used in research on American courts, for example, in recent studies on the Supreme Court of the United States (Carlson, Livermore, and Rockmore 2016) and on state Supreme Courts (Goelzhauser and Cann 2014).

[18] Kandel/Moles scores are calculated based on the following formula:

$$KM = 207 - 1.015\frac{count(words)}{count(sentences)} - 73.6\frac{count(syllables)}{count(words)} \quad (7.1)$$

indicating texts that are very difficult to read, but negative values (suggestive of particularly inscrutable texts) are also possible. Opinions written by the ECJ consistently score below zero, with an overall average of around −57. Readability seems to have improved over time, however. For opinions issued since 2000, the average score is around −20, with around 26% of opinions achieving a score of zero or better. Only 0.5% of opinions achieve a score of more than thirty, which still designates texts which are at best "very difficult" to understand. These results seem to confirm allegations by those critics of the ECJ who lament its unclear style of writing.[19]

Third, the measures described above can be used in an attempt to understand if the person acting as *judge rapporteur* can influence the writing style of an opinion, that is, if judges differ in their writing style. In order to test this hypothesis, the measures displayed in table 7.2 are regressed on a categorical variable indicating the person acting as *judge rapporteur* and a range of control variables.[20] There appear to be two measures which vary widely depending on the person acting as *judge rapporteur*—first, the overall length of opinions and second, the type–token ratio. Differences between the individual judges on the other measures are less pronounced, although on all dimensions at least some judges are depicted as differing significantly from their peers.

---

[19]Note that the score was calculated from measures obtained by means of natural language processing, and it seems possible that the software used was not in all instances able to correctly determine features such as the length of a sentence. Therefore, these scores should be taken with a grain of salt.

[20]Controls consist of dummy variables indicating the year of the opinion, the type(s) of procedure, and the subject matter(s) of the case (both as indicated in the databases of the ECJ), as well as the chamber deciding the case, the latter including the three-year period in which the case was decided. Judges are assigned to chambers anew every three years.

This result is interesting for at least two reasons. The first reason is related to the observation that changes in the second and third measures appear to reflect changes in the writing of the entire court. The fact that most individual judges also do not show substantial differences in this regard might be interpreted as showing that the judges have limited leverage in deciding on the overall structure of reasoning and the vocabulary they use to build their opinions. By contrast, they seem to be able to influence the overall length of the opinion, and the relationship between the *judge rapporteur* and the measures of the variability of language use seems to indicate that they might be able to influence the style of writing as well. In any case, this result suggests that neither the collegiate nature of decision-making at the ECJ nor the involvement of clerks in the drafting process prevent judges from leaving their individual marks on their opinions. However, this conclusion rests, of course, on the assumption that there are no systematic differences in the set of cases dealt with by individual *judge rapporteurs* that are uncorrelated with the stylistic measures described above and are uncaptured by the set of controls used in the regression analysis.

Fourth, it might be interesting to consider whether these measures can be used to uncover the drivers of the changes described earlier in this chapter. The analysis above showed that there were two or potentially three major changes in the development of the writing style at the ECJ. First, in 1979, the ECJ abruptly started to use much shorter sentences. Second, in the late 1990s, the vocabulary used in its opinions seems to have become somewhat more demanding. And third, around 2004, the ECJ partly reversed its previous embrace of shorter sentences. It is interesting to note that all these changes were preceded by different rounds of EU enlargement that brought into office judges from countries with different legal traditions.

One might therefore ask, for example, whether it is possible to determine if the changes observed in 1979 were the result of the accession to the EU of Denmark, Ireland, and the United Kingdom just six years before? In order to explore this possibility, a subset of the data including all decisions between 1973 and 1980 is used in regressions similar to those described above. In fact, it appears that, during the period leading to the shift in the ECJ's way of structuring opinions, all judges from the then-new member states were among those who scored considerably lower on both the number of tokens per sentence and the percentage of sentences with more than twenty tokens. However, these judges are not the only ones whose opinions used shorter opinions than average, and it is impossible to conclude that they influenced other judges to do so.

Evidence From the Distribution of Function Words

The preceding analysis has demonstrated how the writing style of the ECJ has evolved over time. It has also shown that judges seem to differ in how they write. However, based on the simple measures used above, the analysis has been unable to determine if the changes were the result of a loss of influence of the French legal tradition. This second part of the empirical investigation examines whether the distribution of function words as a measure of style can provide additional insight into this question.

The following analysis uses a list of function words which is on display in table 7.4 in the appendix. As the French language often relies on short fixed phrases, the list includes words as well as a number of bigrams, trigrams, 4-grams, and 5-grams.

A Python script was used to count the occurrence of function words in the individual judgments, resulting in feature vectors encoding how often a certain function word is used in an individual judgment. In the empirical analysis, I do not use the feature vectors directly. Instead, following Carlson, Livermore, and Rockmore (2016) as well as Hughes et al. (2012), I use the symmetrized Kullback-Leibler (KL) divergence as a measure for the difference between the distribution of function words in two samples of cases.[21] For this, I normalize all feature vectors so that they add up to one.[22] I then calculate the KL divergence as follows:

$$D_{KL}(P,Q) = \frac{1}{2} \sum_i \left(P(i) log \frac{P(i)}{Q(i)} + Q(i) log \frac{Q(i)}{P(i)} \right) \quad (7.2)$$

The resulting values for D_{KL} are distributed between zero and one, with a score of zero indicating identical distributions and a score of one carrying the interpretation that the distributions are so different that it is impossible to learn anything about one from the other.

Changes Over Time

In a first step, the analysis demonstrates a certain evolution of the writing style of the ECJ over time as manifested by the change in the use of function words. Just as in the context of the US Supreme Court (Carlson, Livermore, and Rockmore 2016) and literature in general (Hughes et al. 2012), we can distinguish between samples of text produced in close succession and samples of text produced at distant points in time. One can describe such an analysis as an attempt to

---

[21] In the remainder of this chapter, KL divergence refers exclusively to symmetrized KL divergence.

[22] Note that because the KL divergence is not defined for distributions in which any item appears zero times, I follow Carlson, Livermore, and Rockmore (2016) in adding 0.0001 to all probabilities after normalizing.

ascertain whether there is a discernible "style of the time" in the writing of the ECJ.

This analysis begins by examining the relationship between the distance in time between two samples of writing and their similarity. For this analysis, in a first step, information about the use of function words is aggregated at the year level. That is, for all years in the time period under observation, all the feature vectors from decisions handed down in this year are added up. The KL divergence is calculated for all pairs of years starting in 1960.[23] KL divergence scores are scaled to obtain a similarity score ranging from one to zero, with one indicating the most similar observation in the resulting dataset and zero indicating the least similar observation. The results are displayed in figure 7.3. The x-axis indicates the temporal difference between two years, while the y-axis indicates the similarity score. Each pair of years is represented by one point. From figure 7.3, it is apparent that the difference between the use of function words in the text produced by the courts increases with the difference in time.

~ 173 ~

Next, the analysis turns to the relationship between the similarity of the texts produced by pairs of judges and the distance between their time in office. For this, the information on the use of function words is aggregated at the judge level. For each judge who authored at least twenty opinions, the median year he or she served at the ECJ is determined. Then, similar to the analysis above, the KL divergence for each pair of judges in the dataset is computed, and the resulting values are scaled to obtain a similarity score similar to the one used above. The results are displayed in figure 7.4. The x-axis indicates the temporal difference between the median year of the judges' tenures at the court, and the y-axis displays the similarity score.

---

[23] Earlier years are omitted because a very limited number of judgments was produced during these years.

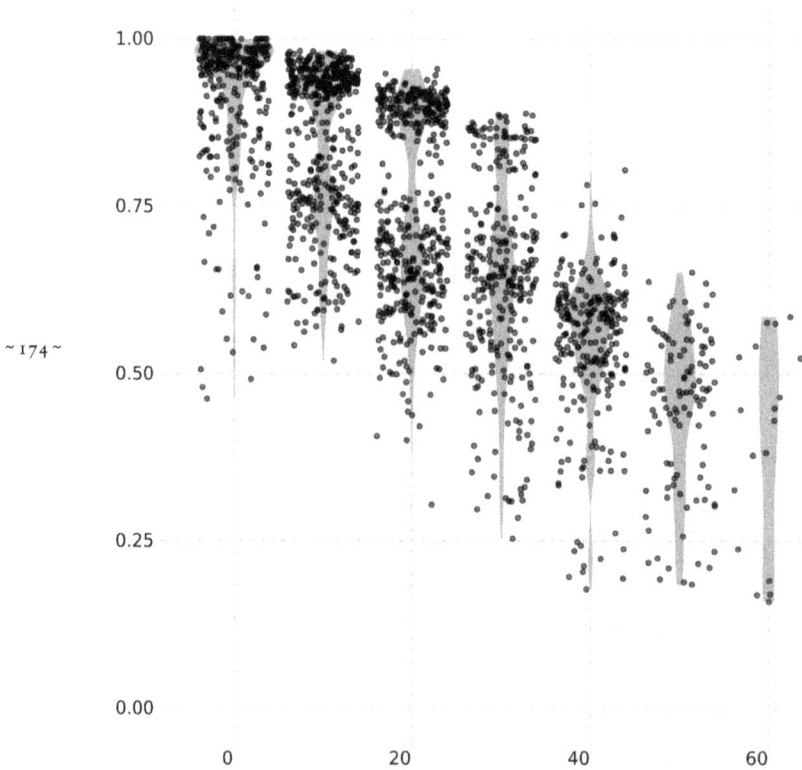

Figure 7.3. Similarity and temporal distance. Distribution represented with mirrored normal density (violin plot). Data points randomly "jittered" on the x-axis and grouped by every ten years.

Again, it can be seen that the further apart two judges' tenures are, the bigger the difference in their use of function words.

Lastly, the analysis uses spectral clustering (von Luxburg 2007) to determine whether judges are, on average, more similar to judges serving at the ECJ during the same period in time or to judges from countries with the same legal tradition. Spectral clustering is an unsupervised algorithm which can be used to sort individuals into groups based on observed similarities between these individuals. If the "style of the time" was the dominant factor in determining a judge's writing

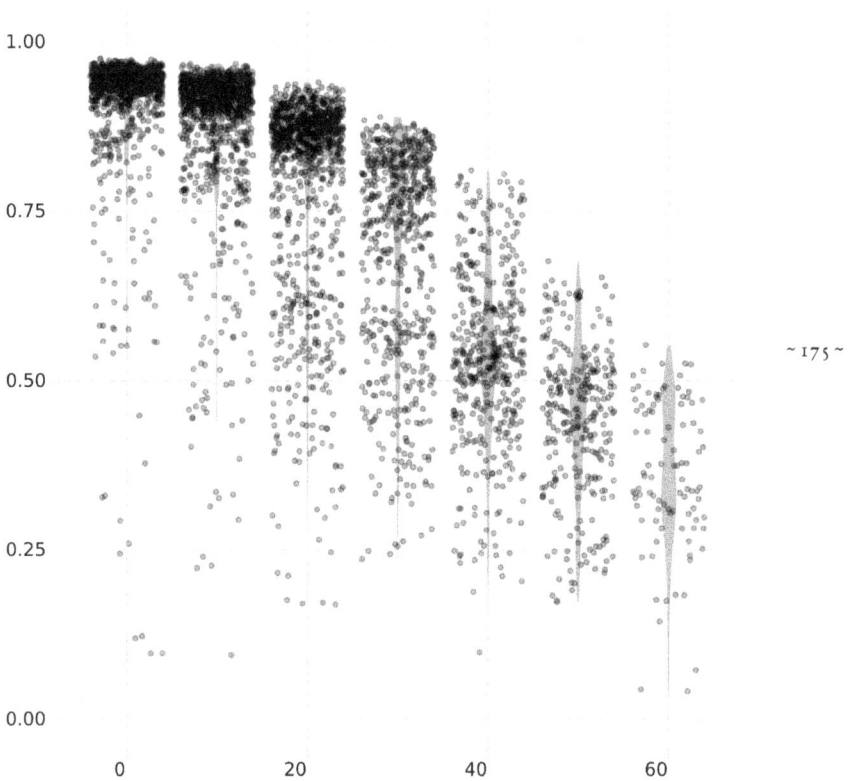

Figure 7.4. Similarity between different judges as a function of time. Distribution represented with mirrored normal density (violin plot). Data points randomly "jittered" on the x-axis and grouped by every ten years.

style, we would expect judges with overlapping tenures to be clustered together. If judges from different legal cultures could be distinguished by their writing styles, we would expect them to be clustered together. For purposes of this analysis, five legal cultures from which judges at the ECJ are drawn are used: French law, German law, (English) common law, Nordic law, and jurisdictions formerly under the influence of the Soviet Union.

This analysis again uses similarity scores obtained from scaling the KL divergence between feature vectors representing the use of function words by individual judges. Judges and similarity scores are then represented as a network with judges as nodes. Edges are placed between the nodes based on the similarity score of the respective judge pair: each judge is connected to the five judges (representing roughly 5% of all judges) with the highest similarity scores in the set. Duplicate edges are treated as one edge. The resulting network graph forms the basis of the analysis using spectral clustering. Spectral clustering classifies nodes in a network as belonging to one of a previously defined number of groups based on the relationship between the nodes. I specify the number of groups to be equal to the number of legal traditions represented at the ECJ (5). The results from this analysis are presented in table 7.1.

One can immediately see that the legal tradition of a judge's home country has at most very limited effects on the cluster he or she is classified into.[24] For example, in cluster 5, judges from all countries are present. Rather, the spectral clustering analysis effectively grouped the judges into clusters according to the time they served at the court. Cluster 1 contains almost all judges appointed to the court before 1967. In cluster 2, almost all judges were appointed between the late 1960s and the early 1980s. Cluster 3 consists of judges appointed in the 1980s and early 1990s, cluster 4 of judges appointed in the 1990s, and all judges appointed after 2002 are allocated to cluster 5. This result suggests that the style of individual judges is more influenced by the "style of the time" than by a style of a particular group of judges.

Overall, the findings in this part provide solid evidence for the hypothesis that the style of the ECJ changes over time in a way that

---

[24]Note that a similar result is obtained by rerunning the spectral clustering analysis on a smaller set of judges including only the judges in table 7.1, cluster 5.

Table 7.1. Results of spectral clustering—All judges.

Cluster 1		
Delvaux (BE, 1952–1967)	Kutscher (DE, 1970–1980)	Rossi (IT, 1958–1964)
Monaco (IT, 1964–1970)	Riese (DE, 1952–1963)	Rueff (FR, 1952–1962)
Catalano (IT, 1958–1961)	Hammes (LU, 1952–1967)	Strauss (DE, 1963–1970)
Lecourt (FR, 1962–1976)	Trabucchi (IT, 1962–1972)	Donner (NL, 1958–1964)
Cluster 2		
de Wilmars (BE, 1967–1984)	Due (DK, 1979–1994)	Capotorti (IT, 1976–1976)
van Kleffens (NL, 1952–1958)	Everlihng (DE, 1980–1988)	Touffait (FR, 1976–1982)
Bosco (IT, 1976–1988)	Koopmans (NL, 1979–1990)	Bahlmann (DE, 1982–1988)
Ó Dálaigh (IE, 1973–1974)	Pescatore (LU, 1967–1985)	Mackenzie Stuart (UK, 1973–1988)
Sørensen (DK, 1973–1979)	O'Keeffe (IE, 1975–1985)	Serrarens (NL, 1952–1958)
Cluster 3		
Joliet (BE, 1984–1995)	Grévisse (FR, 1981–1982)	O'Higgins (IE, 1985–1991)
Kapteyn (NL, 1990–2000)	Zuleeg (DE, 1988–1994)	Galmot (FR, 1982–1988)
Murray (IE, 1991–1999)	de Almeida (PT, 1986–2000)	Rodriguez Iglésias (ES, 1986–2003)
Chloros (GR, 1981–1982)	Schockweiler (LE, 1985–1996)	Slynn (UK, 1988–1992)
Diez de Velasco (ES, 1988, 1994)	Kakouris (GR, 1983–1997)	
Cluster 4		
Jann (AU, 1995–2009)	Sevón (FI, 1995–2002)	Macken (IE, 1999–2004)
da Cunha Rodrigues (PT, 2000–2012)	Wathelet (BE, 1995–2003)	Rosas (FI, 2002–2015)
Mancini (IT, 1988–1999)	Ragnemalm (SE, 1995–2000)	Hirsch (DE, 1994–2000)
Puissochet (FR, 1994–2006)	La Pergola (IT, 1994, 1999–2006)	von Bahr (SE, 2000–2006)
Colneric (DE, 2000–2006)	Ioannou (GR, 1997–1999)	Schintgen (LU, 1996–2008)
Edward (UK, 1992–2004)	Gulmann (DK, 1994–2006)	Skouris (GR, 1999–2015)
Timmermans (NL, 2000–2010)		
Cluster 5		
Berger (AU, 2009–)	Silva da Lapeurta (ES, 2003–)	Kasel (2008–2013)
Toader (RO, 2007–)	Lenaerts (BE, 2003–)	Bonchot (FR, 2006–2018)
Biltgen (LU, 2013–)	Lindh (SE, 2006–2011)	Arabadjiev (BL, 2007–)
Rodin (HR, 2013–2021)	Levits (LEV, 2004–)	Fernlund (SE, 2011–)
Arestis (CY, 2004–2014)	Juhász (HU, 2004–)	Barthet (MT, 2004–)
Ilešič (SI, 2004–)	Malenovský (CZ, 2004–)	Ó Caoimh (IE, 2004–2015)
Prechal (NL, 2010–)	Klučka (SK, 2004–2009)	von Danwitz (DE, 2006–)
Tizzano (IT, 2006–)	Makarczyk (PL, 2004–2009)	Šváby (SK, 2009–)
Larsen (DK, 2006–)	Kūris (LT, 2004–2010)	Safjan (PL, 2009–)
Schiemann (UK, 2004–2012)	Lõmus (EE, 2004–2013)	Jarašiūnas (LT, 2010–)
da Cruz Vilaça (PT, 2012–)	Vajda (UK, 2012–2018)	Jürimäe (EE, 2013–)

makes it possible to attribute a judge to a certain time period in the decision-making of the court. By contrast, this part offers no evidence that judges from countries with similar legal traditions share a common style.

The Drivers of Changes in Style

The analysis above shows that the writing style of the ECJ has changed over time and that the stylistic fingerprints of judges offer only limited hints as to the legal tradition in which the judge was raised. One question that remains unresolved is whether the several rounds of EU enlargements which brought judges from new countries to the ECJ have contributed to the changes in writing style. This part of the analysis explores whether the distribution of function words can be used to shed light on this question.

To set the stage for this analysis, figure 7.5 displays yet another graphical depiction of the development of the ECJ's writing style over time. Figure 7.5 uses a technique called multidimensional scaling (Fischman 2015) to represent the feature vectors for all years of opinion writing at the ECJ in a two-dimensional space based on the KL divergences between all pairs of years. The plot on the top shows the results for all years between 1955 and 2014; the plot on the bottom omits the first fifteen years. The color of the dots shifts from older (dark gray) to newer (light gray) decisions.

The plots confirm, yet again, that the style of the ECJ has evolved over time. Interestingly, two of the three significant changes depicted in figure 7.2 also seem to be reflected in the distribution of function words. In the late 1970s, after years of relative stability, the style of the ECJ appears to change rather quickly before entering a new period of stability. This period lasts during the 1980s and into the second half of the 1990s, when the style of the ECJ again begins to change rather quickly and

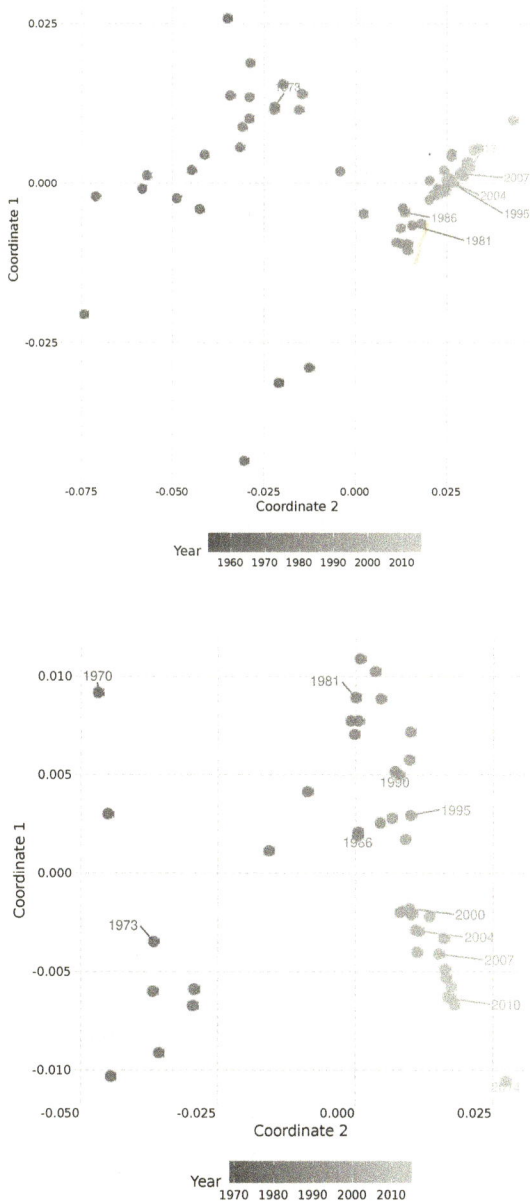

Figure 7.5. Multidimensional scaling of KL divergences between different years.

then stabilizes through the time period under observation. These changes seem to reflect the changes in the general structure of opinions and the length of sentences which occurred in the late 1970s, as well as the change in the complexity of the vocabulary which occurred in the late 1990s.

In the following analysis, two alternative and at least partly competing hypotheses about the development of the ECJ's writing style are tested. The first hypothesis is that EU enlargements have been a driver of the change in style. The second hypothesis is that the French legal culture still dominates the court to the extent that those judges who come from a country with a French legal tradition have more influence on the overall writing style of the ECJ than their peers.

Did EU enlargements change the writing style of judges at the ECJ? The analysis above revealed two clear instances in which the accession of new member states to the court was followed by significant changes in the court's writing style. Both of those rounds of accession were also among those that had the most substantial impact on the composition of the ECJ. In 1973, for the first time, two judges from common law countries entered the court alongside the first judge from a Nordic law country. In 1995, the number of judges from Nordic law countries increased from one to three and the number of judges from German law countries rose from one to two. By contrast, the two rounds of enlargements in the 1980s brought judges from only French law countries to the ECJ. The 2004 EU enlargement increased the number of judges at the ECJ by ten and therefore can be seen as the most consequential change to the composition of the court. However, most judges came from countries that had recently undergone great changes in their legal systems, with many of them attempting to adopt features of the legal systems of countries in Western Europe. Therefore, it seems questionable whether those judges could bring with them a strong sense of their own legal tradition capable of influencing the style of legal reasoning at the ECJ.

To investigate more thoroughly whether the new judges played a role in changing the style of the ECJ following the 1973 and 1995 accessions, the analysis turns to whether the writing of the other judges at the court in subsequent years became more similar to the writing of the new judges during their early years on the court. For this, for each judge actively serving at the ECJ in 1973, the following steps were taken: First, the feature vector representing the use of function words in judgments authored by each new judge in the years 1973–1975 was obtained. Second, the same was done for judgments authored by all other judges during the same time period. Third, this last step was repeated for all three-year periods starting between 1974 and 1988 (i.e., a feature vector was obtained for the writing of all other judges between 1974 and 1976, between 1975 and 1977, all the way to the time period 1988–1990). Finally, the KL divergence representing the difference in the use of function words between opinions authored by the judge between 1973 and 1975 and the opinions authored by other judges in the same and in later time periods was calculated, resulting in a measure for whether the writing of all other judges at the court became more similar or more dissimilar to the writing of individual new judges in the time period 1973–1975.

This setup allows for an exploration of whether the ECJ, in its writing, was more influenced by some judges than by others. If one judge or a group of judges did exercise a stronger influence on the development of the writing style of the entire court, one should expect the distance between the judges' initial writing and the writing of the other judges at the ECJ in later time periods either to decrease over time or at least to increase at a slower pace as compared to judges who were less influential.

The results of this analysis are displayed in the upper part of figure 7.6. It can be seen that the distance between the writing style of all judges in the period 1973–1975 and the writing style of the entire court rapidly increased as the court entered the late 1970s. Again, this likely is a reflection of the change in the overall structure of opinions documented above and confirms the finding that the style of the ECJ changed rather quickly during that time period. However, the differences in style increased at a roughly equal pace for all judges who served at the ECJ in the mid-1970s. In particular, we do not observe that this increase would be less pronounced for the judges from those countries which acceded to the EU in 1973.

The same test was repeated for the 1995 enlargement of the EU. The results of this second test are displayed in the lower part of figure 7.6. For this time period, it seems like the writing style of the ECJ initially became more similar to the writing style of judges from the new member states during their early years: while the lines indicating judges from old member states slope upwards for the years 1996–2000, the lines representing judges from new member states move horizontally, or even slope downward. However, if the ECJ during these years in fact changed because those new judges influenced the writing of their colleagues, this effect did not last long. The writing style of the ECJ fifteen years later was more different from their writing style as from that of their colleagues.

In sum, there is only very limited evidence that it was the writing of the judges joining the ECJ in 1973 and 1995 that caused the writing style of the ECJ to change in the years following these rounds of enlargement.

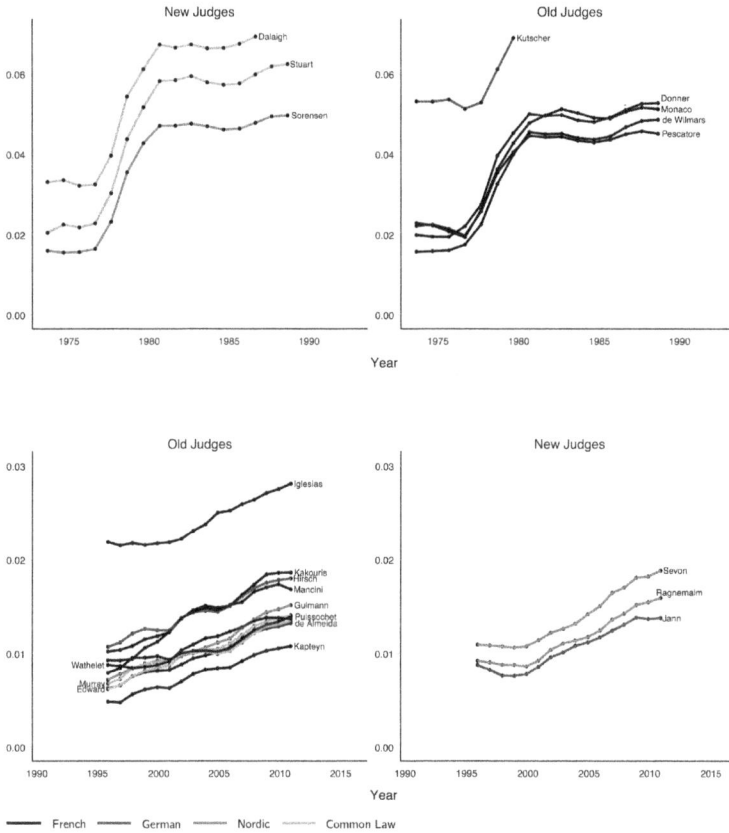

Figure 7.6. Development of ECJ writing style in comparison to the writing of judges between 1973 and 1975.

The analysis now turns to the second hypothesis described above: Did judges from countries with a French legal tradition have more influence on the development of the writing style of the ECJ than their peers from other countries? To answer this question, the analysis conducts what can be considered as a more general version of the test implemented above. Put very simply, it investigates if the writing styles of the other judges at the ECJ became, on average, more similar to the writing style of a judge belonging to a specific group during the first years after his or her taking office as compared to other judges who took office at the same time. This last point is different from the analysis above, in which new judges are compared with all the other judges serving at the ECJ when they took office.

For this test, judges were sorted into groups based on the time of their first appointment to the ECJ. The regular tenure of half of the judges ends every three years (Art. 253(2) TFEU). While some judges enter the court as a replacement for a judge who has left office before the end of his or her official tenure, most judges assume office at the beginning of such a three-year period. The first of these regular changes in the composition of the ECJ took place in October 1958, the second one in 1964, and subsequent changes happened every three years in October. All judges appointed to the ECJ in a given three-year period starting on January 1 (with the first three-year period covering the time period January 1, 1958, to December 31, 1961) are treated as one group of judges.

Table 7.2. Summary statistics.

	Obs.	Mean	Std. Dev.	Min.	Max.
KL_0	68	0.0193	0.0188	0.0057	0.1399
KL_3	68	0.0210	0.0189	0.0065	0.1320
KL_6	68	0.0201	0.0241	0.0076	0.1323
KL_9	68	0.0270	0.0220	0.0079	0.1473
$Trad_{French}$	68	0.4853	0.5035	0	1
$Trad_{German}$	68	0.1176	0.3246	0	1
$Trad_{Common\ Law}$	68	0.1764	0.3451	0	1
$Trad_{Nordic}$	68	0.1029	0.2721	0	1
$Trad_{Soviet}$	68	0.1176	0.3246	0	1
$Trad_{strongFrench}$	68	0.1765	0.3841	0	1

Next, for each judge who took office in 1958 or later, the frequency of function words in texts written by this judge during the first three years of his or her time in office was determined. Then, information on the use of function words in all opinions written by other judges during four time periods was obtained: first, the first three years in office of the respective judge; second, the three-year period three years later; third, the three-year period six years later; and finally, the three-year period nine years later. KL divergences were calculated between the texts attributed to the respective judge during his or her first three years in office and the texts written by other judges in the respective time periods. KL_0 represents the KL divergence between the texts authored by one judge and the texts authored by the rest of the ECJ during the same time period. KL_3, KL_6, and KL_9 contain the KL divergences between the texts authored by one judge and the texts authored by all other judges in the three-year periods three, six, and nine years later. Table 7.2 provides summary statistics for this dataset.

This dataset was then used in a regression analysis to detect whether judges from specific countries have a larger influence on the development of the writing style of the ECJ than others:

$$y_i = \beta_0 + \beta_1 KL_{0,i} + \beta_2 p_i + \gamma X_i + \epsilon_i, \epsilon \sim N(0, \sigma) \quad (7.3)$$

In equation (7.3), y_i is the KL divergence between a judge's writing during his or her first three years in office and the writing of all other judges on the court in later periods of time (either KL_3, KL_6, or KL_9). $KL_{0,i}$ is the KL divergence between the judge's initial writing and the writing of the rest of the court during the same time period. p_i indicates the three-year period in which the judge joined the ECJ. Lastly, X_i is a vector of dummy variables indicating the legal tradition of a judge's home jurisdiction.

Table 7.3 reports the results of this regression analysis. One can first see [rows (1), (4), and (7)] that the different groups of legal traditions as defined above do not differ significantly from common law judges, which is the baseline group in these

Table **7.3.** Ordinary least-squares regression, with models presented as rows. F-test p-values were less than 0.001 for all models. p-values based on heteroscedasticity-robust standard errors in parentheses. Controls for judge time period included in all regressions. $* p < 0.05$, $^{\dagger} p < 0.01$, $^{\ddagger} p < 0.001$.

Dependent variable: KL divergence between opinions by judge and opinions by the rest of the court

		KL_0	$Trad_{French}$	$Trad_{German}$	$Trad_{Nordic}$	$Trad_{Soviet}$	$Trad_{strong\ French}$	_Intercept	Adjusted R2	N
3-year gap	(1)	.9743‡	-.0005	-.0004	-.0020	.0005	-	-.0013	.963	68
		(.000)	(.641)	(.741)	(.279)	(.455)		(.825)		
	(2)	.9767†	.0002	-.0008	-.0024	.0004	-.0021	-.0022	.963	68
		(.000)	(.879)	(.574)	(.202)	(.565)	(.244)	(.725)		
	(3)	.9798‡	-	-	-	-	-.0014	-.0023	.965	68
		(.000)					(.314)	(.573)		
6-year gap	(4)	.9774‡	-.0008	-.0002	-.0013	.0009	-	.0015	.972	68
		(.000)	(.3730)	(.851)	(.210)	(.373)		(.740)		
	(5)	.9820‡	.0004	-.0008	-.0019	.0007	-.0040*	-.0001	.975	68
		(.000)	(.682)	(.4667)	(.066)	(.463)	(.041)	(.982)		
	(6)	.9854‡	-	-	-	-	-.0032*	.0002	.976	68
		(.000)					(.043)	(.995)		
9-year gap	(7)	1.049‡	-.0009	.0009	-.0001	.0007	-	.0043	.975	68
		(.000)	(.524)	(.570)	(.971)	(.580)		(.262)		
	(8)	1.054‡	.0006	.0001	-.0007	.0004	-.0047†	.0024	.979	68
		(.000)	(.000)	(.914)	(.615)	(.730)	(.004)	(.0545)		
	(9)	1.056†	-	-	-	-	-.0043†	.0029	.980	68
		(.000)					(.001)	(.403)		

regressions. This means that there is no indication that judges from countries with a French legal tradition have been more able than others, on average, to influence the development of the writing style of the ECJ.

However, this result might be caused by the fact that the group of countries with a French legal tradition as defined above encompasses countries as diverse as Portugal, Greece, and the Netherlands. Even if judges from France were more able to influence the ECJ's style of reasoning, this effect could be counteracted if, for example, judges from Portugal were, on average, less influential. For this reason, the analysis was rerun using an additional indicator for whether a judge is from a country with a strong French tradition, which is defined as encompassing Belgium, France, and Luxembourg. Different from the other countries, these countries also use French as an official language. Therefore, judges from these countries might have a greater ability to influence the writing style of the ECJ.

Rows (2), (5), and (8) as well as (3), (6), and (9) reveal a relationship for judges from countries with a strong French tradition that is negative throughout and significant in the second and third set of regressions. This effect is also substantial. For example, for an average judge, the writing style of the other judges after nine years is more different from his or her initial writing than it was in the beginning. For judges from French-speaking countries, this effect is reversed: on average, the writing style of the other judges is less different from the initial writing after nine years than it is during the three years after he or she entered the court.

Note that this result is by and large robust across alternative specifications, robust to shortening the period in time under observation, and robust to changing the definition of which country has a strong French legal tradition. In additional regressions, dummy variables indicating a judge's home country are used instead of variables indicating his or her legal tradition. Consistently, French judges are depicted as being more influential than other judges. Belgium and Luxembourg both appear in the group of top five influential countries, while the five least influential countries include four countries that accessed the EU in 2004. This result can also be seen as a confirmation that the method developed here captures a meaningful effect.

This result suggests that judges with a strong French legal tradition are more able than others to influence the writing style at the ECJ. One might interpret this finding as evidence that the French influence at the ECJ is still substantial.

Concluding Remarks

This study is one of the first to explore the use of computational text analysis in an analysis of the development of the writing style of judges at the ECJ. It documents that the writing style of the ECJ has evolved in a way that can be tracked by means of a range of different measures. It also finds that judges serving at the ECJ at a specific point in time use function words in a way that makes it possible to isolate a specific "style of the time," while judges from countries with the same legal tradition do not seem to conform in their use of function words more than other judges. Finally, the study sets out to uncover some of the drivers of the change in style over time. While it fails to provide evidence for the hypothesis that the various rounds of EU enlargements have had an effect on the writing style of the ECJ, its findings suggest that judges from French-speaking countries throughout the years were able to influence the development of the writing style of the ECJ to a higher degree than other judges. This can be interpreted as evidence that the French legal culture still dominates the court to some extent at least.

More generally, the study demonstrates how computational methods can inform debates about judicial institutions. The writing style of the ECJ is a contested topic, with some commentators accusing the ECJ of clinging to an outdated French version of opinion writing, while others defend the court against such criticism by arguing that its writing style has changed considerably over time. So far, most commentators have relied solely on qualitative evidence in support of their arguments. This study provides, apparently for the first time, quantitative evidence of the development of the ECJ's writing style over time. It will be interesting to see how this debate, as well as other debates, will change in the face of a growing availability of evidence gained from analyzing large quantities of legal texts by means of computational methods. 🖛

Acknowledgments

I thank Marion Dumas, Michael Livermore, and Dan Rockmore for helpful feedback. Valuable research assistance by Irma Klünker is gratefully acknowledged. Errors and omissions are solely mine.

Appendix

Table 7.4. Function words and phrases used in the present study.

Single words
a actuellement afin ai ailleurs ainsi ait alors ancien ancienne anciennes anciens apparement apparent apparente apparentes apparents apres assez au aucun aucune aucunes aucuns auquel auquelle aurait auront autant autre autrefois autrement autres aussi aussitot autour autre autrefois autrement autres autrui auxquels auxquelles avant avec aura autant aux auxquels auxquelles avaient avais avait avant avoir avons ayant beaucoup bien bon bonne bonnes bons c ca car cause ce ceci cela celle celles celui cependant certain certaine certaines certains certes ces cet cette ceux chacun chacune chaque chez ci clairement comme comment complet complete completes complets compris comprise comprises conformement consequemment considerablement contre correct correcte correctes   corrects couramment d dans de dedans dehors deja depuis dernier derniere dernierement dernieres derniers derriere des desormais dessous dessus devait devant devraient devrait   dire directement dis dit dite dites doit doivent donc dont du due durant dus effectivement egalement elle elles en encore enfin ensemble ensuite entendu entier entiere entieres   entiers entre envers environ essentiellement est et etaient etant etais etait etc ete etions etre eu eurent eut eux evidemment excepte exclusive exclusivement facilement facon faire faisant fait faite faites faut fois font furent fut generalement grave graves haut hors ici il illegitime illegitimes ils immediatement in inclu inclue inclus inclus inclusive inclusivement incorrect incorrecte incorrects incorrectes injust injuste injusts injustes j jadis jamais je jusque jusqu just juste justs justes l la laquelle le legitime legitimes lequel les lesquelles lesquels leur leurs loin lors lorsqu lorsque lui ma maintenant mais malgre maniere me meme memes mes met mettait mieux mis mise moindre moins mon montant n naturellement ne neanmoins ni nombreuse nombreuses nombreux non normalement notamment nous nos notre nul on ont or ostensif ostensive ou ou outre par pareillement parfois parmi part particulierement partout pas pendant permet permettait permettant permettent permettraient permettrait permis peu peut peuvent plein pleine pleines pleins plupart plus plusieurs plutot pour pourtant pourra pourraient pourrait pouvant premier premiere premieres premiers pres presqu presque principalement probablement prochain prochaine prochainement prochaines prochains propre pu puis puisque puisqu puisse qu quand quant quasi que quel quelle quelles quelqu quelque quelquefois quelques quels qui quiconque quoi quoique raison rapidement recemment rien s sa sait sans sauf saurait se selon sera serait serieuse serieusement serieuses serieux seront ses seul seule seuls seules seulement si sinon soient soit son sont sous souvent subsequemment suis suivant suivante suivantes suivants surtout sur ta tant tantot te tel telle tellement telles tels tes ton toujours tous tout toute toutefois toutes tres trop tu un uns une unes unique uniquement va vers vide vides vient viennent voire vont vos votre vous vraiment y

Bigrams
a bas a part a nouveau a savoir a travers apres cela attendu que au-dela au-dessous au maximum au minimum au moins autant de bien que bien sur celui-ci chaque fois charge de ci-apres ci-dessous combien de comme quoi considerant que d'abord d'ailleurs d'ou d'un d'une de l de la de nouveau denue de depuis peu du coup du moins en apparence en attendant en cela en consequence en bas en effet en fonction en outre en commun en tout entre-temps grace a grace au grace aux hors de hors d la-bas la-dessus lors de meme que nulle part par consequent par exemple par lequel parait-il parce qu parce que peut-etre.   pour cela quand meme sans doute s'agit s'agissait sur quoi tandis que y compris

Trigrams
a cause de a ces mots a cote de a partir de au cours de au lieu de au lieu que au sein de c'est pourquoi dans l'immediat de sorte que de toute facon en cours de en cours d en dehors de en face de en raison de il arrive que n'importe comment n'importe lequel n'importe ou n'importe quand n'importe quel n'importe qui n'importe quoi par la presente par le present pas du tout tous les jours tout a fait tout de suite tout le monde

4-grams & 5-grams
a la place de a la suite de c'est-a-dire par la faute de ou que ce soit s'il te plait s'il vous plait tout a l'heure a la suite de quoi un de ces jours un jour ou l'autre

ᑲ

A COMPUTATIONAL ANALYSIS
OF CALIFORNIA PAROLE
SUITABILITY HEARINGS

Hannah Laqueur, University of California, Davis

Anna Venancio, Primer AI

Introduction

There are over thirty-five thousand inmates in California serving life sentences with the possibility of parole—over a quarter of the entire state prison population. For these inmates, the sentencing phase does not end with a judge in the courtroom. Instead, their freedom depends on an affirmation by the Board of Parole Hearings ("the board") that they are suitable for release from prison. The board holds tremendous discretionary power over inmates' lives, determining whether or not they will die in prison. Yet relatively little is known about the decision-making process. Final deliberations among the parole commissioners are held in private, with only the final decision revealed. Hearings are open only to the victim's family and, with permission, members of the press. However, the *transcripts* from California parole hearings are publicly available.

This chapter uses machine text processing to analyze over eight thousand transcripts (the population of transcripts from hearings held between 2011 and 2014) to provide a window into this relatively invisible and understudied process.[1]

---

[1] There have been two empirical reports on California lifer parole, but they analyze a sample of only 754 hearing transcripts from suitability hearings conducted between October 1, 2007, and January 28, 2010. Our analysis,

The central question before the board is whether the inmate poses an unreasonable risk of danger to society if released from prison. The board has extensive latitude in making this determination. They may consider "all relevant, reliable information available to the panel" (Cal. Code Regs., tit. 15, §2281). The board's assessment includes paper files indicating the offense for which the inmate was committed, the inmate's psychological risk assessment score, their prison record, and letters of support. Beyond these facts on paper, the board also holds a hearing, typically lasting two to three hours. In the words of the Board of Parole Hearings executive officer Jennifer Shaffer, this is the board's opportunity to try to understand, "Who were you then? Who are you today, and what's the difference?"[2]

Inmates and their attorneys view the hearing to be a critical performance. The Post-Conviction Justice Project's guide for California parole attorneys describes the work of preparing the client to testify as "the most important thing you can do to assist them in being found suitable for parole."[3] The hearing is seen as an opportunity for the inmate to narrate a story of redemption: to demonstrate "insight" into the commitment offense, express remorse, and describe efforts to reform.

In this chapter, we use natural language text processing to explore inmate speech during the hearing and assess whether the narration may make a difference. We also identify nonverbal factors that are predictive of parole outcomes, such as the inmate's

---

by contrast, uses the population of hearings from 2009 to 2014, over twelve thousand transcripts, which allows for more current and more extensive analyses (Weisberg, Mukamal, and Segall 2011). We focus on 2011 forward because this represents the current parole administration era.

[2] See J. Ulloa, "More California Inmates are Getting a Second Chance as Parole Board Enters New Era of Discretion," *The Los Angeles Times* (2017), http://www.latimes.com/politics/la-pol-ca-parole-board-proposition-57-20170727-htmlstory.html.

[3] http://uscpcjp.com

psychological risk score, and analyze potential cognitive biases among the parole decision-makers. We find the psychological risk score is by far the most important predictor of whether or not an inmate will be granted parole: inmates evaluated as being a "high risk" for future violent acts and offending are virtually certain to be denied parole, whereas over half of the inmates identified as "low risk" are granted parole. Given the importance of risk, we focus much of our analyses on hearings for low-risk inmates, a setting in which the decision is more likely to be affected by other factors.

The chapter proceeds as follows: We first provide background on the California parole system for inmates serving life sentences with the possibility of parole ("lifers"). This includes a discussion of the legal, political, and administrative changes over the last decades. We then offer a description of the current parole suitability hearing process and describe the construction of the dataset built using computer text parsing of the universe of transcripts of California Parole Board hearings held from 2011 to 2014. Based on these data, we discuss the variables that can be consistently pulled from the transcript text and that are predictive of a grant or denial of parole. We also use an instrumental-variables approach to estimate the causal effect of risk score on the likelihood of a parole grant. The subsequent analysis examines inter- and intra-decision-maker inconsistency in the parole hearing process. This includes estimation of the extent to which key system actors—presiding commissioners, deputy commissioners, and psychologists—affect an inmate's chances of parole. Additionally, the section analyzes decision-maker variation driven by cognitive biases or extraneous factors that should have no bearing on the decisions. We then turn to an analysis of inmate speech during the suitability hearing. We use a bag-of-words approach, focusing on unigrams and bigrams for classification. Finally, we offer some concluding thoughts on our analysis and avenues for future work.

Trends in California Parole

The system of indeterminate sentencing and release through parole reached a high point in California, and nationally, in the 1970s. At the time, more than 95% of prisoners in California were released by the discretionary decision of a parole board, and more than 70% of inmates in the United States as a whole were released by parole (Petersilia 2003). In 1977, California overhauled its indeterminate sentencing system, replacing it with a determinate system for almost all crimes. Indeterminate sentences remain only for those convicted of the most serious crimes and given life sentences with the possibility of parole. Over 90% of inmates serving life terms with the possibility of parole are in prison for first- or second-degree murder. This shift in sentencing reflected broader nationwide changes in criminal justice thinking and policymaking. It was a time of pessimism about the possibilities of rehabilitation, conservative concerns that criminals were being coddled, and progressive worries that the discretionary system of indeterminate release led to discrimination and inequality.

Under the current system, the determination of parole suitability for those serving life sentences with the possibility of parole is made by the Board of Parole Hearings ("the board"), an executive branch agency within the California Department of Corrections and Rehabilitation (CDCR). For decades, the possibility of parole on the books was meaningless in practice— almost no one faced an actual possibility of parole release. Throughout the mid-1980s until the early 2000s, on average, the board granted parole in only 2% of scheduled hearings. In the rare instances in which the parole board determined an inmate suitable for parole, the governor almost always reversed the decision. During Governor Gray Davis's tenure, from 1999 to 2003, he vetoed 98% of the parole recommendations, resulting in a total of only two inmates paroled; between 2003 and 2011, Governor Arnold Schwarzenegger vetoed an average of 73% of the parole recommendations that came before him (Sarosy 2013). This refusal to grant parole contributed to the tripling of the lifer population in California's prisons since the late 1980s. In the late 1980s, an inmate sentenced to a life term with the possibility of parole for second-degree murder served an average of five years; two decades later, he or she would serve an average of twenty-four years (Mullane 2012). There are now roughly thirty-five thousand inmates serving life sentences with the possibility of parole— over 30% of the population in prison. California has the highest proportion of inmates serving life sentences with the possibility of parole of any other state in the nation (Weisberg, Mukamal, and Segall 2011).

There has been a radical shift in the rate of parole release in recent years. As shown in figure 8.1, the number of grants by the board has been increasing steadily since the early 2000s, with a substantial rise since 2007. The *rate* of release by the board has also increased significantly. Because hearings may be scheduled but then postponed or stipulated by the inmate, the best measure of the rate of release is the percentage of cases granted out of the number of hearings actually conducted. Data on this have been available since 2007 and are presented in figure 8.2. In 2007, 8% of hearings resulted in a grant by the board; in 2014, the rate of release was 36%.[4]

The recent change in the board's grant rate tells only part of the story. There has also been a substantial shift in

---

[4]Data from 2007 to 2008 are from the Stanford Criminal Justice Center *Life in Limbo* Report. Data from 2009 to 2014 are from this paper's parole hearing transcript dataset.

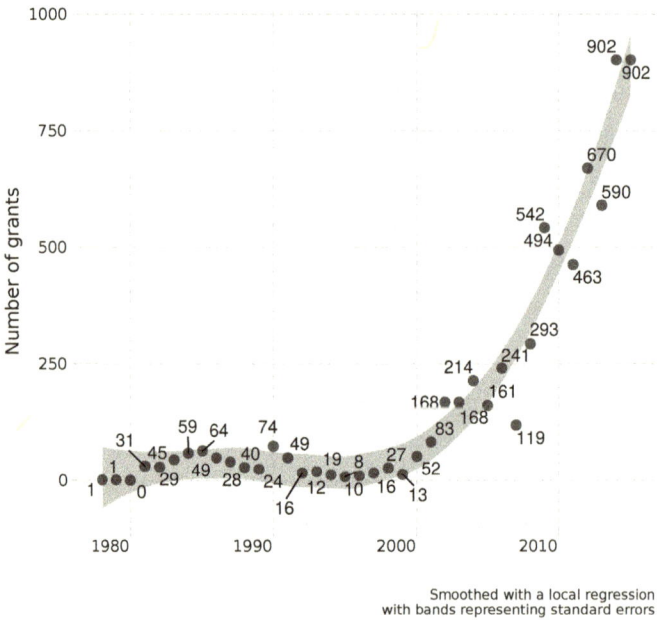

Smoothed with a local regression
with bands representing standard errors

Figure 8.1. Number of hearings resulting in a grant: 1978–2015. Smoothed with a local regression with bands representing standard errors.

The recent change in the board's grant rate tells only part of the story. There has also been a substantial shift in the governor's rate of approving his appointed board's release decisions. California is one of only four states in which the governor has this power of reversal. (Louisiana, Maryland, and Oklahoma are the others.) This additional layer of administrative review was instituted in 1988 with the passage of Proposition 89. The governor has the power to reverse board decisions in all murder cases. For offenders convicted of a crime other than murder, the governor cannot directly reverse the decision but can refer a parole grant back to the board to conduct an *en banc* board review. The current governor, Jerry Brown, has departed from his predecessor's practice of routine reversals, affirming over 80% of

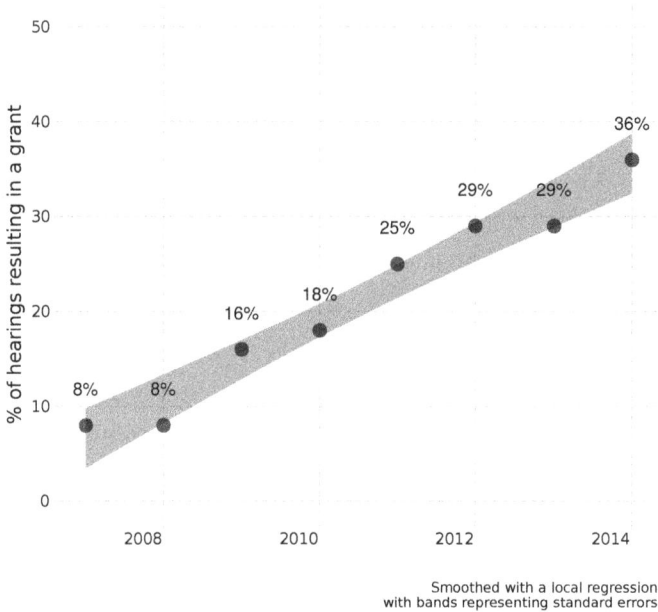

Figure 8.2. Rate of parole grant: 2007–2014. Smoothed with a local regression with bands representing standard errors.

the parole board's release decisions since he took office in 2011. Thus, Governor Brown, rather than compensating for the board's increasing leniency, has re-enforced and affirmed the trend in the board decision-making.

What led to the substantial change in the rate of parole release? And how do we explain the change in the governor's position toward his board's decisions? At least in part, these changes may be explained by recent changes in the law brought about by the California Supreme Court. In 2008, the court held that the board and the governor must reach the release decision based on an inmate's "current dangerousness." Simply citing the heinousness of the inmate's commitment offense is now an insufficient basis on and governor establish some rational nexus between the evidence and a conclusion the inmate still poses a criminal risk, may make blanket denials less likely. At the very least, even if the law does not actually demand a higher threshold, it may offer political cover. Indeed, Brown has explained his lower reversal rate by stating, "I'm bound to follow the law."[6]

The increasing rate of release also came in the shadow of the state budget crisis and the United States Supreme Court's 2011 *Brown v. Plata* decision, which mandated that California reduce its prison population. Despite speculation by many that prison overcrowding encouraged more leniency in parole grants, the Brown administration nonetheless has insisted lifer releases are unrelated to any efforts to reduce the prison population numbers. On the other hand, Brown has recognized changes in public attitudes on crime. "There's still public safety (as a concern), but there's different dominating issues."[7]

Whatever the causes of these changes in California parole, we are at a moment in which, for the first time in decades, the possibility of parole means a real possibility. With this possibility may come actual incentives that encourage the convicted to reform—the putative purpose of parole. Engaging in productive in-prison programming, avoiding in-prison infractions, and narrating an understanding of past wrongs and future ambitions may have a substantive impact on the chances of release. In short, parole hearings now serve a purpose and represent a real process. In what follows, we describe the details of this process and then turn to an analysis of relevant decision variables and inmate speech.

---

[6]See B. Egelko, "Brown Paroles More Lifers than Did Predecessors" (2011), https://m.sfgate.com/crime/article/Brown-paroles-more-lifers-than-did-predecessors-2373800.php.

[7]See B. Egelko, "Brown Paroles More Lifers than Did Predecessors" (2011).

The Parole Hearing Process

California lifer inmates become eligible for parole consideration a year before their minimum eligible parole date, a date set
at the time of sentencing. As a general rule, the initial and all
subsequent hearings are presided over by a commissioner and
deputy commissioner. There are twelve full-time commissioners, each appointed by the governor and confirmed by the senate for staggered three-year terms with eligibility for reappoint-
ment. Deputy commissioners are civil servants who need only
to have a "broad background in criminal justice" (Hopkins and
King 2010). There are approximately seventy deputy commissioners across the state. Hearings are usually conducted at the
prison where the inmate resides and usually take between two
and five hours. Commissioners travel across the state to preside
over hearings, although they are more frequently assigned to
prisons near their residences. The identities of the commissioners assigned to hearings at a given prison in a given week are kept
confidential until the week's hearings begin.

Before every hearing, the board receives the central file, which
includes the inmate's behavioral record in prison, vocational
and education certificates, and the results of the psychological
evaluations assessing recidivism risk. The hearing itself generally
proceeds in three parts: first there is a discussion of the
commitment offense and the inmate's preconviction behavior and
circumstances; then, a consideration of postconviction factors,
including prison behavior, program participation, and the risk
assessment score; and finally, a review of postrelease parole plans
and, if applicable, any statements of support for or against the
inmate's release.

As the California Supreme Court stated in the 2008
In re Shaputis decision "[T]he paramount consideration
for both the Board and the Governor under the governing

statutes is whether the inmate currently poses a threat to public safety."[8] Prior to the 2008 decision, the board, rarely granting parole, would often cite the incarcerating offense as the sole justification for the denial. The board may no longer make decisions based on mere retributive impulses; the central question before the board is whether the inmate poses a risk of future violence and recidivism. In making this determination the board considers a host of factors, perhaps most importantly, the psychological risk assessment score. Until 2013, psychologists from the Board of Parole Hearings' Forensic Assessment Division (FAD) generated a risk score using a combination of instruments known as the HCR-20, LS/CMI, and PCL-R and expressed that score in terms of one of five risk levels: low, low-moderate, moderate, moderate-high, and high. Since 2014, the board has revised its evaluation procedures slightly by ceasing to use the LS/CMI and expressing the risk rating in terms of one of three levels: low, moderate, and high. The psychologist's evaluation may incorporate consideration of the commitment offense, historic risk factors, institutional programming, and the inmate's past and present mental state. Evaluations are valid for five years; however, if an inmate's petition to advance is granted, they are given a new subsequent risk assessment before the next scheduled hearing.

In addition to the risk score, the board may consider factors including whether the inmate has a violent criminal history, exhibits signs of remorse, has plans for the future, has been engaged in institutional activities, vocational and educational certificates, participation in self-help groups such as Alcoholics Anonymous or Narcotics Anonymous, and the extent of his or her misconduct in prison, such as infractions for fights, use of

---

[8] In *re Shaputis*, 190 P.3d 573 (Cal. 2008).

drugs, or possession of a cell phone. The board also considers any letters of support or opposition to the inmate's release and testimony from victims or victims' kin.

If an inmate is found unsuitable for parole, the law requires that a subsequent hearing be set three, five, seven, ten, or fifteen years in the future. At each subsequent hearing the board reviews transcripts from previous hearings along with an updated record. Despite the fact that fifteen years is the default by law, few subsequent hearings are set for fifteen years: in more than 90% of cases in our dataset, the subsequent hearing is set for three years from the time of denial. Under Marsy's Law, the board automatically reviews and may advance a three-year denial one year after the denial was issued. At the same time, an inmate may also request an advanced date regardless of the denial period. If the board determines there has been a "change in circumstances or new information" and there exists a "reasonable likelihood that consideration of public and victim's safety does not require the additional incarceration," a hearing date will be advanced (California Penal Code section 3041.5(b)(3)).

If the commissioners find the inmate suitable for parole, the decision is subject to review by the full board. This rarely happens in practice. Absent action by the full board, the grant of parole becomes final 120 days after the hearing and then goes to the governor for his review. Again, in California, the governor has the unusual power to reverse the decision of the parole board for inmates serving sentences for murder, the vast majority of lifers.

A parole candidate scheduled to go before the board may seek to waive, stipulate, or postpone the hearing. Waivers may be requested for anywhere between one and five years. A request for a short-term delay (postponement) for "exigent circumstances" such as emergencies, illness, or incomplete files may also be made. Both waivers and postponements are granted as a matter of course if made forty-five days before the scheduled hearing, but there are limits as to the number of waivers and postponements an inmate may request. Alternatively, an inmate may stipulate to unsuitability for a period of three, five, seven, ten, or fifteen years in accordance with the current denial periods (California Penal Code, 3041.5(b)). A stipulation is similar in nature to plea bargains, and the board may choose to deny a stipulation. In practice, however, the board appears to grant the vast majority of them.

Finally, all inmates are entitled to a state-appointed attorney at the parole hearing if they have not retained an attorney. These "panel" attorneys are compensated $50 an hour with a cap at $400, which means they spend a maximum of eight hours per case. Advocates and private attorneys with whom we have spoken contend that, not unlike plea bargaining in criminal court, the panel attorneys often try to persuade parole candidates to stipulate to a denial and defer the hearing rather than spend the time going through the hearing process.[9]

The Dataset

Through a public records request, we obtained all parole suitability hearings conducted between 2011 and 2014. The first year in which the universe of transcripts is available electronically is 2009; however, we focus our analyses on 2011 forward because this marks the beginning of the current Board of Parole Hearings administration under Governor Jerry Brown and executive officer Jennifer Shaffer, and the period

---

[9] See J. DeBacco, Bridges to Freedom (2010), https://bridgestofreedom. files.wordpress.com/2014/10/a-b2f-front-pages.pdf.

after the important *In re Shaputis* California Supreme Court decision. We obtained the records as PDFs, which we then converted to text files. They have an intrinsic page layout, which is flattened in the text file. This is strongly reflected in the index at the beginning of the file (see an example in the following table).

Table 8.1. Example of an index.

	Page
Proceedings...................	3
Case factors...................	9
Precommitment factors...	14
Postcommitment factors.	15
Parole plans...................	16
Closing statements..........	20
Recess...........................	21
Decision........................	22
Adjournment..................	24
Transcript certification....	25

The presence of the index underlines the predetermined flow of the parole hearing. This can be very useful in order to extract structured pieces of the information from the raw text (for instance, parole outcome), but also implies that some text cleaning may be necessary to be able to highlight only transcribed speeches. Beyond these architectural elements, the transcripts have another very distinctive attribute to take into account during text analysis: a dialogue structure. In order to focus on the inmate's part of the dialogue, the text must be parsed to selectively retain only the inmate's contribution.

FROM RAW TEXT TO STRUCTURED DATA

We used Python regular expressions to pull key information from each hearing transcript. These variables include the result of the parole hearing, the commitment crime, and the psychological risk assessment score, as well as the identity of the evaluating psychologist, the minimum eligible parole date, the inmate's lawyer, the district attorney (if present at the hearing), the number of victims or victims' next of kin present at the hearing, whether or not an interpreter was present at the hearing, the results of any previous suitability hearings, the inmate's date of entry into prison, the presiding and deputy commissioners, the date and time of the hearing, and the results of the immediately preceding hearings.

Figures 8.3 and 8.4 present snapshots of hearing transcripts, highlighting variables that can be consistently retrieved from the raw text and that might impact the parole board's decision.

PRESIDING COMMISSIONER CHRONES: All right. Good morning, everybody. The time is 9:00 a.m., and this is a Subsequent Parole Consideration Hearing for Leo Robles, CDC Number B-00842. Today is March 13th, 2012. We are located at the California Men's Colony. Mr. Robles was received on February 18th, 1966, from Santa Clara and Marin Counties. The controlling offense for which he was committed include Murder in the First Degree, case number 41303. Two counts of Penal Code Section 187, Assault with Intent to Commit Murder, case number 2780. And a—Penal Code Section 217, and an additional count of Assault in Prison under the same case number as the previous, I believe, with an additional count of 12022 (b) PC. Mr. Robles has a minimum eligible parole date of August 12th, 1977. This hearing is being recorded, and for purpose of voice identification.

Figure 8.3. Structured data in the presiding commissioner's opening statements (example of extracted information).

FROM RAW TEXT TO INMATE DIALOG

During the parole hearing, several key figures participate, asking questions to prompt the inmate and the inmate's attorney to shine light on the inmate's past and current behavior. Because we not only want to extract key variables but also analyze the inmate's participation in the dialogue, this means building a parser able to selectively extract parts of the text, as shown in figure 8.5.

> DEPUTY COMMISSIONER CASADY: Okay.
> PRESIDING COMMISSIONER PRIZMICH: So, let me go through a brief description of 115s. February 21, 1975, Attempted Mail—Subverting the Mail, I guess. He was found guilty on that. February 21, 1975, Possession of a small amount of marijuana, found guilty; [. . .]
> —
> DEPUTY COMMISSIONER TURNER: Okay. Did you get any kind of, like, laudatory chrono from your supervisor saying, you know, you're a good worker and all that?
> —
> Dr. Twohy formed the opinion that you present a low to moderate risk for violence in the free community.

~ 207 ~

Figure 8.4. Structured data in the rest of the text (example of available information).

> DEPUTY COMMISSIONER ZARRINNAM: Ali Zarrinnam, Z-A-R-R-I-N-N-A-M, Deputy Commissioner.
> INMATE ROBLES: I'm Robles, R-O-B-L-E-S. Number B—as in boy—00842.
> ATTORNEY EISENSTAT: Your first name, sir. INMATE ROBLES: Pardon?
> ATTORNEY EISENSTAT: Your first name. INMATE ROBLES: Leo, L-E-O.
> PRESIDING COMMISSIONER LABAHN: And, Mr. Robles, it will be helpful—
> INMATE ROBLES: Oh, okay. I'm sorry.

Figure 8.5. Parsing the text to obtain only the parts of the dialogue spoken by the inmate (examples of sentences to extract).

Preprocessing steps relied on *spaCy*, a free open-source library in Python for the Natural Language Processing Toolkit. We describe them in more detail below.

Nonverbal Factors that Influence Parole

By far the strongest predictor of the release outcome is the inmate's risk score. A rating of high or moderate-high essentially forecloses the possibility of parole: less than 3% of the 1,193 of the inmates in our dataset with a high or moderate-high risk score are granted parole. On the other hand, 56% of the 1,673 inmates given a low risk score are granted parole.

We estimate the causal impact of the risk assessment score using an instrumental-variables approach (Angrist, Imbens, and

Table 8.2. Effect of risk score on grant chances.

	IV Estimate	Naïve OLS Estimate
Risk (1–5)	-0.154 (0.023)	-0.161 (0.004)

Rubin 1996). Risk assessments are completed by one of forty-six psychologists in the Board of Parole Hearings Forensic Assessment Division. The FAD psychologists work exclusively on conducting these assessments for use in suitability hearings. Evaluations are conducted in person using a combination of instruments—HCR-20, LS/CMI, and PCL-R—as well as "clinical judgment." The process of pairing inmates with psychologists is based solely on geography. Thus, within a given prison and year, an inmate's assignment to one of the forty-six psychologists is essentially random. It is thus possible to use the systematic differences in psychologists' penchants for assigning higher (or lower) risk ratings, measured as a cardinal variable from a score of 1–5, with 1 being low risk and 5 high risk. This allows us to estimate the causal effect of risk on the likelihood of parole grant. The results are presented in table 8.2, which shows the estimate of the average effect of moving down a risk score level on the likelihood of parole release. The instrumental-variables estimate is slightly smaller and the confidence interval wider than a simple regression estimate, but the two estimates are remarkably similar: the average of moving up each risk assessment score category (i.e., from low to low-moderate) results in a 15% reduction of prison release chances. We ran several robustness checks, including models with controls for prison and year, and found estimates of similar magnitude and significance.

OTHER POTENTIALLY RELEVANT VARIABLES

Table 8.3 presents other variables that we can consistently computationally retrieve from the hearing transcripts that could impact the parole decision: the inmate's commitment offenses, whether the inmate used an interpreter, and the presence of victims at the hearing. We present the simple difference in mean release rate as well as adjusted estimates controlling for other relevant variables, including the prison where the inmate is housed, their attorney, and the number of previous hearings. The adjusted estimates are made using targeted maximum likelihood estimation (TMLE), a doubly robust estimation technique that incorporates data-adaptive modeling and has been shown to be less statistically biased and more efficient than conventional parametric methods (Van der Laan and Rose 2011). At the same time, even after adjusting for covariates in a rigorous manner, there are too many possible unmeasured confounders—variables that may affect both the variable of interest and the outcome—to allow for these estimates to be interpreted causally. Instead, the measures offer a means of noting important associations that are not easily explained away and thereby raise research questions. For example, we find the presence of an interpreter is strongly associated with the likelihood of release, even after adjusting for an assortment of other variables. Does speaking through an interpreter actually result in a sharp drop in one's chances of getting parole because it is harder for the inmate to connect personally with the commissioners? Or is speaking through an interpreter simply associated with other variables we can't easily extract from transcripts, such as inmate ethnicity?

In most cases, the grant rate does not appear to vary with the commitment offense. One clear exception are offenses involving a sex crime. After controlling for other variables, we find the grant rate is an estimated 13% lower when the inmate's commitment offense included a sex crime. This finding comports with evidence presented in previous research (Weisberg, Mukamal, and Segall 2011) and conventional wisdom among attorneys and advocates

Table 8.3. Likelihood of parole grant: Variable importance measures.

	Bivariate Estimate	Adjusted Estimate
Murder	0.045 (0.015)	0.004 (0.014)
Robbery	-0.001 (0.02)	0.031 (0.019)
Kidnap	-0.033 (0.019)	-0.004 (0.019)
Attempted	-0.012 (0.018)	0.013 (0.018)
Sex Crime	-0.186 (0.028)	-0.125 (0.02)
Interpreter	-0.032 (0.016)	-0.066 (0.015)
0 vs. 1 Victim	-0.015 (0.04)	-0.001 (0.044)
0 vs. 2 Victims	-0.008 (0.03)	-0.02 (0.033)
0 vs. 3 Victims	-0.058 (0.027)	-0.057 (0.028)
0 vs. 4 Victims	-0.073 (0.031)	-0.06 (0.031)
0 vs. 5 Victims	-0.085 (0.039)	-0.073 (0.038)
0 vs. More than 5 Victims	-0.139 (0.033)	-0.141 (0.027)

that suggests the board is particularly unlikely to release an inmate whose crime involved sexual violence, especially if the crime involved a child. As we see in the text analysis in the section "Inmate Expression Analysis" below, the word "sex" is also negatively associated with the chances of a grant of parole.

The role victims play in the parole suitability hearing has been the subject of considerable debate since the passage of Marsy's Law (Proposition 9) in 2008. Marsy's Law, among many things, gave victims and their next of kin expanded rights to receive notice and testify at suitability hearings. It was also under Marsy's Law, ostensibly in an effort to relieve the hardship placed on victims of attending hearings, that the deferral lengths between hearings was extended to the current 3–15 years period.[10] The Stanford Criminal Justice Center Report (Weisberg, Mukamal, and Segall 2011), the first study to offer an empirical portrait of the California parole process, used a sample of hand-coded

[10]See J. DeBacco, *Bridges to Freedom* (2010).

transcripts from 2007 to 2010 and found the overall grant rate when victims attended hearings was 5%, as compared to 14% when victims did not attend. The estimates presented here look at the influence of victim attendance at the hearing ranging from zero in attendance to more than five. This analysis shows that victim attendance at the parole hearing is an important predictor of an inmate's parole chances. While there is minimal difference when one or two victims are at a hearing as compared to zero, the chances of parole drop precipitously as more victims are in attendance: hearings with more than five victims result in a grant roughly 14% less often than hearings with no victims in attendance, even after adjusting for highly predictive variables.

The estimate of the rate of release at hearings conducted through an interpreter is 7% lower than at hearings in which an interpreter was not present. Again, it cannot be concluded that speaking through an interpreter actually causes the board to be less likely to grant parole. It may be that unmeasured factors, such as ethnicity, are associated with having an interpreter and with lower chances of parole. Or perhaps the need for an interpreter represents a broader set of disadvantages faced by inmates who do not speak English and therefore have a harder time engaging in prison programming and planning for postprison release.

Given that essentially no inmate with a psychological risk score of high or moderate-high is granted parole, whereas over half of low-risk inmates are granted parole, we performed the same set of variable analyses on the subset of low-risk inmate hearings. It is among low-risk inmates that there is theoretically the most room for other factors, such as victim or interpreter presence, to impact the parole decision.

Table 8.4 shows a simple difference in mean grant rate among low-risk inmates for the important variables discussed above next to the estimates for the full population. Table 8.5 shows

LAW AS DATA

Table 8.4. Likelihood of grant among low-risk vs. all inmates (bivariate estimates).

	Bivariate Estimate	Adjusted Estimate
Murder	0.036 (0.033)	0.045 (0.015)
Robbery	0.006 (0.048)	0.045 (0.015)
Kidnap	-0.061 (0.045)	-0.033 (0.019)
Sex Crime	-0.215 (0.071)	-0.186 (0.028)
Interpreter	-0.112 (0.031)	-0.032 (0.016)
Victim Present	-0.155 (0.027)	-0.063 (0.014)
0 vs. 1 Victim	-0.089 (0.076)	-0.015 (0.040)
0 vs. 2 Victims	-0.045 (0.059)	-0.008 (0.030)
0 vs. 3 Victims	-0.119 (0.049)	-0.058 (0.027)
0 vs. 4 Victims	-0.200 (0.054)	-0.073 (0.031)
0 vs. 5 Victims	-0.216 (0.069)	-0.085 (0.039)
0 vs. More than 5 Victims	-0.267 (0.060)	-0.139 (0.033)

the adjusted estimates for low-risk inmates versus the adjusted estimates for the full population.

As was the case among the full population, crime type appears to generally make little difference in the parole outcome among low-risk inmates, with the exception of sex crimes. In this case, among low-risk inmates, serving time for a sex crime is associated with a 17% lower chance of parole, controlling for other variables, as compared to 13% among the full population. Both the presence of an interpreter and the presence of victims appear to be even *more* important in the parole decision for low-risk inmates. An interpreter is twice as important among the low-risk group. Hearings with five or more victims present result in a grant roughly 25% less often than hearings with no victims in attendance, after adjusting for relevant covariates. This is an approximately 1.8 times greater difference than among the full population.

Table **8.5.** Likelihood of grant among low-risk vs. all inmates (adjusted estimates).

	Bivariate Estimate	Adjusted Estimate
Murder	-0.054 (0.047)	0.004 (0.014)
Robbery	-0.010 (0.059)	0.031 (0.019)
Kidnap	-0.108 (0.067)	-0.004 (0.019)
Sex Crime	-0.165 (0.088)	-0.125 (0.02)
Interpreter	-0.113 (0.031)	-0.066 (0.015)
Victim Present	-0.174 (0.026)	-0.127 (0.013)
0 vs. 1 Victim	-0.129 (0.075)	-0.001 (0.044)
0 vs. 2 Victims	-0.065 (0.058)	-0.02 (0.033)
0 vs. 3 Victims	-0.144 (0.048)	-0.057 (0.028)
0 vs. 4 Victims	-0.224 (0.054)	-0.06 (0.031)
0 vs. 5 Victims	-0.248 (0.068)	-0.073 (0.038)
0 vs. More than 5 Victims	-0.248 (0.059)	-0.141 (0.027)

Inconsistency in the Parole System

"Justice," the saying goes, "is what the judge ate for breakfast" (Kozinski 1992). The problem of inconsistency in judicial decision-making has been documented in a number of contexts. Recent research indicates, for example, that the outcome of a football game (Chen 2016c), the results of the immediately preceding case (Chen, Moskowitz, and Shue 2016), and the time of day (Danziger, Levav, and Avnaim-Pesso 2011a) can substantially affect legal decisions. Also startling are the wide between-judge disparities found in domains including immigration asylum (Ramji-Nogales, Schoenholtz, and Schrag 2007; Fischman 2014), social security disability (Nakosteen and Zimmer 2014), and criminal sentencing (Abrams, Bertrand, and Mullainathan 2012).

In what follows, we offer estimates of the extent to which differences between major decision-makers—presiding commissioners, deputy commissioners, and FAD psychologists—affect parole outcomes. The results suggest a process that contains some real elements of chance. The presiding commissioner or deputy commissioner can alter whether an inmate is released or remains in prison. On the other hand, there is no evidence of *intra*-commissioner inconsistency. The time of day of the decision is uncorrelated with outcomes, a finding at odds with a highly publicized study of the Israeli parole board (Danziger, Levav, and Avnaim-Pesso 2011a). We also found no effect from the so-called "gambler's fallacy," a psychological effect documented in other judicial decision-making contexts (Chen, Moskowitz, and Shue 2016). In sum, the findings, although they cannot speak to the inherent quality of the release decisions themselves, do suggest that individual officials operate in a relatively consistent manner (at least with respect to the studied variables).

INTER-DECISION-MAKER INCONSISTENCY

The following analysis of inter-decision-maker inconsistency provides two measures of decision-maker differences: average and extreme, building on the approach introduced in Fischman (2014). The average estimate provides an overall measure of how inconsistency affects outcomes. This estimate of average difference may be interpreted as the probability that a hearing would be decided differently by two randomly selected decision-makers due to one of them being systematically more likely to grant parole than the other. The estimate of extreme inconsistency is the difference in grant rates between the harshest and most lenient commissioner and deputy commissioner. We can think of this as the percentage of cases that could come out differently if they were assigned to the most lenient decision-maker rather than the harshest.

ESTIMATION PROCEDURE

Estimates are made using prison–year combination fixed effects with dummy variables for actor identities. Commissioners are assigned to hearings based largely on geography (they are more likely to decide hearings at prisons near their residence), and different security level prisons will have inmates with different characteristics. Commissioners in one year may hear very different types of cases than a commissioner hearing cases in another year. Within prison and year, however, commissioners should see the same type of inmates on average, and thus, at least with an infinite sample, estimates should provide reasonable quantification of actors' relative impacts on the probability that an inmate will be granted parole. We conduct a randomization check and find support for the assumption that decision-makers are randomly assigned to inmates once prison and year are controlled for. We use the presence of a district attorney at the hearing and whether a hearing is an initial hearing or subsequent as test outcomes. If cases are randomly assigned, district attorney attendance and hearing number should not be related to the decision-maker, and indeed, we find no evidence of a relationship.

A second stage is needed in the estimation procedure to correct for finite sample bias and to properly estimate confidence intervals. In a finite sample, estimates will be biased upwards because there will almost always be differences between decision-makers due to chance, even if they would make the same decision in each case. For the measure of average difference in grant rates, we estimate and correct for finite sample bias via subsampling (Fischman 2014). Our estimates of finite sample bias are slightly responsive to the percentage of the dataset we subsample: for presiding commissioner inconsistency, as the percentage subsampled increases from 25% to 55%, our estimate of bias decreases by about 0.5%; for deputy commissioner inconsistency, the estimate of bias decreases by almost 1%. We report bias-corrected estimates derived from a subsampling percentage of 40%. Measures of extreme differences in grant rates are especially susceptible to finite sample bias because we would expect the extremes to regress toward the mean with larger samples. Thus, taking a particularly conservative approach to the estimates of extreme differences, we permute outcomes within prison–year combinations to estimate bias under the null distribution that assumes no actual differences between decision-makers. For both extreme and average measures of inconsistency, we also guard against finite sample bias by limiting our analysis to decision-makers with at least one hundred observations.

ESTIMATES OF INTER-DECISION-MAKER DIFFERENCES

It is worth stressing that these estimates represent lower bounds for inconsistency in decision-making. Insofar as commissioners grant parole differently in different types of cases, the grant rate differentials may understate the degree of inconsistency. For example, one commissioner may decide to grant in 20% of hearings and another in 30% of hearings. While we know that the commissioners decide at least 10% of cases differently, the number could be considerably higher. Imagine, for example, that the first commissioner's grants are all first-degree murder cases while the second commissioner's grants are all second-degree murder cases, and the first commissioner denies parole in all first-degree murder cases. The inconsistency between the two commissioners would actually be 30%, but our estimate would only be 10%.

Nonetheless, even the lower bound estimates of inconsistency in the decision-making are illuminating. As the estimates presented in table 8.6 show, at least 11% of cases would be decided differently based on the presiding commissioner that happens to be assigned to an inmate's case, and at least 15% of cases could be decided differently depending only on the deputy commissioner assigned to hear the case. An evaluating psychologist can affect an inmate's chances of parole by at least 6%, despite the fact that they are not involved in the actual parole decision. As shown in the preceding section, an inmate's risk score is an important determinant of the likelihood they will be released; consistent with *In re Lawrence*, risk is the primary consideration for the board. This raises two questions. First, given the importance of risk, how accurate are the current clinical risk assessment instruments used by the board? Second, are the risk tools being applied consistently? If two clinicians administer a tool to an inmate, the score should not reflect the idiosyncratic tendencies of a given clinician. At the same time, the average differences in parole rates are, as expected, substantially smaller and relatively small: at least 3% of cases could be decided differently if they were randomly reassigned to a different psychologist, at least 6% of cases would be decided differently if they were randomly reassigned to a different presiding commissioner, and at least 7% of cases would be decided differently if assigned to a different deputy commissioner.

Table 8.6. Inconsistency in decision-making: Differences in grant rates.

Decision-Maker	Average Difference	Extreme Difference
Presiding Commissioners	6.1% (5.0% - 7.2%)	11.3% (7.8% - 14.8%)
Psychologists	3.3% (1.2% - 5.4%)	5.6% (3.3% - 9.7%)
Deputy Commissioners	7.0% (5.7% - 8.3%)	14.6% (10.3% - 18.7%)

Given a substantial literature devoted to the study of how judicial decisions are affected by judicial ideology, one might surmise commissioner characteristics could account for the differences in the grant rate. One might expect, for example, that commissioners appointed by Governor Schwarzenegger, a governor known for being "tough on crime," would be less likely to grant parole than commissioners appointed by

Governor Brown (Ball 2011). Yet despite the well-powered dataset, there is surprisingly little difference in grant rates between commissioners with different backgrounds and characteristics: neither appointing governor, gender, prior employment with the California Department of Corrections and Rehabilitation (CDCR), military experience, nor prior service as a parole panel attorney appear to matter in the likelihood the hearing will result in a grant.

INTRACOMMISSIONER INCONSISTENCY

A growing body of work in the judicial decision-making literature has documented extraneous factors and cognitive biases that can make judges internally inconsistent in their decision-making. Most relevant to the present study of California parole, a highly publicized study of Israeli judges found the time of day in which the decisions were made in relation to breaks and meals had a significant impact on the parole decision (Danziger, Levav, and Avnaim-Pesso 2011a). Danzinger et al. found the rate of parole grants dropped from roughly 65% to almost zero as a session neared its end and then rose again to 65% after a session break. The authors speculate that the effect is likely driven by mental depletion. However, more recent work has challenged the results, arguing that cases are not randomly distributed throughout the day and that, because favorable rulings take longer than unfavorable ones, judges may be strategic in the order in which they hear cases (Weinshall-Margel and Shapard 2011).

The California Parole Board does not take such clear and scheduled recesses. But it is possible to capture and assess the effect of the time of day on release outcomes. Table 8.7 shows the differences in grants for each start hour relative to 8:00 a.m., the earliest time in which hearings are scheduled. Start times are rounded to the nearest hour. The model includes prison-commissioner fixed effects. That is, for each presiding commissioner within a given prison, the model measures differences in release decisions that depend on the time in which the hearing is conducted. The results suggest no such significant differences in the outcome regardless of the time of day the hearing begins. One possible exception is late-starting hearings. There is some evidence that hearings starting between 3:00 p.m. and 6:00 p.m. are more likely to end with a grant, but the sample sizes at those times are too small to allow for robust conclusions.

Table 8.7. Effect of hearing start time on grant rate.

Start Time	Grant Rate (relative to 8:00 a.m.)
9:00 a.m.	-0.029 (0.018)
10:00 a.m.	0.005 (0.022)
11:00 a.m.	-0.010 (0.018)
12:00 p.m.	-0.010 (0.010)
1:00 p.m.	0.007 (0.017)
2:00 p.m.	-0.002 (0.021)
3:00 p.m.	0.070 (0.028)
4:00 p.m.	0.062 (0.043)
5:00 p.m.	0.102 (0.088)
6:00 p.m.	-0.029 (0.018)

The "gambler's fallacy"—the mistaken idea that the chances of something occurring increases (or decreases) depending on recent occurrences, despite the fact that the probability of the occurrence is fixed—is another cognitive bias scholars have speculated could impact judicial decision-making (Tversky and Kahneman 1974). In the parole context, this would be the tendency to respond to streaks of grants (or denies) by becoming more likely to deny (or grant) in the next hearing. Chen, Moscowitz, and Shue (2014) document such negative autocorrelation in the decisions made by asylum court judges and find growing effects as the length of a streak of decisions in one direction or another increases. We find no evidence for such effects in the California parole hearing process. Running a model of lagged release decisions regressed on the present hearing outcome does generate statistically significant lagged coefficients, *prima facie* evidence that a streak of previous grants (or denials) increases the probability that a commissioner will grant (or deny) the next inmate's request for parole. For every sequential grant, commissioners are about 3% less likely to grant parole to the next inmate. However, recent work suggests that standard techniques for analyzing the gambler's fallacy are subject to finite sample bias (Miller and Sanjurjo 2018). Indeed, in our dataset, further analysis reveals the result is *not* driven by commissioner psychology but is instead an artifact of finite sample bias. Permuting hearing outcomes, thereby randomly reassigning a grant or deny to each hearing, and repeatedly running the same lagged models on the permuted data generates estimates of negative correlation that are essentially equivalent to the initial estimate.

In summary, there is no evidence that the gambler's fallacy affects parole decisions. The current administration has made efforts toward increasing professionalization—commissioners are sent to national judicial college for training, and efforts have been made to reduce workloads to guard against decision fatigue and to allow for more deliberate decision-making.[11] It is possible that this has had an impact on parole decisions. Another explanation is simply that the cognitive biases documented in previous studies have been overstated. In fact, statistical properties of the underlying data may explain phenomena that have been attributed to decision fatigue or the gambler's fallacy.

---

[11] Personal communication with Howard Mosely and Jennifer Shaffer, December 2014.

Inmate Expression Analysis

Many factors, explicit or not, are naturally taken into account when the board makes parole determination. Here we propose to examine the relationship between the parole outcome and the words the inmates used during the parole suitability hearing.

PREPROCESSING

The parser described above extracts out the parts of the dialogue spoken by the inmate. The resulting text is then tokenized to obtain unigrams (single words) and bigrams (two consecutive words), a standard process in the "bag-of-words" approach to text analysis. This approach discards word order and uses a simple count of the number of times the word appears in the text. The tokenization is performed using *spaCy* in the Python Natural Language Processing (NLP) toolkit and includes partial preprocessing of the text, such as the removal of a subset of punctuation and of all *spaCy* stopwords, i.e., common words such as "a," "and," "like," "this," "that," "so," and "on." Unigrams are also normalized further by applying *spaCy*'s lemmatization algorithm, which reduces all words to their base form. For example, "talk," "talked," "talks," and "talking" are reduced to the verb lemma "talk."[12]

Finally, we zoom into word functions within sentences by relying on *spaCy*'s part of speech (POS) tagger to identify verbs, nouns, adjectives, adverbs, and interjections. This allows us to pay particular attention to more coherent subcategories of the vocabulary, for instance, by selectively retaining only verbs or nouns used by the inmates.

---

[12]Research indicates for simple and common tasks like measuring sentiment, topic modeling, and classification, *n*-grams do little to enhance performance (Hopkins and King 2010).

TEXT-BASED PREDICTION

To examine the relationship between the parole outcome and the words the inmates used, we established a vocabulary (a list of words) and transformed each transcript into a sparse vector by looking at the counts of each word in the preprocessed text. To investigate the impact of words on the outcome of the hearing, we favored explanatory models and found logistic regression to satisfy our needs. Discounting interaction effects, which are hard to capture for words, logistic regression produces coefficients whose magnitude reflects the importance of the features in the model.

We focus the analyses on suitability hearing transcripts for inmates identified as low risk. As described above, almost no inmate with a psychological risk score of high or moderate-high is granted parole, while over half of low-risk inmates are granted parole. We therefore focus our attention on this subset for whom the hearing itself could theoretically make a difference in the parole decision. It is among low-risk inmates that there is the most room for commissioner discretion and room for subtle and subjective factors to impact the decision.

Finally, we restricted the dataset to parole hearings resulting in a clear "deny" or "grant" outcome. Although several outcomes, including "postpone," "waive," or "stipulate" are possible,[13] retaining deny or grant results gives us a clearer signal. The positive class is taken to be the grant outcome.

We begin with an analysis of lemmatized unigrams with stopwords removed. The vocabulary is taken to be the top thousand words from both granted and denied hearings. The features for each transcript are a straightforward word count vector. The dataset is split into training/validation sets, with

---

[13] See http://www.cdcr.ca.gov/BOPH/pshResults.html for a detailed description of the various hearing results.

an 80/20 ratio. A logistic regression classification model is trained using Scikit Learn's LogisticRegressionCV, with the L2 regularization constant determined with cross validation to avoid overfitting.

The accuracy of the model, obtained on the withheld validation set, is 66%. The baseline accuracy, obtained with a simple prediction of majority label, is 56%. Diving a little deeper into the performance of the model, we investigated the overall precision and recall, as well as the performance on a per class level. Overall precision is at 65% and recall is at 66%, giving an F_1 score of 65%. Precision offers a measure of the quality of the label: it is computed as the ratio of correctly predicted positive observations to the total predicted positive observations. Recall gives an idea of the quantity of data points of each class captured: it is computed as the ratio of correctly predicted positive observations to all observations in the actual class. The F_1 score is used to capture both measurements and is computed by obtaining the harmonic average of precision and recall. This underlines that there is some signal captured by the model. Looking at per label performance shows that recall for the minority class is what is keeping the overall performance metrics down, as it is at 47%. The precision of the minority class is at 61% and that of the majority class is at 68%.

Table 8.8 presents the ten most important unigram features of grants and denials. "Thank" shows up as the most important feature for parole grants. This is consistent with the interpretation that inmates who are polite and show respect and deference are more likely to be granted parole. As the Post-Conviction Justice Project's manual for attorneys states: "It is crucial that they not seem hostile, disrespectful, or defensive. The Panel generally looks most favorably on

Table 8.8. Top ten most important unigram features for parole grants and denials.

Grant		Denial	
thank	0.107	would've	-0.045
sponsor	0.032	guess	-0.043
service	0.032	mistake	-0.042
situation	0.032	responsibility	-0.040
turn	0.032	115	-0.036
commissioner	0.03	respect	-0.036
emotion	0.027	sell	-0.032
world	0.027	class	-0.032
alcoholic	0.027	cell	-0.031
care	0.026	lie	-0.031

clients who are respectful, penitent and truly appear to be remorseful."[14]

Much of what we see are words suggestive of factors that we would expect to influence the parole decision, as compared to words expressing sentiment or affect. For example, "sponsor" is likely an indicator of an inmate who has participated in substance abuse programming such as Alcoholics Anonymous, which the board looks upon favorably as an indicator of rehabilitation and reform. "Alcoholic" is also among the ten most important features of a parole grant among low-risk inmates.

With respect to denials, the conjunction "would've" is the most important unigram. To fully understand its meaning, we would, of course, need to look at complete transcript text. But often, "would've" is part of the expression of past unreal conditionals—phrases used to express wishes about the past,

---

[14]http://uscpcjp.com/wp-content/uploads/2016/11/
Parole-Manual-and-Case-Law-Chart.pdf.

often expressions of regret: "Had I known A, I *would have* done B." We know the parole board looks for expressions of remorse and regret, but perhaps this past hypothetical conditional appears to avoid full responsibility.

"Would've" is followed in importance by the word "guess." This may point to the parole board's desire for the inmate to claim ownership of their lives. For example, as shown in the snapshot below of a suitability hearing transcript, the commissioner pushes the inmate on the question of his honesty and ownership of his past transgressions: dealing drugs and pimping. "Were you a pimp at any time in your life?" the commissioner asks. "I guess—I guess I was," the inmate responds.

> And you would deal drugs. Do you recall selling drugs?
> INMATE MITCHELL: Yes, ma'am.
> PRESIDING COMMISSIONER FRITZ: You do. Okay. I guess with the clinician you initially denied selling drugs. Then noted, "I think I may have sold drugs or something like that." And you were vague about having pimped. Were you a pimp at any time in your life?
> INMATE MITCHELL: I guess—I guess I was. I was, like, what do you call a watcher, you know.
> PRESIDING COMMISSIONER FRITZ: Now what does that mean?
> INMATE MITCHELL: When you—if you got somebody working, and you look out for them and stuff like that.
> PRESIDING COMMISSIONER FRITZ: Okay. So, someone, maybe a prostitute, working for you.
> INMATE MITCHELL: Yeah.

Figure 8.6. Transcript excerpt: "Guess."

"Lie," also an important feature of denial, points to the centrality of truthfulness in the board's assessment of insight, remorse, and rehabilitation. "If you're *lying* to the Panel today, what do you think that says about your rehabilitation?" a commissioner asked inmate #P-05816 at his suitability hearing. In another case, arguing against parole, the district attorney states: "He is still trying to present his crime as an insurance fraud that turned into a robbery without his prior knowledge. The prisoner either has no idea why he committed the robbery or he's hiding the truth. Either way, that would make him a risk for future violence if paroled . . . The prisoner has committed perjury before on numerous panels, and some have expressed concern about his truthfulness."

Finally, we also find a number of words predictive of denial that are indicative of negative behaviors; "115" represents a rules violation report. This can be for serious violations (e.g., fighting, possessing escape paraphernalia) or administrative violations. The snapshot below shows an example of dialogue around an inmate's 115 citation for a relatively minor infraction of cursing to a guard after being asked to tuck in his shirt.

PRESIDING COMMISSIONER LABAHN: Okay. So, the 115 indicates that after being told to tuck your shirt in, that you responded fuck you; is that true?
INMATE COLEMAN: That was a while back.
PRESIDING COMMISSIONER LABAHN: That was ten years ago.
INMATE COLEMAN: Yes.
PRESIDING COMMISSIONER LABAHN: I'll tell you what, I'm a little younger than you, not that much, but if I told somebody in a uniform fuck you, I'd probably remember it even ten years ago. I think I'd remember saying that. Did you say that?
INMATE COLEMAN: If they said I said it, I said it, sir.
PRESIDING COMMISSIONER LABAHN: You don't remember it.
INMATE COLEMAN: Sometimes my mind gets a little foggy.
PRESIDING COMMISSIONER LABAHN: All right.

Figure 8.7. Transcript excerpt: "115."

EXAMINING PARTS OF SPEECH

We also examine building classification models on parts of speech: nouns, verbs, adverbs, and adjectives. Table 8.9 presents the model accuracy, precision, recall, and F_1 score for the four classification models. We again use a logistic regression model with cross validation; performance metrics are obtained on the withheld validation set. We find verbs and nouns perform much better than do adjectives or adverbs. Adjectives and adverbs are often taken in natural language processing to be indicators of subjectivity; they are parts of speech used to convey sentiment. That they perform much less well in our classification task offers some indication that the parole decision is, indeed, grounded in the facts as the board sees them rather than in the more subjective way in which the inmate narrates them.

Table 8.9. Comparing parts of speech.

NOUNS: 64% Accuracy			
	Precision	Recall	f1-score
0	0.57	0.45	0.5
1	0.68	0.77	0.72
Avg./Total	0.63	0.64	0.63
VERBS: 64% Accuracy			
	Precision	Recall	f1-score
0	0.57	0.49	0.53
1	0.68	0.75	0.71
Avg./Total	0.64	0.64	0.64
ADVERBS: 60% Accuracy			
	Precision	Recall	f1-score
0	0.54	0.27	0.36
1	0.6	0.83	0.7
Avg./Total	0.58	0.59	0.55
ADJECTIVES: 57% Accuracy			
	Precision	Recall	f1-score
0	0.45	0.1	0.17
1	0.58	0.91	0.71
Avg./Total	0.53	0.57	0.48

BIGRAMS

The inherent drawback of the unigram model is its inability to capture relationships between two words (e.g., a word and its modifier, a word and its negation, etc.) because it treats each word in isolation. In what follows, we consider bigram features. Table 8.10 presents the top ten features again using a logistic regression classification model.

Table 8.10. Top ten most important bigram features of grant and denial.

Grant			Denied		
thank	thank	0.105	I'm	trying	-0.042
thank	yes	0.041	don't	remember	-0.039
's	going	0.034	cell	phone	-0.036
's	house	0.031	don't	recall	-0.031
feel	like	0.03	don't	don't	-0.028
yeah	Yes	0.027	don't	know	-0.028
thank	sir	0.025	know	going	-0.028
sir	thank	0.025	yes	don't	-0.027
don't	care	0.025	I'm	saying	-0.025
don't	need	0.025	ask	forgiveness	-0.024

Many of the bigrams most predictive of grants are expressions of thanks, as we saw with the unigram analysis: "thanks yes," "thanks sir," "sir thank." There are also several affirmations: "thanks yes," "yeah yes," This again comports with advice given in the Post-Conviction Justice Project guide for parole attorneys: "It can be extremely effective to have on record something as simple as your client testifying 'Yes, I did the crime, I'm responsible and I'm sorry.'"

On the other hand, we see a number of negative bigrams that are predictive of denial: "don't remember," "don't recall," "don't don't," "don't know." This difference in positive versus negative sentiment in grants versus denials is illustrative of one of the advantages of using bigrams: we are able to see this negative (and positive) syntactic structure, which would not have appeared when looking at unigrams alone.

Besides these bigrams expressing negative sentiment, several of the expressions are ones regarding memory: "don't remember," "don't recall," "don't know." This is consistent with the idea that the board is looking for inmates to have "insight" into their past misdeeds. Insight is, of course, impossible if you can't remember the set of events in question.

The snapshots below offer two examples from suitability hearings of the inmate telling the board he doesn't remember. In the first case, he tells the board he doesn't remember committing the crime for which he is incarcerated: "I probably did it.... I *don't remember* picking her back up again.... No, I have no idea why I did it, if I did it." Studies of parole release decision-making have consistently demonstrated that an inmate's willingness to acknowledge their culpability and express remorse for the commitment offense is a vital component of the parole decision calculus (Medwed 2007).

> I probably did it. I remember being with her the whole evening. I remember taking her home and the child, too. I don't remember picking her back up again. No, we didn't have an argument. No, I have no idea why I did it, if I did it.

Figure 8.8. Transcript excerpt: "Don't remember."

Worse than not remembering, claims of innocence are essentially never looked upon favorably by the board and substantially diminish the chances of parole. The attorney manual referenced above advises: "The Panel generally looks most favorably on clients who are respectful, penitent and truly appear to be remorseful. . . . If your client maintains their innocence, it is possible for them to be remorseful about what happened to the victim even if they are not responsible for it." While the parole board cannot require the inmate to admit guilt as a condition of parole (Cal. Penal Code 5011(b).), it is a much more challenging task to express remorse for an event in which the inmate claims he had no part.

> INMATE STEPHENS: I don't remember saying I still have the occasional violent fantasies. Now, as far as, you know, objectification of females, sometimes, yes. I mean, Playboy magazine objectifies females, but I try to look at them as—
> DEPUTY COMMISSIONER CHAMBERS: (Inaudible.)
> INMATE STEPHENS: Yeah, I try to look at them as humans and people because I got a little thing that I do. I imagine them with a little stuffed animal, and all of a sudden, they quit being a thing or an object and you can see the person, you can see the warmth, you can see the glow.
> DEPUTY COMMISSIONER CHAMBERS: So, you view them as objects of sexual, for your sexual gratification.
> INMATE STEPHENS: Sometimes, long ago in the past, yeah.

Figure 8.9. Transcript excerpt: "Don't remember."

In the snapshot of a transcript shown in figure 8.9, the inmate is not claiming innocence or even denying the commitment offense. The exchange is with regard to his memory of an admission of sexual and violent fantasies.

Conclusion

Parole suitability hearings for those sentenced to prison in California for life with the possibility of parole was for decades a process without a purpose. Hearings were held, but the outcome was almost never in doubt: virtually no one was released. Things have changed. Parole now represents an important release valve in the California criminal justice system, yet it has received relatively little attention.

With a unique dataset built using computer text processing, this chapter has offered a rigorous account of release decisions. The results suggest a system that contains some element of chance. The evaluating psychologist or commissioner can alter whether an inmate is released or remains in prison for many more years. Further, the analysis of variables suggests that, at least to some degree, factors that, by law, should not matter in the parole decision, such as the presence of victims at the hearing, may well have a substantial impact on how the board decides. On the other hand, the psychological risk score, a measure of an inmate's dangerousness and risk of reoffending, which is the central criteria upon which the board is required to make its decision, is indeed of great importance in the decision. And there is no evidence that the California parole commissioners exhibit the psychological effects that researchers have discovered in other adjudicators: the gambler's fallacy and decision fatigue appear to play little to no role in the California Parole Board's decision-making.

Our analysis of inmate speech during the suitability hearing helps to illustrate and confirm some of the conventional wisdom with respect to what matters at the hearing: the idea that the parole board looks most favorably on clients who are respectful, clearly state admission to the crime, and appear rehabilitated, for example. Our classification models do capture some signals suggesting inmate narration at the hearing may be of some importance. On the other hand, that the models are not more predictive suggests that inmate narration is likely not a critical determinant of the parole outcome. This is as we would hope for a system that strives to be rational, consistent, and objective. ❧

ﬁ

ANALYZING PUBLIC COMMENTS

Vlad Eidelman, Fiscal Note
Brian Grom, Fiscal Note
Michael A. Livermore, University of Virginia

The public comment process is one of the hallmarks of the American administrative state. As the informal notice-and-comment rulemaking procedure has grown into one of the most important national policymaking venues, the public comments process has become a forum for both organized interest groups and ordinary individuals to engage in public deliberation and political debate. In recent years, as both the ease of participation and interest in rulemaking have grown, there has been an explosion of public participation, and agencies now receive millions of comments from the public each year concerning proposed agency actions. These comments are voluntarily generated by individuals and organizations representing a vast diversity of interests—from large industrial trade associations representing businesses with billions of dollars at stake to individual citizens who have an interest in a particular regulatory outcome.

At the same time that agencies find themselves deluged in public comments, recent advances in machine learning and natural language processing have made powerful text analysis tools more broadly available. Both commercial enterprises and academic researchers have recently begun to put these tools to use in a variety of settings, from tracking employee morale based on email communications to testing the relationship between online blogging and political opinions. Computational text analysis of public comments, however, is relatively rare, leaving largely untapped a substantial resource for both scholars and policymakers.

Public comments are a valuable source of data that can be used to empirically examine how bureaucratic institutions interact with the public. As a form of political participation that is unique to the bureaucratic setting, commenting behavior is an interesting and important phenomenon in its own right and provides information on how agencies and their actions shape and are shaped by the publicly expressed views of individuals and groups. In recent years, a small number of political scientists and others interested in bureaucratic behavior have begun to take advantage of public comments to study agencies—work that can be substantially facilitated by leveraging new tools in computational text analysis.

In this chapter, we describe an analysis of over three million public comments received by administrative agencies over the course of the Obama administration. Applying a basic, replicable procedure of sentiment analysis to these comments and comparing those results to information on agency ideology from the political science literature, we find that agencies with more moderate ideological leanings tend to receive comments that contain more positive language. This analysis indicates, as a threshold matter, that political characteristics of agencies are correlated with comment characteristics. Future work can build on this insight to inform subsequent research into the relationship between agencies' behavior and the public comments that they receive.

Moving from the descriptive to the normative, we examine how agencies and agency oversight institutions can use computational text analysis of public comments to improve agency decision-making and accountability. In the era of mass commenting, agencies face both a "needle-in-the-haystack" problem (i.e., identifying the most substantive comments) and a "forest-for-the-trees" problem (i.e., extracting overall trends or themes in large, unstructured collections of documents). To examine the usefulness of text analysis techniques to address these challenges, we carry out a case study of the comments received by the Environmental Protection Agency (EPA) in response to its proposed rule to limit greenhouse gas emissions from the electricity generating sector, called the Clean Power Plan. We apply two tools: a measure of the *gravitas* of a public comment and a topic model that we use to identify overall semantic trends in the comments. We find that, although not perfect, these techniques already have value for agencies and can be further refined to improve on their current performance.

The Deluge

Over the past several decades, informal rulemaking has become one of the most important policymaking forums in American politics (DeMuth 2016). Partisan rancor and divided government have often inhibited the ability of Congress to pass meaningful legislation. Legislative gridlock on the major issues of the day, including immigration and climate change, has led to an ever more active executive branch. Informal rulemaking is perhaps the executive's preeminent tool for setting domestic policy, and administrations of both political parties have wielded it to great effect (Kagan 2001). This state of affairs has had several consequences for the public comment process. Comments have taken on greater importance as a means to influence major policy decisions and, at least occasionally, serve as a preliminary step in litigation over high-stakes rulemaking. In addition, the public comment process is sometimes incorporated into broader advocacy efforts to influence public opinion and politicians. Advocacy campaigns around rulemaking increase public attention, leading to a higher volume of public comments.

At the same time as the stakes of the public comment process grew, information technology lowered the costs of participation. In the past, there were fairly substantial barriers to learning about a rule, engaging in research on the public policy choices involved, and submitting comments. Now, rather than attempting a trip to the local library for a copy of the Federal Register, interested individuals can quickly access a diverse array of information about proposed regulations online. From websites of individual agencies or the comprehensive government-wide regulatory portal at Regulations.Gov, interested persons can now easily and inexpensively identify ongoing regulatory proceedings, access relevant documents, and submit comments. In addition to extensive explanatory regulatory preambles, agencies typically include a great deal of additional substantive information on their rulemakings on agency websites, and any official supporting documents are also made available on Regulations.Gov. Advocacy organizations also publish their own analyses of proposed rules, and journalists and other content authors (i.e., bloggers, opinion writers, academics, etc.) often provide additional information for free. Within the time it once took to drive to the library and find a proposed rule's text, a relatively well-informed and conscientious researcher can amass a substantial amount of information about any rule of interest.

With lower costs and higher stakes, participation in the notice-and-comment process has ballooned. Several recent high-profile rulemakings have generated what might be called megaparticipation, with comments numbering well over a million. The State Department Keystone XL oil pipeline decision received more than 2.5 million comments; the Federal Communications Commission received over 1.25 million comments on its original net neutrality rules; the EPA received over four million comments on its proposed Clean Power Plan. Within these voluminous submissions are form comments that have been circulated by advocacy groups; detailed, well-researched submissions by nongovernmental organizations, industry, and academics; and comments from other interested groups and individuals, including local organizations, states and municipalities, and members of the general public.

In terms of simple administrative manageability, opening up the floodgates of participation in the rulemaking process presents clear difficulties. Agencies are obligated to consider and respond to substantive comments; having to review many millions of comments to even determine their substance is an extraordinary burden. Even if many of the comments are repeats of a form submission, agencies must still separate out the unique comments and give at least some cursory examination to them. Given the lower cost of acquiring information, agencies may also face a higher volume of substantively meaningful comments. Even if these comments contain valuable information, processing them can require substantial commitment of agency resources.

From the perspective of expanding citizen involvement in administrative decision-making, recent innovations create obvious opportunities. More comments allow agencies to collect more information on matters such as the interaction of a rulemaking with technological innovation or business practices. In addition, more people participating in the comment process may increase the legitimacy of a rulemaking. Just as a higher voter turnout is often interpreted as a sign of a more robust democracy, a larger number of public comments indicates a more inclusive and participatory administrative process.

But translating the promise of mass participation into a public comment process with enhanced value has proven to be no easy challenge. One approach has been to attempt to improve the quality of comments. For example, Cynthia Farina and her colleagues facilitated a project called the Regulation Room, an an online deliberative portal. The idea behind the Regulation Room was for human facilitators to help potential commenters identify the kinds of information that was most likely to be of value to agency decision-making (Farina et al. 2011).

An alternative approach to addressing mass participation attempts to extract more meaning from comments in their current form. The idea is to use advanced information processing and text analysis techniques to extract as much usable information as possible from the public comments that are submitted to agencies. In the subsequent discussion we will focus on how the data that already exist in the form of agency comments can be put to use, first to better understand how agencies interact with the public and then to improve those interactions to ultimately increase the value of the public comment process.

Studying Bureaucratic Politics

There is a considerable body of empirical and theoretical literature on bureaucratic politics (Aberbach and Rockman 2006; Brehm and Gates 1999; Baekgaard, Blom-Hansen, and Serritzlew 2015; Bendor and Meirowitz 2004; Fiorina and Noll 1978). Some of the fundamental contributions in this literature have helped illuminate the heterogeneous nature of agencies and the importance of institutional design and administrative procedure in affecting relationships between agencies and political institutions (Balla 1998; Devins and Lewis 2008; Epstein and O'Halloran 1996; McCubbins, Noll, and Weingast 1987). Even among US federal agencies, there is great variation in agencies' missions, in the interest group environment in which they operate, in their institutional structure and procedures, and in their internal cultures. Taken together, the collection of these differences can be thought to contribute to distinct profiles that remain at least somewhat consistent over time, inclining some agencies toward certain behaviors while inclining other agencies toward other behaviors.

Among the differences that appear to matter for agency decision-making is the makeup of the career personnel and, specifically, their policy preferences (Baekgaard, Blom-Hansen, and Serritzlew 2015; Brehm and Gates 1999; Gailmard and Patty 2007; Prendergast 2007). Although policy is at least sometimes responsive to the desires of political principals, the substantial policy discretion given to career personnel and their role in structuring and informing the decisions made by principals gives them substantial ability to shape outcomes toward their preferred policies (Imbeau, Pétry, and Lamari 2001; Knill, Debus, and Heichel 2010; Wood and Waterman 1991; Carpenter 2014, 2001; Livermore 2014). This fact does not imply bad faith on the part of career staff—they may genuinely intend to serve the public interest, as they understand it, and have little at stake, in terms of personal satisfaction, for the policy decisions made by agencies. Nevertheless, the values, perspectives, and beliefs of career personnel can (perhaps appropriately) influence the choices that agencies make.

There is a subfield within the bureaucratic politics literature that attempts to estimate the policy tendencies of agency personnel. This work builds on earlier efforts within political science to examine the role of *ideology*—understood as a consistent set of preferences over policy outcomes—in other decision-making contexts, such as Congress and the courts (Poole 2005; Martin and Quinn 2002). These ideal point models enable a data-driven method for estimating preferences via construction of a low-dimensional latent space that captures similarity among individuals. This same notion of ideology has been applied to agencies and agency personnel, and a variety of methods have been used to estimate this "latent variable," including prior agency decisions, the opinions of outside experts, the political moment of an agency's formation, and the campaign contributions and survey responses of agency personnel (Lewis 2003; Clinton and Lewis 2008; Clinton et al. 2012; Lewis 2007; Nixon 2004; Bonica, Chen, and Johnson 2012). These studies tend to come to relatively consistent results, confirming that agencies have something like an ideological profile that persists over time.

Agency ideology has been found to have several interesting consequences for the relationship between political and administrative decision-making. For example, presidents treat agencies differently in their political appointment decisions in light of agency ideology: some agencies are targeted for patronage while other agencies are targeted for more intensive policy supervision, depending on how well agency ideology tends to align with the governing philosophy in the White House (Lewis 2008).

An important but understudied question in the bureaucratic politics literature is how agency ideology affects interactions between agencies and the broader public. In addition to making official decisions—promulgating rules, issuing licenses, initiating enforcement actions—agencies carry out a variety of public engagement activities, which include not only those that are required by law (such as the notice-and-comment process) but also through voluntary initiatives that include public meetings, publications, media relations, and social networking. These actions speak to the importance of agencies in managing public perception, not only for direct reputational benefits but also as part of a broader effort to influence the oversight activities of actors that are more directly accountable to the public.

One of the difficulties of studying agency–public interactions is that data and methods have not been as fully developed as in other areas of the bureaucratic politics literature. Some studies have focused on the identity of commenters as the primary explanatory variable to test whether interests that commenters have submitted in comments tend to influence the regulatory process (Balla 1998). Survey techniques have also been used to examine how participants perceive their role in the regulatory process (Furlong and Kerwin 2005; Yackee 2015). Recently, researchers in political science, public administration, and law have begun to exploit public comments to study agency–public interactions (Boustead and Stanley 2015; Golden 1998; Krawiec 2013; McKay and Yackee 2007). Susan Webb Yackee, in particular, brought attention to the value of public comments in understanding agency–public interactions with several studies that relied on hand coding a large number of comments (McKay and Yackee 2007; Yackee 2005). Two recent papers published in law journals also engage in human-coded analysis of public comments to gain insight into how the public perceives highly salient agency rulemakings (Boustead and Stanley 2015; Krawiec 2013).

One underexplored feature of comments is their sentiment. Similar to consumer reviews on websites such as Amazon.com or Yelp.com, which express satisfaction or dissatisfaction with a product or service, public comments express agreement (satisfaction) or disagreement (dissatisfaction) with a rulemaking or provisions thereof. The sentiment carried within public comments provides a valuable lens into attitudes and perceptions held by the public on agency decisions. Of course, comments are not submitted by a random sample of the population and should not be thought of as replacing public opinion surveys, which are carefully designed to provide insight into the general public. But the self-selection process itself conveys information—public comments express the views of the interested public, a number that can exceed over a million individuals for mass-comment-volume rules. This large group of interested individuals may be more likely to vote, contribute to or volunteer for political campaigns, or act as opinion leaders within local social networks. Whatever their status *vis-à-vis* these other political activities, the group of public commenters is sufficiently large that it is a useful object of study in its own right, so extracting and analyzing the sentiment in their comments has the potential to provide worthwhile information for the study of how agencies and the public interact.

Ideology and Comment Sentiment

In a simple model of participation in the public comment process, a proposed regulation can be understood as representing a point within a one-dimensional, left–right ideological space.[1] A liberal proposed regulation (for example, one that increases regulation of financial products) would occupy a position to the left of a conservative proposed regulation (for example, deregulating financial products). Both agencies and commenters in this model have "ideal points" within this ideological space that represent their preferred policy outcomes. Agencies will tend to propose regulations that are close to their ideal point. The distance from

[1] For background on ideal point estimation see Bateman and Lapinski (2016).

a commenter's ideal point and the proposed regulation affects the sentiment of the comment. A proposed regulation that is relatively close to a commenter's ideal point will spur positive sentiment, and proposals that are relatively distant will generate more negative sentiment. The likelihood of commenting may also be related to the distance between a potential commenter's ideal points and proposed regulations, as well as other factors such as the salience of the underlying issues affected by a rulemaking. Individuals are more likely to comment on a proposal they strongly favor or disfavor and on rulemakings that involve highly salient issues.

This model of commenting involves strong assumptions about the nature of ideological space and the motivations of commenters. In reality, people's ideological dispositions may vary on multiple dimensions and be poorly represented by a simple left–right spectrum. Even on a single dimension, we are unable to directly observe the distribution of ideal points of potential commenters, which may be relatively normally distributed and clustered toward the center or, alternatively, may be bimodal, with relatively few moderates and separate groupings oriented toward the extremes or even uniformly distributed across the space. We also have little information on how individuals learn about regulatory proposals and decide to comment, and what mix of salience, satisfaction with the policy choices made by an agency, and other factors influence commenting behavior.

What simple models lack in nuance, sophistication, and complexity they can sometimes offset with parsimony. Despite the limitations in the microlevel account of commenting behavior offered above, it can nonetheless help motivate some macrolevel predictions about the relationship of sentiment and the ideological characteristics of agencies. At the aggregate level, certain features of comments may be predictable, even if the behavior of individual commenters is highly stochastic, unchanging, or affected by a wide range of unobserved variables.

The macrolevel prediction that we test is that agencies with more moderate ideological tendencies will receive comments with relatively more positive sentiment. This result is consistent with the simplified model in which personnel ideology is correlated with the likely policy choices made by agencies, the desire to comment is relatively uniform across the ideological spectrum, and there is some centralized tendency to the distribution of ideal points within the pool of potential commenters.

Given the number of potential variables in play, there are other microlevel accounts that would also be consistent with a positive correlation between ideological moderation and sentiment. For example, under a bimodal distribution, moderate proposals might be distant from all potential commenters' ideal points, but many unhappy potential commenters may be demotivated by milquetoast proposals and decide not to comment at all. There may be other causal stories as well, such as the possibility that negative sentiment in comments causes ideologically moderate individuals to avoid seeking work at an agency. If this is the case, the location of regulatory proposal in ideological space does not cause negative sentiment in comments; rather, some other variable causes negative sentiment (e.g., an unpopular mission), which then affects the type of personnel attracted to an agency, which in turn affects ideology estimates. Or there may be unobserved variables—such as the influence of a special interest group or congressional oversight committee—that affect both agency ideology and comment sentiment. The current analysis will not be able to distinguish between these alternatives. Rather, its value lies in investigating empirical associations in the data and opening the door for additional research to develop and test alternative theories of commenter behavior.

DATA

For our analysis, we exploit the extensive dataset of public comments compiled by FiscalNote, a Washington, DC–based government analytics firm. FiscalNote scraped all publicly available comments for all agencies during the study period from Regulations.Gov whenever possible and from the individual agencies' websites otherwise. It is worth noting that not all comments are publicly released; for example, agencies sometimes release only unique comments. Every comment that was available was analyzed. Our study period covers the two terms of the Obama administration. During that time, more than two million public comments were received and released in response to solicitations by US federal administrative agencies concerning pending actions (primarily, but not exclusively, rulemakings). Our analysis is based on 1,461 rules adopted during the study period for which one hundred or more comments were received. The data cover 106 administrative agencies. The three rules with the largest number of comments during the study period were a rule by the Department of Health and Human Services implementing provisions of the Affordable Care Act, an Internal Revenue Service rule concerning candidate-related political activities by tax-exempt organizations, and a State Department permit application concerning the TransCanada–Keystone Pipeline.

Our predictor variable of interest is agency ideology. We will test the relationship between ideology and sentiment with the basic hypothesis that agencies with more "extreme" ideologies will receive public comments with lower mean sentiment. To construct the ideology variable, we rely on a measure based on responses to a 2007–08 survey conducted as part of the Survey on the Future of Government Service project.[2] That survey was sent to over seven thousand senior-level officials within the federal government, both career and political appointees. Completed surveys were received from over two thousand respondents (with an overall response rate of 34%). To develop an ideology measure that was commensurable with Congress, the survey included several questions on how the respondent would have voted on fourteen questions that were subject to House and Senate votes in 2006 (including the confirmation of Justice Alito and a bill to increase the minimum wage). Using congressional actions allowed the researchers to compare the hypothetical votes of survey respondents to the actual votes of senators and representatives, placing them in common ideological space.

The Survey on the Future of Government Service researchers reported separate ideology scores for career officials and appointees; for our analysis we will use the ideologies for the careerists only, disregarding the political appointees. Given the relatively low level of turnover within ranks of career civil servants, a greater degree of interadministration ideological consistency can be expected. Use of only the career respondents may also allow us to examine more persistent agency-level effects that are less likely to be subject to variation as the White House changes between presidents and political parties.

The FiscalNote data include fairly granular-level specification of the issuing agency, which is more finely grained than the estimates of agency ideology. In those cases, we identify a parent agency and tag an ideology score accordingly. For example, the Fish and Wildlife Service is a bureau within the Department

---

[2] We use data from an earlier working paper version that reported more detail on the survey responses (Clinton et al. 2012).

of the Interior—since we have no specific ideology score for the Fish and Wildlife Service, all of its rules are attributed to Interior. This lack of granularity adds a fair amount of noise to our analysis; it is quite possible that the ideology of personnel associated with the Fish and Wildlife Service (which is charged with administering the Endangered Species Act) is different than personnel at the Bureau of Oceans Energy Management (which administers offshore oil drilling), even though both fall within the Department of the Interior. The coarseness of our ideology measure can be expected to attenuate our results rather than lead to spurious correlations.

Our ideology variable is not simply a location on the left–right spectrum. Rather, we estimate the degree to which an agency departs from the middle of the ideological distribution. To do this, we calculate an ideological midpoint by taking the mean ideology for agencies in our sample and then using the distance between that midpoint and an agency's score as an estimate of the degree of ideological polarity.

We analyzed the sentiment of all of the comments in our dataset using the publicly available sentiment analysis model in the pattern Python library. The text was preprocessed using TextBlob and NLTK to tokenize, stem, and remove stop words. Every comment is given a sentiment score based on the occurrence of lexical items indicating positive and negative polarity within the document; these comment sentiment scores are then categorized by the corresponding rule to create a per rule distribution, and a mean sentiment value is determined for every rule. This mean sentiment is used as the dependent variable in our analysis. For all rules in our dataset, the distribution of mean sentiment is roughly normal.

RESULTS

There are eighteen agencies in our dataset for which we have both ideological scores and rules. Many of these agencies have bureaus, departments, or offices that are not represented separately. Figure 9.1 presents the relationship between agency ideological polarity and mean sentiment. Recall that each agency's ideology score is not a location in left–right ideological space but rather a measure of the distance between that agency's ideology and the midpoint in the space. Most of the agencies have a unique score, but several of the agencies share ideological scores with each other (there are thirteen unique scores for eighteen agencies).

The horizontal axis presents a measure of agency polarity that is calculated by taking the absolute difference between an agency's ideology and the mean ideology. Sentiment falls on

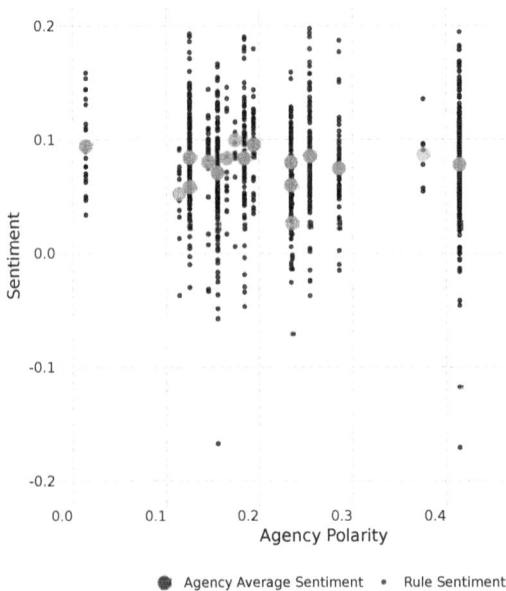

Figure 9.1. *Agency polarity and comment sentiment.*

the vertical axis, with the scores representing the percentage of positive words minus the percentage of negative words. Each dot represents a single rulemaking, and gray circles are the sentiment scores aggregated by agency.

Visually, it is clear from figure 9.1 that there is a substantial amount of variation at the agency level; the same agencies issue rules that generate comments with very different levels of positive or negative sentiment. Because ideology is assigned at the agency level, its usefulness in predicting sentiment at the rule level is relatively low, and at first glance it is not clear that there is a relationship at all. The large number of rules in our sample, however, allows us to detect even fairly weak signals in these noisy data. When sentiment is averaged at the agency level, the relationship between ideology and sentiment becomes somewhat more visually apparent.

To examine the relationship's size and test for significance, we estimate a linear model in which sentiment is treated as a dependent variable and agency ideological polarity is treated as a predictor variable with controls for the number of comments (which may be correlated with both agency ideology and sentiment) and year fixed effects. An ordinary least-squares regression shows that agency ideological polarity is significant at the 1% threshold with a coefficient of −0.03 and an adjusted R-squared of 2%. This coefficient can be interpreted to mean that agencies with highly moderate ideologies can expect, other things being equal, comments that include roughly 1.5% more net positive words compared to agencies at the extreme end of our sample. The relatively low R-squared value is consistent with the limited predictive power of ideology for sentiment at the rule level. We also estimate a separate model in which agency average sentiment replaces rule sentiment with the same controls for total comments received and year fixed effects. In that model, ideological polarity is again significant at the 1% threshold and R-squared increases to 21%.

To check whether any specific agency accounts for the relationship, we conducted an additional round of analysis. We first analyzed whether EPA rules were particularly controversial, adding an indicator for whether a rule was issued by the EPA—there was no change in the results. We then constructed a new model with the issuing agency as an indicator variable and identified seven agencies with significant correlations with mean sentiment (Commerce, Education, Interior, Transportation, Housing and Urban Development, Nuclear Regulatory Commission, and State). We gave each of these agencies the same treatment as the EPA, adding them individually to the third model. In none of these specifications did the significance of agency polarity fall below the 1% level.

The upshot of this analysis is that there is a relationship between agency ideology and the sentiment that is expressed in comments to the agency. More ideologically moderate agencies tend to receive comments that are, on average, more positive. It is worth emphasizing that these results are detectable even though the measures of both sentiment and agency ideology are quite noisy. The sentiment analysis scores are based on publicly available technology that is not tailored to the regulatory context, and the document-level aggregation obscures the content of the sentiment within the comments. Agency ideology is assigned at a very high level up the hierarchical ladder within the executive branch, which misses out on the potentially substantial ideological differences between bureaus and offices within agencies. Nonetheless, in the aggregate, there appears to be an identifiable relationship.

This result is consistent with the simple model of commenting introduced above, but as discussed, there are also other plausible stories that might account for this relationship—a fuller reckoning with the underlying causal mechanisms will require further research. Our limited understanding of how agency ideology interacts with the comment process also reduces our ability to draw normative conclusions. Under a very strong traditionalist understanding of agencies (Goodnow 1900) in which politics should be kept entirely separate from administration, our result would raise red flags. But on a more nuanced understanding that allows for some interaction of politics and administration, it is at least potentially unproblematic for agencies to occupy different places along the ideological spectrum and, accordingly, generate differing public responses to their rules.

~251~

Certainly, it would be overly hasty to conclude from our analysis that the more ideologically distinctive agencies are poorly performing their regulatory tasks. It is not obvious that agencies should seek to avoid negative sentiment and encourage positive sentiment—unlike in the case of consumer products, immediate satisfaction simply may not be the best interpretation of agency performance. Furthermore, there is no reason to think that the comments that are submitted to agencies are representative of the views of the American public in general; they may well better reflect the interests of organized groups or ideological activists, in which case negative sentiment from commenters may be associated with rules that better protect the public interest.

FURTHER RESEARCH

Although both the empirical and normative conclusions that can be drawn from our results are murky at this stage, the potential directions for future research are clearer. We propose three research areas that are raised by the proceeding analysis that could be addressed by leveraging existing data sources and that have the potential to generate a richer understanding of the public comment process and interactions between public bureaucracies and the public more generally.

The first research area would be to develop better measures of comment characteristics as well as measures of rule characteristics based on regulatory texts. The basic sentiment measure used in our analysis is very rough and is not tailored to the corpus or question of interest. Better measures could capture more relevant features of comments that are estimated with greater precision. In addition, we do not analyze the content of regulatory texts and relate that content to agency ideology or comment content. Some simple analyses of regulatory texts have been carried out, but more sophisticated measures that reveal the policy content or ideological valence of regulatory text could serve as the basis for more nuanced analysis of the relationship between proposed regulations and the comments that they generate.

The second research area concerns commenter ideology. In the analysis above, identifying information was not used, and commenters were treated as an undifferentiated mass. As with other public filings, however, commenter identity is not hidden and can, in theory, be extracted from comment texts. Generally speaking, there are two types of commenters: organizations and individuals. Once commenter identity is revealed, that information can be used to categorize comments by the authoring organization or individual and can be linked via that information to other sources of data, such as campaign donation information for individuals or lobbying filings for organizations. This information can provide insight into the distribution of ideological tendencies within the pool of commenters and how commenter ideology relates to agency ideology and comment characteristics.

A third research area would involve interactions of comment characteristics with political oversight. Our analysis is limited to the Obama presidency and so does not capture variation in the party that occupies the White House. As the Trump administration unfolds, these new data will become available. Nor do we mine information on congressional oversight activities or the institutional design characteristics of agencies (such as whether they submit rules to OIRA for clearance) to examine how oversight from political bodies might affect regulatory text or the content of comments. Data from media sources might also be used to estimate the public salience of regulatory matters, which may interact with political oversight or affect the propensity of interested individuals to comment.

Overall, the initial exploratory analysis presented in this section indicates that there is substantial potential to make use of public comments to understand the relationship between agencies and the public. Of course, as with any source of data, public comments do not provide a complete picture, but they nonetheless are a valuable lens into the thoughts, concerns, and reactions of interested parties to the proposed actions of administrative agencies. As new generations of computational text analysis tools become available, they have the potential to provide substantial new insights that challenge existing understandings and create opportunities for new avenues of research that enrich and deepen our understanding of questions at the heart of the social scientific study of government.

Enhancing Participation

The most salient risks posed to agencies when faced with a large number of comments is the failure to identify and respond to substantive comments that are subsequently used as the basis for litigation. Courts enforce the Administrative Procedure Act's requirement for agencies to collect and consider comments, occasionally striking down rules when agencies fail to adequately respond to comments.

A judicial reversal imposes major costs. Agencies spend considerable resources on rulemaking, including personnel hours and funds to hire consultants. The most significant category of rulemaking costs, however, may be political. Each agency's rulemaking agenda is extremely constrained, with the potential to issue only a small number of rulemakings each presidential term. The opportunity cost for any rulemaking includes all of the other policies that could have been pursued. Agencies also expend general political capital during rulemakings, inevitably courting controversy and opposition even in contexts where regulation is broadly popular. Given the costly investment that a major rulemaking represents, when a rule is struck down it is an enormous disappointment for the agency. When agencies fail to account for highly substantive comments, not only do they undermine their own decision-making process, they also open themselves up to these litigation risks.

When conforming to judicial requirements to respond to comments, agencies are not under a general obligation to "discuss every item of fact or opinion" offered in a comment.[3] But they are required to include in the record sufficient response to comments to show to a reviewing court that the "major issues of policy were ventilated" during the rulemaking deliberations.[4] Courts have emphasized the importance of considering comments that raise relevant scientific or technical

---

[3] *Automotive Parts & Accessories Association v. Boyd*, 407 F.2d 330, 338 (DC Cir. 1968).

[4] *ibid.*

information, illuminate undesirable consequences of the proposed rule, or offer alternative courses of action for the agency to consider.[5]

In addition to meeting their legal obligations, agencies may also consider comments for the purpose of improving their own decisions. Although there is some disagreement over whether and how to consider less sophisticated mass comments (Farina, Newhart, and Heidt 2012; Mendelson 2011), there is broad consensus that there are some highly substantive comments that have obvious information value.

Although agencies have both internal reasons and external incentives to identify highly substantive comments, the low cost of submitting comments means that more substantive submissions can be buried under a mountain of less substantive comments. We refer to this challenge as the *haystack problem*— agencies must find the proverbial needle in the haystack. Substantive comments must be unearthed from within a large number of comments that are highly unlikely to pose any litigation risk. Failure means that the agency loses the opportunity to improve its rulemaking or head off a judicial challenge. When there are a large number of unique comments of a less substantive nature and a relatively small number of sophisticated comments, the task becomes more difficult. Agencies that find themselves inundated with tens of thousands or even millions of comments have an especially difficult chore.

The second challenge agencies face is how best to respond to the less sophisticated comments offered on behalf of small stakeholders or interested individuals. These are not the technocratic, jargon-laden comments that are submitted by large law firms on behalf of major industry but are nonetheless genuine expressions of concern or support. The welfare stakes

---

[5]See, e.g., *Pub. Citizen Inc. v. FAA*, 988 F.2d 186, 197 (D.C. Cir. 1993).

can be quite real for the small business owner who faces a rule that would cut into already slim profits or the parent commenting on an air quality rule out of concern for a child with asthma. Although there is some disagreement among experts about the information value of these informal comments, the reality is that they are regularly submitted, and they may, if properly analyzed, provide insights that are of use to regulators.

This second challenge is much greater than the haystack problem because it is not just a matter of screening out the informal comments. Since the information in these comments is diffused, there has to be some mechanism to distill and aggregate meaning that might be spread over many tens of the thousands of comments. It is relatively easy to sit down and extract the content from a set of well-researched, persuasively written comments that have been prepared by professionals whose expertise is exactly in communicating to agency officials. It is an altogether different task to face perhaps thirty thousand comments submitted by individuals from a huge range of backgrounds, many of whom have no experience communicating to agencies, lack familiarity with the governing jargon, and have little time to devote to researching the relevant issues, and attempt to extract any kind of collective meaning from those.

This difficult task we refer to as the "forest problem." For a major rulemaking with tens of thousands of unique comments, it is impossible for any person to gain a sufficient vantage point from which to view the entirety. Any single person can read only a small share and attempt to summarize them; others may read the summaries and try to detect broader trends, but as the level of resolution decreases, important details can blur. The limits of human cognition (not to mention hours in the day) require a trade-off, and the process of communication between the fine-grained level (of individual comments) and higher-order meaning (trends within the group of comments) is highly imperfect.

GRAVITAS

At the heart of the haystack problem is the need to identify the most useful comments within a large, unstructured corpus of documents. In essence, agencies would like to be able to separate comments that have substantive weight, or gravitas, from the mass of less substantial informal comments. Farina, Newhart, and Heidt (2012) argue that agencies can focus exclusively on the comments with greater gravitas, while Mendelson (2011) argues that agencies should analyze even the less substantial comments. Regardless, the treatment given to comments will vary depending on their gravitas—even if informal comments are not ignored altogether, the types of analysis that they will be subject to and the information that will be extracted will be very different. When a major law firm representing a multinational company submits comments to an agency on a proposed rule, agency lawyers must comb through the document to identify potential legal lines of attack, and economic, engineering, and policy personnel will examine the comments for data or arguments that inform potential changes in the rule. The information in informal comments will often be of a very different sort and is more likely to bear on issues such as risk communication, framing, or the political interpretation of the proposal.

There are some obvious approaches to addressing the haystack problem that do not require advanced computational techniques. For example, comments that are filed by major companies or organizations, or their representatives, are likely to receive attention. The comments of established experts may also be more likely to be flagged for special treatment. Other relevant features of comments that can be identified fairly easily using basic computational tools include the length of the submission, the sophistication of the vocabulary, the relevance of the comment to the rule (perhaps based on shared vocabulary or phrases), the amount of citation, and basic identifying information on the author (such as whether the comment was from an institution or an individual, or whether the commenter has participated in other rulemakings).

We conduct an initial assessment of the viability of this technique using comments from a recent rulemaking by the EPA to reduce greenhouse gas emissions from existing electricity-generating units. This rule is one of the most consequential proposed environmental regulations in the nation's history, and the number of public comments received by the agency reflected the historical status of the rule: the EPA received over 4.3 million comments. Given this massive volume of comments, it is clear that the agency cannot give each one a great deal of individual attention, and so some automated means of directing the agency's focus would be particularly helpful in this context.

A first step is simply to remove duplicate comments. The EPA appears to have done so; although the agency reports over four million comments received, less than 1% (34,388) are posted online. Even with the duplicates removed, however, there are still tens of thousands of comments for the agency to review, so additional steps to prioritize responses and categorize comments remain helpful.

The following analysis relies on a gravitas measure developed by FiscalNote. This measure ranks comments based on several identifying features, including comment length, attachment count, the complexity (or coarseness) of the language that is used, whether the author is an organization, key person, or ordinary individual, the number of cogent arguments expressed, and other cues that together serve as a proxy for sophistication. According to the FiscalNote gravitas measure, the single most sophisticated comment was submitted by the environmental organizations Sierra Club and Earthjustice. This comment was truly behemoth, clocking in at 274 pages with an additional seventy-three exhibits (an example, "The History of Energy Efficiency") and thirteen appendices (including "Literature Survey, Efficiency Improvements through Upgrades of Existing Plants"). Other gravitas leaders include environmental group the Center for Biological Diversity and the Ameren Corporation (an energy holding company).

Some other points on the gravitas spectrum give a sense of the wide diversity of comments received by the agency. At the top of the bottom quintile of the FiscalNote measure are two short comments sent by individuals, one of which appears to be a form comment:

- Please reduce carbon pollution from existing power plants to protect public health. Set strict limits on carbon pollution from power plants. (forwarded via the American Lung Association)

- Thank you, thank you for your courage and foresight. Change must happen immediately to save the planet.

At the top of the second quintile there are two short comments from individuals that remain fairly unsophisticated:

- This sounds good. Up front at least. It's been a while since this country has passed any major environmental movements, and I definitely agree with one like this.

- Support american energy & not the marxist agenda.

The comment with the median gravitas score was submitted by an individual and reads as though it might be a mixture of a form comment along with some additional personalized commentary:

- I strongly support the EPA's effort to limit industrial carbon pollution from existing power plants. These new clean air standards will protect public health, fight climate change, and create jobs through innovation in cleaner, safer energy technology. It's our obligation to protect our children and future generations from the effects of climate change—and that means moving forward with these clean air protections now. I have been screaming about this for fifty years! What the hell are you waiting for, the end of life on earth?

At the top of the third quintile is the following:

- It is absolutely imperative that this plan be enacted. As an American citizen, I am ashamed by the near-total lack of action on climate change in this country. Climate change is a very real threat to the health and well-being of my children and their future children. If we do not take decisive action now, there will be no world for my grandchildren to inherit. This plan makes economic sense and promotes public health and national security. It would be a complete disgrace if this plan was not enacted because of the short-sighted special interests of a small group of industry executives and politicians that are beholden to the coal industry. It is time for the United States to begin taking actions to reduce our disproportionate share of global greenhouse gas emissions.

At the top of the fourth quintile is a comment that may be or have elements drawn from a form comment:

- Please keep in mind all of us who support clean renewable energy, which is very do-able in the USA. We vote and we spread information. Along with all 2.4 million Sierra Club members and supporters, I want to see strong positive climate action. I'm encouraged with the framework put forward by EPA's proposed safeguards against carbon pollution from existing power plants but want to see the standard strengthened. We need states to be encouraged to choose clean energy and energy efficiency. It's not the appropriate time to build more dangerous nuclear power plants and invade our fragile water supplies with fracking. We must dedicate resources to create more productive jobs in an industry that doesn't pollute our air and water or disrupt the climate. (forwarded via the Sierra Club)

~ 261 ~

It is, of course, the most sophisticated comments that are likely to receive the most agency attention. Even focusing at the top, there are still a fairly large number to consider. For example, the comment with the one-hundredth highest gravitas score was from a group called the Small Business & Entrepreneurship Council, a northern Virginia–based "nonpartisan advocacy and research organization." These comments are fairly long, at 2,300 words, and include comparatively sophisticated arguments in support of the three criticisms of the rule summarized by the group:

- After a thorough review of the CPP, [we] believe the wisest course of action is for EPA to withdraw the proposed rule and abandon its costly agenda to regulate carbon dioxide under the Clean Air Act. The reasons for our position are straightforward. EPA's proposed rule: Is illegal, stretching far beyond the narrow boundaries of Section 111(d) of the Clean Air Act; Imposes high costs for no meaningful benefits; and Threatens the reliability of the nation's bulk electric power system, which raises the prospect of blackouts and brownouts, which can in turn increase operating expenses and uncertainty, as well as reduce output and revenues.

The one-thousandth most sophisticated comment was submitted by the West Virginia Community Action Partnership, a "statewide membership association" of local community groups that serve over forty thousand "low-income families annually in all of our state's disadvantaged communities, providing [them] with multiple services to promote their economic security." These comments clocked in at 1,600 words and offered support as well as detailed criticism for the rule. The group summarized its view as follows:

- We are in strong support of the general goal and ultimate result of the 111[d] rule. We are enthusiastic about a framework that allows a building block of efficiency and clean energy policies as part of the solution. However, we believe EPA has failed to establish: The needed framework for participatory planning; The criteria for establishing that each state's plan is equitable with respect to access to clean energy; Protection from environmental degradation; and, Protection of human health.

Any automated measure will sometimes fail, and one form comment broke the top five most sophisticated because it was delivered in bulk as a set of images (rather than individually as text), creating a very large file. The comment, which appears to have been individually hand signed by over thirty thousand electricity customers in South Carolina, urged that agency to give "equitable treatment for [South Carolina's] nuclear units under construction." But even if imperfect, the ability to conduct a rough triage using automated tools can have substantial savings. Assuming that it would take an employee ten seconds per comment to group the comments into rough categories, the human resource cost to conduct this simple operation on all of the comments received on the Clean Power Plan would be over ten thousand person hours—a considerable avoided cost.

Agencies also have access to information that could build on the simple approach used above. FiscalNote's gravitas measure is based on a handful of very easily identifiable characteristics that are used to generate an intuitive measure, such as comment length and the sophistication of the language used. These factors make sense, but they are not independently verified. Agencies, on the other hand, have access to metadata that could be used to validate a constructed measure, or perhaps even more promising, to train a supervised model for classification. Specifically, agencies have data from past rulemakings on which comments were considered serious enough to warrant a substantive response—either in terms of revision to a proposed rule, a response in a preamble, or a response-to-comments document. In essence, the agencies have tagged certain documents as worthy of a deeper consideration. This information could be used to generate a training set for a supervised machine learning experiment. It is possible that a properly constructed and trained algorithm could predict, with a fair degree of accuracy, those comments that are most likely to warrant additional attention. At the very least, such an approach could take a first cut to identify high-substance comments, with the remainder subjected to the existing human evaluation procedure.

Identifying Emergent Meaning

At the heart of the forest problem is the potential for emergent meaning that arises at the level of the corpus (i.e., the collection of comments) that is not apparent when the component documents (i.e., the comments) are analyzed in isolation. This emergent meaning exists at the aggregate level in the form of patterns and regularities that are determined, at least in part, through the relationship of documents to each other, as well as the internal relationships between concepts, arguments, and ideas within documents. Addressing the forest problem requires aggregated representations of a large number of comments that can be interpreted by agencies and released to the public.

As it stands, in major rulemakings, agencies release comments in an entirely unstructured fashion. In response-to-comments documents or in regulatory preambles, agencies will detail how they have considered substantive issues that were raised during the public comment process. Sometimes, agencies will include some sweeping language concerning informal comments, characterizing the grounds for general support or opposition for the rule. When they identify common themes that are the subject of many comments or when they flag some comments as duplicates that have been generated by mass email campaigns, agencies already engage in some rudimentary efforts to identify emergent meaning. Both of these observations require agencies to compare comments to each other, in addition to taking each on its own terms. But frequently, the response-to-comments document is not indexed at all to the pool of comments, so it is impossible to track which comments match which revisions or responses (other than, theoretically, reading all of the comments themselves).

Ultimately, although these rudimentary efforts are useful, there is very little nuance, and the diversity of the individual voices that participated in the commenting process and their relationship to each other is lost.

The volume of comments, their low information density, and the relative resource scarcity of agencies make more sophisticated attempts to develop higher order meaning difficult without the aid of computational tools. Topic modeling is one possible approach (Blei 2012). A major advantage of topic models is their ability to radically reduce the number of dimensions for analysis, from a vocabulary of tens of thousands of words down to a small number of topics (set by the user, typically to between ten and one hundred). With this information, researchers can examine how documents relate to each other and how semantic content varies over time.

Topic models also create a representation of a textual corpus that is highly amenable to additional analysis that can provide further insight into a corpus. For example, topics can be used to construct a network that is based on the document in a corpus that illustrates how topics relate to each other, showing, in essence, how often subject matter categories are mixed within a document. Topics can also be used as a way to estimate the semantic distinctiveness between documents or groups of documents, which allows a way for analysts to determine, for example, the degree of relatedness between authoring institutions.

From the perspective of an agency, topic modeling comments and carrying out associated analyses could help identify relationships between portions of the rule that are not obviously apparent from the face of the regulation. These emergent relations could provide the agency with insights into how revisions could be made that mitigate a shared concern through a pathway that the agency did not anticipate. This type of representation might also help the agency target its revisions toward areas of the rule that are of concern to many diverse stakeholders by addressing issues that tend to be central within a topical network (i.e., are linked with many different clusters of issues) rather than dealing with concerns that are on the outskirts of the network (and therefore have few relations with other issues).

To illustrate the usefulness of this tool, we applied a standard topic model to the nearly forty thousand publicly released comments received by the EPA in regard to the Clean Power Plan. We used the Latent Dirichlet Allocation implementation provided by Genism for building the topic model. The text was tokenized, stemmed, stop words removed, and the top ten thousand unigrams (single-word expressions) and five thousand bigrams (two-word expressions) were retained.

As discussed in detail above, there was a broad diversity in terms of the sophistication of comment, with many simply offering words of support or opposition and others that engaged in detailed analysis. Because topic models are based, at a very general level, on word cooccurrence, sophisticated and unsophisticated comments are given the same weight. That being said, if more sophisticated comments tend to use similar language, that may be reflected in the topics that are generated by the model. This actually seems to be the case. In a twenty-topic model of the Clean Power Plan comments, one of the topics includes a number of two-level phrases that are very specific to the regulatory context at issue and seem quite likely to be associated with fairly sophisticated players. The top words (or words indicated by word stems) for this topic include "stationary source," "electric utility," "generating unit," and "emissions guidelines." All of these are terms of art specific to the Clean Air Act context and so are likely a flag for sophisticated comments.

In that same twenty-topic model, there is a fair amount of overlap in the vocabulary, with words like "EPA," "energy," "carbon," "clean," and "rule" showing up across multiple topics. Looking at the top words that are not common across topics, some interesting results emerge. For example, one of the topics includes "carbon tax" as a top unit when bigrams were included. That same topic included relatively high in the distribution the words "fee," "dividend," and "citizen." To an observer of climate policy, this topic conforms to arguments forwarded by those who favor an approach to greenhouse gas reductions that favors a revenue-neutral price on carbon that is accompanied by a per capita "dividend" of the collected funds.

Other topics in the same model include a topic that appears oriented toward the gravity of the threat from climate change, which includes several bigrams within the topic words; these include "children enact," "today threaten," "problem greatest," "challenge face," and "undeniable problem." Two topics appear to focus on the question of how nuclear power should fit into climate change policy: one with top words like "nuclear power," "waste," "reactor," "sustain," and "radioactive"; the other with top words such as "nuclear," "clean," "current," "construct," "calculate," "credit," and "baseline." The role of environmental groups in organizing commentators also shows up, with one comment substantially devoted to words associated with the Sierra Club and another with words associated with the Environmental Defense Fund.

On the Horizon

From a content perspective, the analysis of the Clean Power Plan comments provides some interesting insights, but its primary utility is in illustrating the ability to extract emergent content from a large unstructured corpus of comments. At the same time, there are even more ambitious possibilities for a new kind of public deliberation that is created when agencies provide a feedback mechanism within the group of commenters rather than a closed, one-way flow of information from commenters to agency.

One can imagine a technologically enhanced notice-and-comment process that is superior to the current approach in many ways. Natural language processing tools could be embedded directly into the public comment interface and allow commenters easier access to information on a rulemaking that could include automated summaries of rulemaking text or links to relevant background documents, which themselves could be automatically summarized and distilled for ease of consumption. Prompts could be used to help commenters understand the types of information that they might want to include, and automated argument analysis could help facilitate more persuasive and well-supported comments.

These tools could help improve the baseline quality of comments, but it is in facilitating interaction between commenters that there is even greater potential value. For example, an agency could engage in a two-step public comment process that allows for an initial round of comments that close by a certain date, followed by a pause and a second public comment window. During the pause, a specifically designed content analysis engine that combines elements of topic modeling, sentiment analysis, and perhaps other computational text analysis tools could work on the first round of comments to develop an interactive representation of their content. During the second-round comment period, automated tools could guide participants through the first-round comments and flag areas of interest and points of agreement or disagreement—even, perhaps, facilitating up-voting or down-voting particular comments or ideas. An automated approach could be used to identify higher quality comments that could be presented to commenters to inform their own understanding of the rule, or could be integrated into an annotation engine that would allow commenters to voice agreement or disagreement with the arguments presented by others.

~ 269 ~

Once the final set of comments is collected, the agency (or a third-party agency that might have a more neutral attitude) could analyze the digital traces of the conversation, examining points of agreement and disagreement, collating opinions or sentiment with commenter characteristics (such as industry affiliation or region), scraping through for new bits of technical information, and developing a final aggregate representation that captures the state of deliberation on the rulemaking. Because it would be designed to be highly automated, this more interactive notice-and-comment process could be used for rulemaking at many different levels of broader public salience, from highly charged rulemakings that generate millions of comments, to the more run-of-the-mill rulemakings in which only several hundred comments are submitted.

This technologically enhanced public comment process would likely have substantial value, as participants would be given a much richer environment in which to engage in public deliberation, with the associated legitimacy-enhancing effects. There would also be greater information value, as higher quality comments are generated and more of the information that is embedded within them is extracted. The final aggregated representation of the comments could then serve as the foundation for revisions to the proposed rule. The agency could also use the comments as a starting place for a final substantive discussion that grapples with both the technical decisions and value-laden choices that it made.

Conclusion

We hope to have illustrated how new tools in natural language processing, combined with the massive (and publicly available) corpus that is generated by the notice-and-comment rulemaking process, create opportunities for both researchers as well as government officials. Scholars of public bureaucracies can use the information embedded within public comments to test theories on the interaction of agencies and the public that they serve. Agencies can use advanced computational techniques to respond to the challenges of the era of mass commenting, specifically by identifying the most substantive comments that require more sustained attention and by aggregating and analyzing comments to identify emergent content that is only apparent when comments are understood in relation to each other and not simply read as individual, atomized responses to a regulatory proposal. Although computational text analysis is certainly not a panacea for all that ails the regulatory process, or American democracy more generally, these tools can be usefully applied both to enhance understanding and ultimately to improve public deliberation over pressing and highly contested policy decisions. ❧

Acknowledgments

This chapter draws from Livermore, Eidelman, and Grom (2018). Our thanks to the editors of the *Notre Dame Law*

Review for their excellent editorial support on that article. We also thank Kevin Quinn, Jed Stiglitz, and participants at a faculty workshop at the University of Virginia School of Law and the 2016 Political Economy and Public Law Conference at Cornell Law School for helpful comments.

EXPLORATION

USING TEXT ANALYTICS TO PREDICT LITIGATION OUTCOMES

Charlotte S. Alexander, Georgia State University
Khalifeh al Jadda, Georgia State University
Mohammad Javad Feizollahi, Georgia State University
Anne M. Tucker, Georgia State University

This chapter describes the goals, methodologies, and preliminary results of an ongoing litigation outcomes-prediction project conducted by the Legal Analytics Lab at Georgia State University. Drawing on the lab's experience with the project as a case study, the chapter offers guidance for researchers engaged in or contemplating research in a similar vein: predicting trial court outcomes in a particular doctrinal area.

As background, the lab was established in 2017 as a site for collaboration between the university's data science and analytics faculty housed within the business school and the faculty at the law school. The lab grew out of the business school's data analytics unit, which is home to computer scientists, engineers, and statisticians, and whose faculty have developed particular expertise in the use and analysis of unstructured data, including text, imagery, and audio. Over time the business school's data science faculty began working with domain experts in other fields, and these ad hoc collaborations developed into four subject matter–specific research labs: the Legal Analytics Lab, the FinTech Lab, the Social Media Intelligence Lab, and the Operations Analytics

Lab. The design and goal of each lab is similar—to leverage the business faculty's data science expertise across other departments and schools of the university and to explore the application of analytics-based techniques across a wide variety of use cases and contexts. Faculty affiliated with the four labs have also developed interdisciplinary coursework, experiential learning opportunities, and degree and certificate programs to involve students in their work.

Faculty collaborators in the Legal Analytics Lab pursue a variety of grant-funded research projects at the intersection of data science and law. In addition, teams of faculty, graduate analytics students, and law students conduct what are known as "sprints": semester-long, short-term projects commissioned by outside sponsors (law firms or companies) that focus on discrete real-world, law-related data problems. The project described here began as a sprint conducted during the spring 2018 semester on behalf of a plaintiffs' side employment law firm in Atlanta, and has continued as a stand-alone research project within the lab.

The project took as its subject all employment law cases filed and closed in the US District Court for the Northern District of Georgia in the period 2010 to 2017. This included 5,111 cases for which we had approximately 8,600 court documents (complaints, magistrates' reports and recommendations on summary judgment, and district court judges' summary judgment decisions) in PDF form and all docket sheets in a CSV file (about 200,000 text entries). Each document type is defined and described in more detail in the sections below.

The law firm that sponsored the sprint was interested in answering a set of descriptive and predictive questions.

Descriptive:

- What was the frequency and distribution of case-ending events: settlement prediscovery, settlement after discovery had begun, dismissal, granted motion for summary judgment for plaintiff, granted motion for summary judgment for defendant, and trial?

- In summary judgment decisions, what were the legal doctrines used and cases cited most frequently by judges?

- Could we identify defense lawyer and judge "playbooks"?

Predictive:

- What features of a lawsuit, observable at different points in the intake and litigation processes, predicted its case-ending event?

The law firm envisioned three uses for the deliverables that the project would generate. The first was to improve the firm's intake process by identifying characteristics of successful cases, defined as those that settle prediscovery or withstand a defendant's motion for summary judgment. The second was to gain intelligence on the litigation strategies used by opponents and favored by judges. The third was to develop an empirical portrait of judges' summary judgment behavior on the Northern District of Georgia given other researchers' findings of relatively high summary judgment grant rates in that district (Eisenberg and Lanvers 2008).

The research team, in turn, had its own set of goals: to develop code that would classify all entries on a docket sheet into stages in a "life cycle" model of litigation; to write code that would extract all case law citations from a document and count their frequency (including short- and long-form citation formats); to test techniques such as topic modeling in extracting actionable or useful information from fact-heavy documents such as summary judgment decisions; and, more generally, to experiment with whether litigation outcomes are susceptible to predictive modeling.

The remainder of this chapter describes the data assembly process, methodologies employed to extract features from our text and build a predictive model, preliminary results, and areas of continuing work. Along the way, the chapter offers observations about the challenges inherent in applying data science techniques to legal text. This chapter is thus primarily a methods description in the context of a particular type of research project rather than a discussion of substantive results, although preliminary results are reported briefly throughout.

Data Assembly

The project began with a set of docket sheets and court documents that we received from the law firm sponsor. The firm had originally assembled the materials from the US Courts' Public Access to Court Electronic Documents (PACER) system and paid the associated download fees. Together, this material was associated with 5,111 lawsuits that had opened and closed in the US District Court for the Northern District of Georgia within the study period and bore one of PACER's four employment law-related Nature of Suit (NOS) codes: 442 Civil Rights- employment; 445 Americans with Disabilities Act- employment (ADA); 710 Fair Labor Standards Act (FLSA); and 751 Family and Medical Leave Act (FMLA). In rough terms, these codes cover lawsuits concerning employment discrimination of all types (442 and 445), wage

Table 10.1. NOS code distribution.

NOS Code	Frequency	Percent
442 Civil rights—employment	2596	51
710 FLSA	1934	38
445 ADA—employment	429	8
751 FMLA	152	3
Total	5111	100

and hour violations (710), and disputes around employees' family and medical leave from work (751).

The NOS code is assigned by the plaintiff or his/her lawyer at the time a case is filed, chosen from a list on a required document known as a civil cover sheet.[1] Plaintiffs may choose only one NOS code; they are instructed to "select the most definitive" code if more than one could apply. Table 10.1 shows the distribution of NOS codes across cases in our dataset.

Notably, relying on the self-reported NOS code as a threshold filtering device likely produced results that were both under- and overinclusive, sweeping in cases that did not, in fact, contain claims made under that statute and excluding cases that may have contained those statutory allegations but others were deemed more "definitive" by the plaintiff. Therefore, this project did not actually, as stated in the Introduction to this chapter, "take as its subject all employment law cases filed and closed in the US District Court for the Northern District of Georgia in the period 2010–2017" but instead all cases in which the plaintiffs or their lawyers deemed an employment law claim to be "most definitive." Christina Boyd and David Hoffman have recently explored this issue and proposed a series of smart

---

[1] US Courts, Services, and Forms 2018, "Civil Cover Sheet," https://www.uscourts.gov/forms/civil-forms/civil-cover-sheet.

reforms to the NOS code assignment process that would greatly improve the quality of PACER's case-filing data for researchers and policymakers (Boyd and Hoffman 2017).

In addition to the NOS code, the plaintiff or plaintiff's attorney also assigns each case a "cause of action" classification, which could be used as an additional or alternative filter. With respect to the cause classification, the civil cover sheet instructs, "Cite the U.S. Civil Statute under which you are filing (Do not cite jurisdictional statutes unless diversity)," and directs the plaintiff to provide a "brief description of cause."[2] Within PACER's data, the values of these fields appear as something like "29:201 Denial of Overtime Compensation" or "42:2000 Job Discrimination (Race)" or "42:2000 Job Discrimination (Sex)." This text identifies the specific statutory provisions under which claims are being made: "29:201" refers to 29 U.S.C. § 201, or the citation for the FLSA, while "42:2000" refers to 42 U.S.C. § 2000, the citation for Title VII of the Civil Rights Act of 1964, the main federal employment discrimination statute.[3] The rest of the cause classification offers some detail about the type of claim being made under that statute—for example, an overtime claim under the FLSA (as opposed to a minimum wage claim), or a race and sex discrimination claim under Title VII (as opposed to discrimination on the basis of religion, color, or national origin). These cause classifications would therefore seem to nest within the NOS codes, providing more granularity than the broader Fair Labor Standards Act and Civil Rights-employment NOS categories.

---

[2] See PACER Service Center, "US Courts, Services, and Forms, Civil Cover Sheet," http://www.uscourts.gov/forms/civil-forms/civil-cover-sheet.

[3] Title VII prohibits employment discrimination on the basis of "race, color, religion, sex, or national origin." Civil Rights Act of 1964 § 7, 42 U.S.C. § 2000e-2(a) (1964) (unlawful employment practices).

However, examining each case's NOS code and cause classification together revealed some strange pairings. A manual review of the 2,596 cases with "442 Civil Rights: employment" as their NOS code identified 4% with seemingly unrelated cause classifications (e.g., "Qui tam False Claims Act," which allows private plaintiffs to sue federal contractors for defrauding the government, or "Libel, assault, slander") or causes that had their own, separate NOS code (e.g., "29:2601 Family and Medical Leave Act" or "42:12101 et seq. Americans with Disabilities Act of 1990"). The other NOS code categories fared better: 1% or fewer cases with FLSA and ADA NOS codes had mismatching cause classifications, and no cases with the FMLA NOS code were mismatches.

There could be multiple explanations for NOS code–clause classification mismatches and for the higher rate of mismatches within the 442 NOS code category specifically. Rather than treating the cause classification as providing additional granularity about the same claim captured by the more general NOS code, the plaintiffs or their attorneys may have treated the cause classification as a way to record separate additional claims present in the case, beyond the single claim captured by the NOS code. It is also possible that the plaintiffs or their attorneys may have been mistaken or confused.

Further, with respect to the higher mismatch rate within the 442 Civil Rights-employment NOS code, cases with employment discrimination claims may be particularly likely to include additional nondiscrimination employment law claims. As explored in previous work (Alexander, Eigen, and Rich 2016), plaintiffs' lawyers report adding nondiscrimination employment law claims to their employment discrimination lawsuits to increase those cases' viability in the face of perceived hostility toward discrimination claims by the federal courts. The cause classification mismatches within the 442 NOS code may therefore be capturing these add-on claims based on plaintiffs' lawyers' strategic choices.

Thus, while it is possible to assign meaning to the NOS code–cause classification pairs associated with each lawsuit, these PACER fields are problematic filtering devices if the goal is to construct a comprehensive set of docket sheets and court documents in a particular doctrinal area. They may also be unreliable indicators of the statutory allegations and claim types that are actually present in any given lawsuit. However, this approach may be the best of a relatively bad set of data assembly options. An alternative strategy would require downloading all docket sheets and complaints for all NOS codes from PACER for a given court within a given time period and then parsing the complaint text to classify a case according to its statutory allegation and claims made. Because PACER charges ten cents per page downloaded, such a cast-a-wide-net strategy would likely be cost prohibitive in most situations.[4]

---

[4]The charge applies to docket sheets and party-filed documents. There is no charge for judges' opinions accessed via PACER's Written Opinions Report (WOR), which, according to the US Judicial Conference, is supposed to contain "any document issued by a judge or judges of the court sitting in that capacity that sets forth a reasoned explanation for a court's decision" (PACER Service Center 2005). However, our own work has revealed that the coverage of the WOR is woefully inconsistent across judges and districts. For example, in a separate project examining judges' summary judgment decisions in employment law cases, two districts— Wyoming and the Southern District of Iowa—had zero WOR entries in the period 2008–2016, despite having hundreds of summary judgment decisions in employment law cases available on Westlaw (Alexander and Feizollahi, 2019 (forthcoming)).

In addition, the number of WOR entries per civil case filed in a district (obtained from Federal Judicial Center data)—normalized measure of WOR activity—ranged from a high of 0.55 for the Northern District of California to a low of 0.0007 in the District of South Dakota. The courts' ranking by WOR ratio roughly tracks the underlying number of civil cases filed, suggesting

Other sources for assembling docket sheets and court documents include the legal research products offered by Westlaw, LexisNexis, and Bloomberg Law. These services allow keyword searching, which would seem to avoid the NOS code and cause classification problems identified above and instead go straight to the text of the documents to identify the relevant set. However, these services do not make their search algorithms public and sometimes produce dramatically different result sets. As an example, in a separate project an identical search across all three vendors produced 9,712 results in Westlaw, 7,261 results in LexisNexis, and 5,644 results in Bloomberg Law.

In the end, the project team took the NOS code-filtered dataset downloaded from PACER and provided by the sponsor as its corpus but did not rely solely on the plaintiff-provided NOS codes and cause classifications to identify the statutory allegations and claim types at issue in each case. Instead, as part of the feature extraction process described below, the team attempted to identify the full set of statutory allegations and claim types in each lawsuit from the text of the complaints themselves when that text was machine readable.

Feature Extraction

Our dataset contained four document types from which we attempted to extract a set of features that described each case. All of the 5,111 cases had an associated docket sheet and a complaint. Some cases also had a magistrate's report and recommendation on summary judgment and/or a

---

that higher-volume courts issued more opinions or perhaps designated more of their opinions as free via the WOR. However, the Northern District of Georgia, which was the eleventh highest-volume court during the time period, was thirty-ninth with respect to its WOR ratio (Alexander and Feizollahi, 2019 (forthcoming)).

district court judge's ultimate decision on summary judgment. The discussion that follows describes the methodologies employed to extract from each document type the features, or characteristics, of the plaintiffs, defendants, lawyers, judges, claims, and litigation. These sections also provide preliminary summary results, addressing the three descriptive questions set out in this chapter's introduction, and describe the predictive model into which we fed the extracted features as independent variables in an attempt to predict any given lawsuit's case-ending event.

Docket Sheets

A docket sheet is a chronological index of all activity in a case, listing all documents filed by the plaintiff and defendant and all actions taken by the judge. Figure 10.1 provides a snippet of a docket sheet downloaded from PACER's website and then converted to CSV format for purposes of illustration.

The text of the docket sheet entries (the rightmost column shown in figure 10.1) provides information about the "players" in each case: the district court judge's and magistrate judge's names, the number and names of the plaintiffs' and defendants' lawyers, and whether the plaintiff filed the case *pro se* (without a lawyer) or moved to proceed *in forma pauperis* (to be relieved of the requirement to pay a filing fee).[5] In each instance,

---

[5]A note on judges: Each case filed in US district court is assigned to a US district court judge who is a federal judge with life tenure, nominated by the president and confirmed by the US Senate. The federal courts also employ a corps of "magistrate judges," who, despite their "judge" title, are employees of the court rather than appointed judges. District court judges may "refer" certain discrete issues or decisions within a lawsuit to magistrate judges to handle on a preliminary basis. In those circumstances, a magistrate considers the parties' arguments and filings, and then issues a "report and recommendation" or "R&R" to the district judge. The parties may file objections, and then the district court judge makes the final decision, adopting or rejecting the magistrate's R&R in

Case Number	Activity Date	Activity Number	Docket Text
1:15-cv-02410-TWT	2015-07-06	1	COMPLAINT with Jury Demand filed by...
1:15-cv-02410-TWT	2015-07-06	2	Electronic Summons Issued as to Souto...
1:15-cv-02410-TWT	2015-07-13	3	Return of Service Executed by Marvin A...
1:15-cv-02410-TWT	2015-07-13	4	NOTICE Of Filing Reissued Summons...
1:15-cv-02410-TWT	2015-07-14	5	Electronic Summons Issued as to Souto...
1:15-cv-02410-TWT	2015-07-28	6	ANSWER to 1 COMPLAINT with Jury...
1:15-cv-02410-TWT	2015-07-28	7	Certificate of Interested Persons by...
1:15-cv-02410-TWT	2015-07-29	NA	Clerks Notation re 7 Certificate of...
1:15-cv-02410-TWT	2015-08-13	8	Application for Leave of Absence for the...
1:15-cv-02410-TWT	2015-08-14	NA	Clerks Notation re 8 Leave of Absence...
1:15-cv-02410-TWT	2015-08-26	9	JOINT PRELIMINARY REPORT AND...
1:15-cv-02410-TWT	2015.08.27	10	CERTIFICATE OF SERVICE of Discovery...
1:15-cv-02410-TWT	2015.08.27	11	SCHEDULING ORDER: re: 9 Joint...
1:15-cv-02410-TWT	2015.08.28	12	Initial Disclosures by Souto Foods, LLC...
1:15-cv-02410-TWT	2015-11-02	13	Joint MOTION to Approve Settlement by...
1:15-cv-02410-TWT	2015-11-12	14	ORDER GRANTING 13 Joint Motion to...
1:15-cv-02410-TWT	2015-12-07	15	STIPULATION of Dismissal by Marvin A...
1:15-cv-02410-TWT	2015-12-08	NA	Clerk's Entry of Dismissal APPROVING 15...
1:15-cv-02410-TWT	2015-12-08	NA	Civil Case Terminated. (adg) (Entered...

Figure 10.1. Docket sheet extract.

the research team wrote relatively simple code to extract the keywords or names associated with each feature from the docket sheet text.

The docket sheet text also provided a wealth of information about the litigation itself, requiring both simple and more sophisticated methodological approaches. On the simple end of the spectrum, the project team calculated the number of days each case was open by extracting the first and last date on each docket sheet and counted the number of docket sheet entries. Both features could be used as rough proxies for a case's level of activity and/or complexity. The team also identified whether a case had been removed to federal court from state

---

full or in part. The magistrate's role within that particular case is then over. In other circumstances, both parties may agree to have their entire case adjudicated by the magistrate rather than the district court judge. In those cases, the magistrate, rather than the district court judge, makes all relevant decisions and the district judge has no role. See rules 72–73 of United States Courts, "Federal Rules of Civil Procedure," https://www.federalrulesofcivilprocedure.org/frcp/.

Table **10.2.** Litigation life cycle stages.

Stage number	Stage description
1	Complaint
2	Answer
3	Motion to dismiss
4	Motion to dismiss decision
5	Discovery
6	Motion for summary judgment
7	Motion for summary judgment decision
8	Settlement
9	Trial

court by identifying the presence of the trigram, "Notice of Removal" in the first docket sheet entry, and counted the number of depositions taken by each party by identifying "Notice of Deposition" in proximity to the name of a lawyer for the plaintiff or defendant. The "removal" variable might stand in for defendant sophistication, as more savvy defendants might choose to pull cases from state into federal court in search of the most favorable jurisdiction. Similarly, the deposition count feature could be used as a rough proxy for each party's investment in the case, as well as an indication of factual complexity and the level of development of the record during the discovery period.

More sophisticated work was required to capture the stages of litigation through which each lawsuit progressed before concluding with one of six case-ending events: settlement prediscovery, settlement after discovery had begun, dismissal, granted motion for summary judgment for plaintiff, granted motion for summary judgment for defendant, or trial.

Here, the research team created a set of stages through which a lawsuit might progress, shown in table 10.2, and

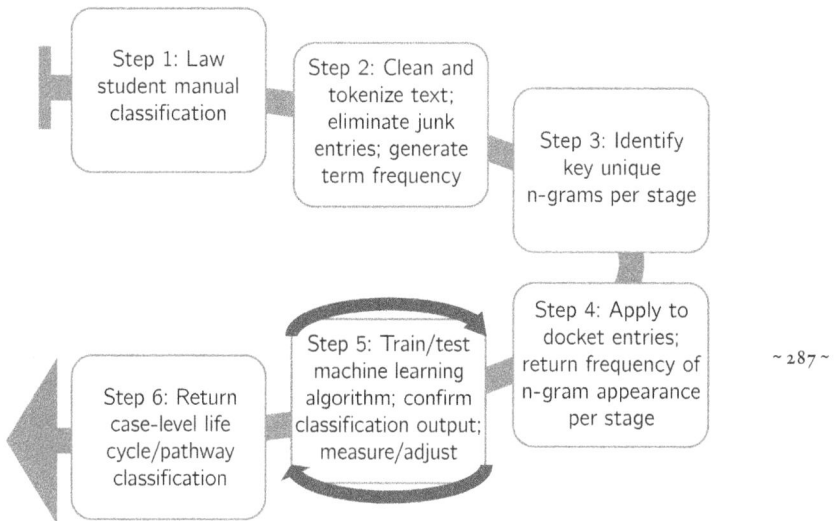

Figure 10.2. Docket entry classification procedure.

then experimented with using a random forest algorithm to assign each docket entry to a single stage after first excluding some docket entries as junk. (Note that, although "Settlement" appears as stage 8 in table 10.2, it could occur at any point in litigation; the remainder occurred in sequence, though not every lawsuit reached every stage.) Law students manually classified approximately one thousand docket entries into the life cycle stages as a training set, and the predictive model then identified the key *n*-grams that were unique to each stage and persisted across all docket entries that were manually assigned to that stage. Using those *n*-grams, the model iterated across all docket entries. The docket entry-level classifications were then rolled up to a rougher set of case-level classifications, generating a picture of each case's pathway through litigation. Figure 10.2 illustrates this process.

During the initial phases of the project, we were unable to identify whether a case-ending motion for summary judgment was granted in favor of the plaintiff or defendant. We were also unable to identify trials successfully, as they were exceedingly low-occurring events within our dataset and in civil litigation more generally. Therefore, our initial results grouped case-ending summary judgment decisions into a single event and omitted trials. Moreover, although precision and recall were quite high for certain classifications, such as complaints and answers, the model performed poorly in identifying discovery and settlement correctly. This is likely because there is no single

Table 10.3. Pathways through litigation.

Pathway	Frequency	Percent
Complaint | case-ending dismissal	197	4
Complaint | discovery | case-ending motion for summary judgment	57	1
Complaint | discovery | non-case-ending motion for summary judgment | settlement	100	2
Complaint | discovery | settlement	1248	24
Complaint | non-case-ending dismissal | discovery |case-ending motion for summary judgment	603	12
Complaint | non-case-ending dismissal | discovery |non-case-ending motion for summary judgment | settlement	618	12
Complaint | non-case-ending dismissal | discovery |settlement	1152	23
Complaint | non-case-ending dismissal | settlement	97	2
Complaint | settlement	233	5
Unknown	806	16

docket entry called "discovery" or "settlement" but rather collections of candidate entries that indicate that discovery is underway or a settlement has happened.

In the final section of this chapter, we describe the ongoing refinement of our techniques, including distinguishing between granted motions for summary judgment for plaintiffs and for defendants, adding more granular litigation stages (e.g., default judgment and *sua sponte* dismissals), experimenting with additional layers of preprocessing to better identify and exclude "junk" docket entries, and using other techniques, including a neural network trained on a much larger set of docket sheet entries, and the various text classification tools within the fastText library, to achieve better classification performance.[6]

Nevertheless, after an initial set of passes through the data, the team was able to classify 84% of the cases into one of nine pathways, ending in settlement prediscovery, settlement after discovery had begun, dismissal, or a granted motion for summary judgment (for either party). Table 10.3 shows the pathways and their frequency; figure 10.3 provides a visualization; table 10.4 shows the frequency of each case-ending event for which we were able to generate data. These results remain preliminary; they are presented here only to illustrate the goals of this phase of the project.

It is tempting to compare these results to the work of other researchers who have studied employment law case outcomes, with a particular focus on the way that employment discrimination cases end. In the Eisenberg and Lanvers study mentioned above, for example, the authors found that the judges on the Northern District of Georgia (NDGA) granted summary judgment at a higher rate than judges from other districts (Eisenberg and Lanvers 2008). As they state, "The

---

[6]See https://fasttext.cc.

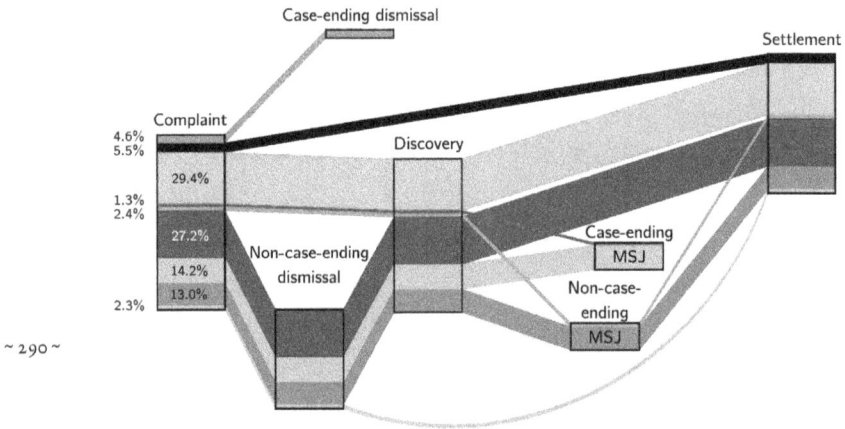

Figure 10.3. Pathways through litigation stages—visualization.

most striking effect was the approximate doubling—to almost 25%—of the NDGA summary judgment rate in employment discrimination cases and a substantial increase in the NDGA summary judgment rate in other civil rights cases." In another study focused exclusively on the outcomes of employment discrimination lawsuits, Nielsen, Nelson, and Lancaster (2010) found that 40% of plaintiffs lost their Title VII claims on dispositive motions or at trial, 58% of cases settled, and 2% of plaintiffs ultimately won at trial. Finally, in a study of employment discrimination cases filed in federal courts between 1979 and 2006, Clermont and Schwab report an overall plaintiff win rate—regardless of stage—of 15% (Clermont and Schwab 2009).

All of this research provides a useful baseline against which to measure our project's outcomes, and particularly our findings with respect to employment discrimination lawsuits' case-ending events. Because of the high number of "unknown" pathways and unspecified case-ending events, however, as well as the underperformance of our classification methodology,

Table 10.4. Case-ending events.

Event	Frequency	Percent
Settlement	3448	67
(Pre-Discovery)	(552)	(16)
(Post-Discovery)	(2896)	(84)
Case-ending motion for summary judgment	660	13
Case-ending dismissal	197	4
Unknown	806	16
Total	5111	100

our results are not yet reliable at this stage. The other features extracted from the docket sheets—e.g., number of days open, number of docket entries, and number of depositions taken by plaintiffs and defendants—are more reliable and able to be included in the final predictive model described later in the chapter.

Moreover, this initial work on litigation pathways and case-ending events has created a foundation that will facilitate continued refinement of our docket sheet classification code. This work will also allow us to answer the first and last of the descriptive questions listed above: the frequency and distribution of case-ending events, and lawyer/judge playbooks. By overlaying judge and lawyer identifiers atop the lawsuits' litigation pathways, in addition to other filters such as claim type, we will be able to determine whether certain lawyers' or judges' cases follow certain patterns through the stages of litigation.

Complaints

In addition to docket sheets, the second document type analyzed by the project team was the complaint filed in each case. A complaint is a case-initiating document written by the plaintiff or his/her attorney that identifies the law(s) that the plaintiff claims the defendant violated and describes what happened to trigger the lawsuit. Here, our goal was to extract from the complaints' text the allegations of statutory violations, claims nested within those allegations, defendant industry, and plaintiff characteristics, including occupation and—for discrimination claims—race and/or national origin.

It soon became apparent, however, that the complaints would not be the rich source of information that we had hoped. Of the 5,111 complaint PDFs, only 3,263, or about 64%, were machine readable. The unreadable complaints were either hand-written or scanned images, and were largely filed by *pro se* plaintiffs, or parties who were not represented by an attorney.

Using the readable 3,263 set, the first task was to extract all references to statutes from the text. The research team developed a lexicon of all relevant statutes' names and abbreviations, and variations thereof. This list was not limited to the options available as NOS codes or cause classification choices on the civil cover sheet, but resulted from law students' initial reading of a sample of complaints with the goal of identifying the range of statutory violations that were alleged. The team then wrote code to generate a frequency count of those key terms' appearance within each document and decided on an acceptable frequency threshold for classifying a complaint as containing an allegation under one or more statutes.

For the complaints that could not be classified using these methods, the team created a citation finder that extracts citations to the US code and the Code of Federal Regulations from the text. This piece of code relied on regular expressions to find all citations to statutes and regulations, which tend to appear in a consistent format in court documents with relatively little variation. For example, if a plaintiff alleged that the defendant violated the overtime provision of the Fair Labor Standards Act, and if the process described in the previous sentence failed to identify the usage of "Fair Labor Standards Act" or "FLSA" in the text, then the citation finder would extract "29 U.S.C. § 207," which the team would identify as a citation to the FLSA. Throughout, law faculty and students reviewed the output to manually classify the remaining complaints according to the statutes and regulations cited therein.

Table 10.5 reports the total number of allegations made under Title VII of the Civil Rights Act of 1964 (Title VII), the Age Discrimination in Employment Act (ADEA), the ADA, the FMLA, and the FLSA. These statistics capture whether a complaint alleged a violation of any of those five statutes. They therefore represent allegation counts, not lawsuit counts, so the total number exceeds the 3,263 machine-readable complaints. For the same reason, these totals are not directly comparable to the NOS code totals shown in table 10.1, which capture only one NOS code per case. Nor do the statutes align perfectly with the NOS codes. However, for purposes of rough comparison, Title VII and ADEA together would fall under 442 Civil Rights-employment, and the remaining statutes would align with their own, stand-alone NOS codes (445 ADA, 710 FLSA, and 751 FMLA).

In theory, the classifications generated by the methodology described here could be validated against either or both the NOS code or the cause classification derived from PACER for any given case. In fact, the research team's text analytics identified the presence of an ADA, Title VII, or FLSA violation

Table 10.5. Allegations of statutory violations extracted from complaint text.

Statute	Frequency	Percent
FLSA	1600	45
Title VII	1089	30
FMLA	377	11
ADA	258	7
ADEA	258	7
Total	3582	100

allegation in a complaint in 77% of the cases with either the relevant NOS code or that statute listed in the "cause" field. The figure for FMLA cases was 97% and 82% for ADEA cases. However, given the known problems with both the NOS codes and cause classifications reported by PACER, it is difficult to know whether these figures are reliable. Without reading every complaint, it is hard to determine whether the lower match rates are driven by problems with our text-based classification system or problems with plaintiffs' assignment of NOS codes and cause classifications on the civil cover sheet. Moreover, this protocol was highly supervised. Another, less manual, approach might have used machine learning to train an algorithm to recognize the statutory allegations within a complaint or topic modeling to identify the statutory allegations at hand.

For example, using a machine learning approach, a team of law students might be assigned to read and manually classify a set of complaints according to the allegations made. This would function as the training set. The team would then build a type of machine learning algorithm known as a multilabel classifier, which could "learn" from the manual classifications and assign one or more labels to new, unclassified complaints. The team would check the accuracy of the labels assigned to this test set and, if necessary, proceed through additional training-test iterations until the algorithm is able to label complaints' statutory allegations with the desired level of accuracy.

As an alternative or complementary strategy, using a topic modeling approach, the team could build an algorithm that would identify commonly co-occurring clusters of words or topics across the entire corpus of complaints. Reviewers would then determine whether any of these topics corresponded to identifiable sets of statutory allegations. If so, the team would next classify each complaint according to the topics—here, statutory allegations—that were present within the text. Topic modeling is discussed further below in connection with the project team's analysis of summary judgment decisions.

In addition to classifying each case by the statute(s) mentioned in the complaints, the project team created an additional, more granular classification: claim type. Here again, the law student team members read a sample of complaints and constructed a lexicon of keywords and phrases that identified the particular claim being made under the statute that was alleged to have been violated. For most statutes, this required one step down in granularity. For ADA allegations, for example, the team identified keywords that were associated with the following claims: hostile work environment; retaliation; reasonable accommodation; and discriminatory hiring, promotion, discipline, transfer, demotion, termination, and constructive discharge. The team wrote code to tally the frequency with which these terms appeared in each complaint. Next, as in the statutory allegation classification process, the team established an acceptable keyword appearance frequency threshold, which was applied in order to classify complaints by claim type, nested within statutory allegations.

For Title VII allegations, we needed to go two steps down in granularity. First, the team identified a set of claim-type keywords, similar to the ADA list above, e.g., hiring, promotion, transfer, demotion. Next, the team identified keywords that were associated with the particular protected class that was at issue: race, sex, national origin, religion, and color. Finally, the team constructed a set of rules that identified all combinations of claim-type keywords and protected class keywords in close proximity to one another within the text and established relevant frequency thresholds. This protocol allowed any complaint alleging a Title VII employment discrimination violation to be classified further as alleging hiring discrimination by religion, for example.

Applying this methodology to complaints that contained allegations under only one statute was relatively simple: the team simply ran the statute classification code and then the claim classification code, which produced both levels of classification, with claim types nested within statutory allegations. However, complaints that contained allegations of violations of more than one statute presented a trickier challenge, as some claims could be nested within more than one statute. For example, a single complaint might make both Title VII and ADA violations as well as discrimination in the form of harassment and demotion. Because harassment and discriminatory demotion are actionable under both statutes, the task was to sort the claim types properly into their statutory buckets.

To solve this problem, the teams wrote code that first identified the statutes, as described above, and then created a roughly two-sentence window before and after the statute identifiers as they appeared in the text. The code then searched for the claim-type keywords within that window. Here, the code searched only for the keywords that were relevant to the particular statute at hand—for example, searching for overtime-related keywords only within FLSA windows and not in the windows around any of the discrimination-related statutes. Once the code identified keywords within any given window, it moved to the next instance of a statute in the text and drew a new window but did not permit the windows to overlap. Through this process, and by again applying acceptability thresholds, the team was able to assign to each of the 3,263 machine-readable complaints one or more statute-claim classification pairs.

Returning to the question of validation, it is, again, difficult to compare the results of this process to the PACER-provided fields because granular claim classifications may or may not appear in PACER's cause classifications. For example, all 152 cases with an FMLA NOS code also had "29:2601 Family and Medical Leave Act" as their cause classification, providing no additional detail on claim type. In contrast, the team was able to classify FMLA cases into those that made "interference" and/or "notice" claims—two types of FMLA violations—introducing a greater level of granularity than what is available via PACER.

The clause classification field for cases with the 442 Civil Rights–employment NOS code provided some detail as to Title VII cases, indicating the relevant protected class that we could then map onto our claim classification output. Here, the best match rate was for religious discrimination; our code identified religious discrimination claims in the text of 77% of the complaints that carried a PACER cause classification that indicated religious discrimination. The match rate for other protected classes hovered between 30% and 60%. Generally, mismatches resulted both from instances where our code failed to find allegations that were present in the text and identified by the PACER fields, and where manual checks revealed that the PACER fields seemed themselves not to match the claims we found within the text.

One potential pathway to improve this claim classification process is through the use of less supervised methods such as topic modeling. The challenge there, however, is the large amount of case-specific factual detail that appears in complaints, which can result in topics that are not illuminating as to the types of allegations being made. Perhaps more pressing is the task of validating the output of the code by performing manual statute- and claim-type classifications. Because the PACER-provided NOS code and claim classification data are flawed in all of the ways described above, they are an unreliable metric against which to measure code performance. Increasing the amount of high-quality human-coded training data may be the only method of substantially improving model performance.

Separate from the statute and claim classifications just described, the project team also constructed an additional set of classifications using the text of the available 3,263 complaints. First, for the cases that alleged race or national origin discrimination, the team wrote code that attempted to identify those plaintiffs' specific race or national origin by extracting keywords from windows of text. This exercise produced the results shown in figure 10.4, which also illustrates the Title VII claim types extracted from the text. The chart on the right can be read as providing more detail for the "Race" and "National Origin" slices of the chart on the left.

Setting aside whether these text-derived distributions map onto the distributions of the PACER NOS codes and cause

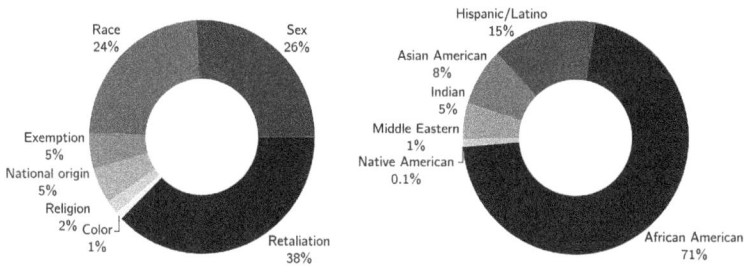

Figure 10.4. Title VII claim types and plaintiffs' race or national origin extracted from complaint text.

classifications, it is interesting to note that the results shown in figure 10.4 are consistent with other research on the frequency of different types of Title VII claims. The chart on the left shows retaliation as the largest slice, followed by race discrimination claims; this squares with research that shows that the largest category of discrimination charge filed with the Equal Employment Opportunity Commission (a predecessor step to filing a Title VII lawsuit in court) has been retaliation, followed by race discrimination, since 2010.[7]

Finally, apart from race and national origin characteristics, the team began work on identifying the occupation of the plaintiff and industry of the defendant, using both a keyword-centric approach that draws on the US Bureau of Labor Statistics' industry and occupation lexicons and a less supervised natural language processing approach involving tools such as part-of-speech tagging within the *spaCy* library, a prepackaged set of text analytics tools.[8]

In summary, the team's work on the complaints is incomplete but promising. Yet it is important to remember throughout that

---

[7]See PACER Service Center, "US Equal Employment Opportunity Commission, Charge Statistics, FY 1997 through FY 2017," https://www.eeoc.gov/eeoc/statistics/enforcement/charges.cfm.

[8]See https://spacy.io.

the results presented here were derived from only 64% of the complaints associated with the full set of 5,111 cases in the study set. Further, the non-machine-readable complaints were not randomly distributed across case and plaintiff types: large numbers were filed by *pro se* plaintiffs who filed lawsuits on their own, without representation by a lawyer. This means that the preliminary results presented here are incomplete in another important way, as they do not capture all plaintiff types. In particular, *pro se* plaintiffs may differ in systematic ways from their represented counterparts, in terms of claim types, socioeconomic status, or some other characteristics. Their absence skews the data, with results describing relatively better-financed litigants rather than the pool as a whole. Manual work may be needed to convert those complaints into a readable format; options include transcribing them in full by hand or manually extracting the relevant features.

Stepping back from the particulars of this project, this issue points to a much larger challenge that many legal analytics projects must face: that of data quality and access. If many law-related insights are locked up in unreadable PDFs, then the conclusions that can be drawn from only their readable counterparts are necessarily limited and the potential for misapplication is great. In particular, research will exclude populations with limited resources for representation, whose complaints are hand-written or so idiosyncratic that they are not susceptible to analysis (Noble 2018). Although great strides have been made in the areas of image processing and handwriting recognition in recent years, these remain challenges on the frontier of the digital deployment of natural language processing techniques. Until inexpensive and broadly available software to process precisely these types of data exist, such texts will continue to be less than accessible to researchers.

Summary Judgment Documents

The final two types of documents from which the team extracted features of the lawsuits were the reports and recommendations (R&Rs) issued by magistrate judges on the parties' motions for summary judgment and the district court judges' final summary judgment opinions and orders (SJs).

This portion of the research was, in a sense, its own separate miniproject within the larger effort, intended to answer the second descriptive question set out in the Introduction to this chapter: In summary judgment decisions, what were the legal doctrines used and cases cited most frequently by judges? To this initial question we added a question about the dynamics of judging: To what extent did district court judges adopt the magistrates' R&Rs in ruling on motions for summary judgment? In other words, what might these results suggest about the distribution of the decision-making function between these two levels of judging?

These questions swept in both case-ending summary judgment decisions, which had already been captured by the litigation life-cycle work described above, and non-case-ending summary judgment decisions. Thus, the unit of analysis here became the summary judgment decision, not the lawsuit, which was the unit of analysis in the docket sheet and complaint portions of the research.

The team used a combination of strategies to answer these questions. First, we tried to use topic modeling on the corpus of R&Rs and SJs to get at the legal doctrines upon which judges were relying, experimenting with a variety of approaches: singular value decomposition (SVD), non-negative matrix factorization (NMF), latent Dirichlet allocation (LDA), and the hierarchical Dirichlet process (HDP) (Chen et al. 2018). We generated ten topics initially and then explored aggregating topics in order to identify fewer but more useful topics. We also explored using the term frequency-inverse document frequency (TF-IDF), which identifies the most "important" words within a document by assigning weights to terms in relation to the frequency within which they appear within a document or corpus of documents (Manning, Raghavan, and Schütze 2008).

Even after substantial cleaning, including the removal of all proper nouns, and grouping documents by statutory allegation and claim type, however, the results that were produced were too specific to the individual facts of the cases to provide much insight into the legal rules that judges were deploying. For example, one typically fact-specific topic generated using the LDA approach from the set of R&Rs and SJs alleging FLSA (wage and hour) violations consisted of the following terms: "FLSA, dancer, club, wage, docket, driver, plan, discrimination, fee, agent." This result does reveal that a subset of FLSA cases concerned the question of whether exotic dancers were properly classified as independent contractors rather than employees by their employers. However, it does not reveal how the judges resolved the dancers' claims or the legal rules or doctrines upon which the judges relied in doing so. Thus, while not completely unhelpful, topic modeling seems better suited to discovering common fact patterns than legal rules.

We then shifted to an approach that combined supervised and unsupervised techniques in which the law students first identified keywords that were associated with doctrines that might be used by a judge in deciding a summary judgment motion in any case, as well as in the particular types of employment law cases at issue in this research. The team generated a simple frequency count for each keyword and selected a threshold for inclusion. Next, the team used a Word2Vec model to pull additional important terms from the context in which the keywords were used in the set of R&Rs and SJs (Mikolov, Sutskever, et al. 2013). Specifically, the team used continuous skip-gram architecture, which can predict a window of context words in which a given word appears.[9] Law faculty reviewed the context output to identify any additional terms that should be added to the keywords list, and the team then reran the frequency table. From the table, the teams created a word cloud in Tableau with filters by judge and additional filters for statute and claim types, year, and other case features.

The team also wrote code that extracted all citations to case law, in both long and short form, from each R&R and SJ and displayed citation frequency across the dataset in a similar Tableau word cloud dashboard. Ongoing refinements to the citation dashboard include displaying citations on a per-opinion basis and creating greater filtering ability, as well as perhaps combining the keyword and citation dashboards into a single interface.

Finally, the team turned to the additional question about the balance of summary judgment decision-making between magistrates and US district court judges. The team began by writing code that would extract the magistrates' recommendations from the text of the R&Rs (grant, deny, partial), on the one hand, and the district court judges' action on the R&Rs (adopt, reject, partial) from the text of the SJs, on the other. However, this approach was stymied by the wide variation in the language that judges, and particularly the

---

[9] Possible next steps might include further exploration of vector space models. Such approaches allow the simultaneous representation and exploitation of multiple aspects of each word in a document, including frequency, context, and sequence. These approaches embed words within their surroundings and structures and could perhaps generate superior classification results by picking up on latent patterns within text (Turney and Pantel 2010).

magistrates, used. The team briefly explored a less supervised approach but lacked the number of documents necessary to train a machine learning model to classify R&Rs and SJs accurately by outcome.

The team then turned to the text of the docket sheet entries associated with R&Rs and SJs as an alternative. After first writing code that paired each R&R with its relevant SJ order, the students were then able to identify the frequency with which a district court judge adopted the magistrate's recommendation. The students also determined which party filed the initial motion for summary judgment and whether the SJ order ended the case as an alternative way to get at the same summary judgment–related case-ending events explored

Case Number	Docket Text	Paired Order Text	Filer	Order Adopted	Case Ending
1:10-cv-00007-JEC	FINAL REPORT AND RECOMMENDATION recommending GRANTING 22 MOTION for Summary Judgment as to all of Plaintiff's claims. Signed by Magistrate Judge Gerrilyn G. Brill on 06/14/11. (Attachments: # 1 Order for Service (fap) (Entered: 06/15/2011)	ORDER regarding Magistrate Judge's Final Report and Recommendation. IT IS HEREBY ORDERED that the Court ADOPTS the Magistrate Judge's Final REPORT and Recommendation 28 GRANTING defendants' Motion for Summary Judgment as to all of Plaintiff's claim 22. The Clerk is directed to close this action. Signed by Judge Julie E. Carnes on 7/5/11. (cem) (Entered: 07/06/2011)	D	Adopt	TRUE
1:10-cv-00383-ODE	REPORT AND RECOMMENDATION that Defendant's 42 MOTION for Summary Judgment be GRANTED IN PART AND DENIED IN PART. Specifically, the undersigned RECOMMENDS that the motion for summary judgment be DENIED with respect to Plaintiffs' S. 1981 claims, and GRANTED with respect to Plaintiff Lewis, Power, and Johnson's claims under the Equal Protection Clauses. Signed by Magistrate Judge Alan J. Baverman on 1/31/2012. (rej) (Entered: 02/01/2012)	ORDER the City of Kennesaw's Objections are 56 SUSTAINED and the 55 Report and Recommendation is ADOPTED IN PART AND REJECTED IN PART. The City of Kennesaw's 42 Motion for Summary Judgment is GRANTED in accordance with Federal Rule of Civil Procedure 56(a) because Plaintiffs have failed to establish a prima facie case of discrimination. Signed by Judge Orinda D. Evans on 3/30/2012. (anc) (Entered: 03/30/2012)	D	Partial	FALSE

Figure 10.5. Docket sheet analysis output extract.

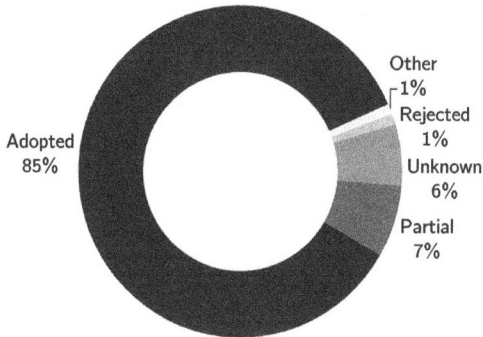

Figure 10.6. District court judges' actions on magistrate judges' report and recommendations on motions for summary judgment.

above. Figure 10.5 gives two examples of the data structure produced by these protocols. Preliminary results are shown in figure 10.6.

In 85% of cases with both an R&R and an SJ order, the district court judge adopted the magistrate's recommendation wholesale. This finding sheds light on where and how summary judgment decision-making was actually happening during our study period.

Predictive Model

Taken together, the processes described above generated a set of features that we could then feed into a random forest model as explanatory or independent variables to attempt to predict cases' termination in one of four case-ending events: dismissal, motion for summary judgment, prediscovery settlement, and postdiscovery settlement. Additional features that could be added in the future include the successful party on summary judgment as well as other events, including trial, default judgments, and *sua sponte* dismissals (as distinguished from rulings on a motion to dismiss). These additional features hold out the promise of substantially improving the predictive models.

Using our preliminary data, we constructed four models using only the information that would be available to a plaintiff's attorney at four points in time: (1) the prefiling intake stage, (2) early in litigation, just after a case is filed, (3) at the close of discovery, and (4) in a state of "omniscience," our term for an all-in model involving the full set of known case features. We assigned features as independent variables only to the version of the model in which that information could be known. As an example, the judge assigned to a case is a feature that would not be known at prefiling intake; therefore, that variable was included only in models 2–4 and omitted from 1.

The model was the richest at the omniscience stage (4), incorporating twenty-eight case features; when it simulated intake (1), only ten features were available.[10] Using each relevant set of available features, our four models attempted to bucket cases into the four case-ending events.

The preliminary results of our modeling are shown in figure 10.7, along with the four most important features in each. Accuracy is listed; this refers to the model's success in bucketing cases by outcome, as described above.

Though this output is extremely preliminary, one interesting result is the predominance of the attorneys' own prior case-handling patterns as important predictors of case outcomes. Notably, this echoes other researchers' findings on the importance

---

[10]Features available in the omniscience stage included, but were not limited to, the number of plaintiffs and defendants, the number of attorneys on both sides, whether the plaintiff was *pro se*, the type and number of statutory allegations made, the year of case filing and termination, the number of days a case was open, and dummy variables representing the judge and his or her characteristics. At intake, by contrast, no judge-related variables were available, nor were any defendant attorney features.

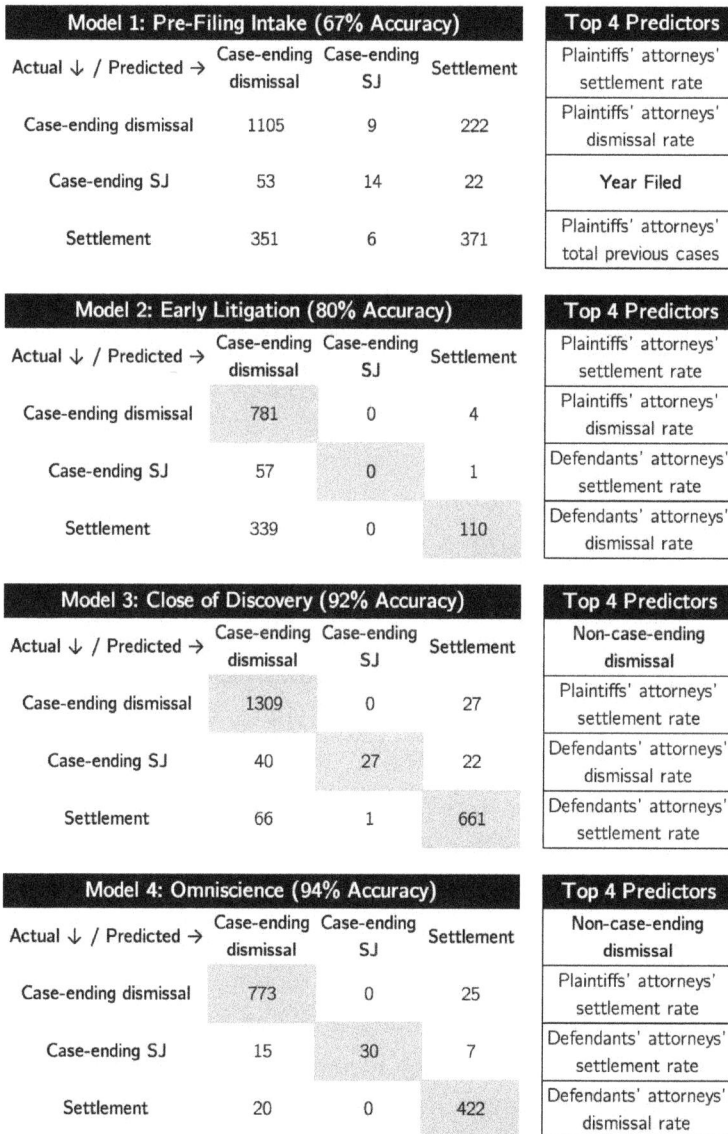

Model 1: Pre-Filing Intake (67% Accuracy)			Top 4 Predictors
Actual ↓ / Predicted →	Case-ending dismissal	Case-ending SJ	Settlement	Plaintiffs' attorneys' settlement rate
Case-ending dismissal	1105	9	222	Plaintiffs' attorneys' dismissal rate
Case-ending SJ	53	14	22	Year Filed
Settlement	351	6	371	Plaintiffs' attorneys' total previous cases

Model 2: Early Litigation (80% Accuracy)			Top 4 Predictors
Actual ↓ / Predicted →	Case-ending dismissal	Case-ending SJ	Settlement	Plaintiffs' attorneys' settlement rate
Case-ending dismissal	781	0	4	Plaintiffs' attorneys' dismissal rate
Case-ending SJ	57	0	1	Defendants' attorneys' settlement rate
Settlement	339	0	110	Defendants' attorneys' dismissal rate

Model 3: Close of Discovery (92% Accuracy)			Top 4 Predictors
Actual ↓ / Predicted →	Case-ending dismissal	Case-ending SJ	Settlement	Non-case-ending dismissal
Case-ending dismissal	1309	0	27	Plaintiffs' attorneys' settlement rate
Case-ending SJ	40	27	22	Defendants' attorneys' dismissal rate
Settlement	66	1	661	Defendants' attorneys' settlement rate

Model 4: Omniscience (94% Accuracy)			Top 4 Predictors
Actual ↓ / Predicted →	Case-ending dismissal	Case-ending SJ	Settlement	Non-case-ending dismissal
Case-ending dismissal	773	0	25	Plaintiffs' attorneys' settlement rate
Case-ending SJ	15	30	7	Defendants' attorneys' settlement rate
Settlement	20	0	422	Defendants' attorneys' dismissal rate

Figure 10.7. Predictive model preliminary results (random forest).

of attorney-related variables in predicting litigation outcomes (Ashley 2017).

This result is not useful as a business matter: a law firm is not helped at intake by its own historical case outcome rate. Nevertheless, these attorney-centric results may be acting as proxies for some filtering or case selection process performed by plaintiffs' attorneys, indicating that attorneys' own nonquantitative predictive modeling may be quite effective. Further work is needed to unpack these results and to attempt to extract further case features from the docket sheets and court documents that might themselves be the subject of the attorneys' own filters.

Here, the gap between a purely predictive approach to an analytics problem and an inference-based, causation-focused, data modeling approach becomes clear. If a given variable, such as attorneys' records, predicts with high accuracy how a case will end, that may be enough to satisfy some business goals but not others where knowledge about the causal consequences of interventions is needed. In his insightful commentary on what he calls "the two cultures" within statistical modeling, Leo Breiman puts it this way: "Approaching problems by looking for [an inference-based, causation-focused] data model imposes an *a priori* straight jacket [*sic*] that restricts the ability of statisticians to deal with a wide range of statistical problems" (Breiman 2001). On the other hand, techniques such as the random forest model used here "are A+ predictors. But their mechanism for producing a prediction is difficult to understand."

The research team is continuing its modeling work, considering different regression model specifications (consistent with the former, data modeling approach) and continuing to refine its decision tree modeling (consistent with the latter). By jumping between the two cultures and deploying each one's tools, we hope to come closer to answering the predictive question posed above: What features of a lawsuit, observable at different points in the intake and litigation processes, predicted its case-ending event?

Continuing Work

Lab researchers remain engaged with ongoing research on these data. In general terms, researchers are testing approaches that are less supervised and less reliant on expert-generated keywords. These endeavors require more data than our 5,111 subject lawsuits can provide. As a solution, the team is accessing the RECAP archive of 2.2 million docket sheet entries and millions more court documents from all ninety-four US district courts available via Court Listener, a nonprofit, free legal search engine run by the Free Law Project.[11] This larger dataset will enable the use of a neural network or other deep learning techniques (Goodfellow, Bengio, and Courville 2016) to process the docket sheet entries and more accurately classify them into a more complete set of litigation stages. The team also continues to experiment with various machine learning approaches to classifying complaints by statutory allegations and claim types, and R&Rs and SJs by outcome.

This work has opened promising new avenues for research that seeks to understand the mechanics of civil litigation and the day-to-day operation of trial courts as they adjudicate disputes. As Pauline Kim and her coauthors have observed, scholarship in law, political science, and related fields too often focuses exclusively on the text of judges' written decisions, neglecting the rich trove of data that resides in docket sheets and party-filed court documents (Kim et al. 2009). This neglect is unsurprising, as the tasks of assembling this text and then analyzing it are

---

[11] See https://www.courtlistener.com.

daunting. However, as this chapter has described, advances in computation now allow researchers to mine masses of unstructured legal text, beyond just written opinions. These computational tools are developing in parallel with, and are perhaps encouraging, a growing movement to increase access to court data of all types (Schultze 2018).

It is too early to claim that litigation pathways are now predictable, however, or that judges' decisions can be easily classified, forecast, or understood in bulk. Legal text remains a bramble bush in many ways, to borrow Karl Llewellyn's characterization of the law as a whole (Llewellyn 1951). Indeed, as one reviewer recounted, Llewellyn "tells us that the law is not a self-contained set of logical propositions; that rules of law do not explain results at law; that the stated reasons for decision regularly mask the inarticulate major premise; that facts are slippery things with a nasty habit of changing shape and color, depending on who is looking at them; [and] that judges are not automatons who announce the law but human beings" (Gilmore 1951).

If these propositions are true—and the project described in this chapter suggests that they are—then legal analytics has its work cut out for it. Nevertheless, as researchers continue to experiment with the application of computational methods to legal text, it is likely that the leaves and stalks may start to become visible within the thicket. A computational approach holds the promise of revealing hidden patterns and structures within the work of the courts; we will then be left to reckon with what we find, thorns and all. ✦

Acknowledgments

We thank the student members of the sprint team, Harry Alex, Ayushri Bhargava, Colt Burnett, Vivian Chew, Chris Cirelli,

Pearson Cunningham, Brad Czerwonky, Nathan Dahlberg, Fei Drouyor, Hayden Hillyer, Ziying Huang, Amanda Iduate, John Lesko, Xiaotong Li, Siddhant Maharana, Ojasvi Maleyvar, Vahab Najari, Babak Panahi, Lucas Perdue, Kyle Price, TJ Sizemore, Pallavi Srinivas, Renate Walker, Zhe Wang, Chad Williams, and Caroline Xu; and the sprint sponsor, Amanda Farahany of Barrett and Farahany. Thanks also to Michael Livermore and Daniel Rockmore for their helpful feedback.

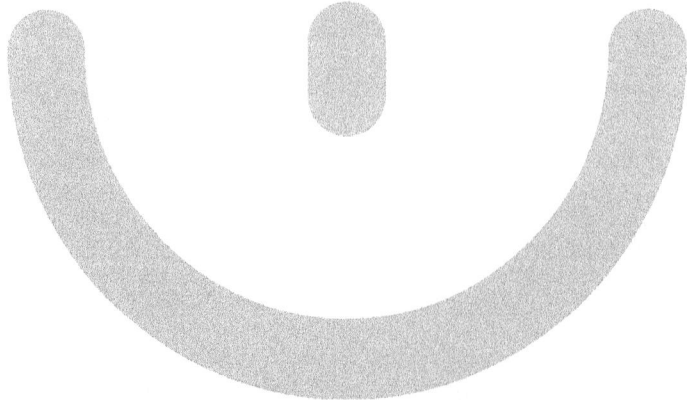

ᯭ

CASE VECTORS: SPATIAL REPRESENTATIONS OF THE LAW USING DOCUMENT EMBEDDINGS

Elliott Ash, ETH Zurich
Daniel L. Chen, University of Toulouse

Law is an artifact of language. In this chapter, we ask what can be gained by applying to the law new techniques from natural language processing that translate words and documents into vectors within a space. These vector representations of words and documents are information-dense—in the sense of retaining information about semantic content and meaning—while also being computationally tractable. This combination of information density and computational tractability opens up a wide potential realm of mathematical tools that can be used to generate quantitative and empirically testable hypotheses about the law.

This new approach to legal studies addresses the shortcomings of existing methods for studying legal language. At a theoretical level, even the best formal models of legal decision-making require strong simplifying assumptions that treat the law metaphorically. The case-space literature, for example, assumes the language of law to be a function over an idealized geometric space, where the law separates the fact space into "liable" and "not liable" or "guilty" and "not guilty."[1] Case-space models give us some insight into the legal reasoning process, but they

---

[1] Cameron and Kornhauser (2017) provide a recent review of this literature.

have been somewhat limited because it has been unfeasible to empirically realize the legal case-space model in any formal mathematical way.

Likewise, because law consists of text, the standard empirical research methods are somewhat limited in the questions that can be asked. Traditionally, text-based empirical legal studies research has relied on small-scale datasets, where legal variables are manually coded (e.g., Songer and Haire 1992). Hand-coding legal documents is labor-intensive and requires subjective and simplifying decisions.

However, new computational approaches to the study of text are enabling new kinds of large-scale quantitative text-based projects across a wide range of disciplines. For example, recent work in computational linguistics has made breakthroughs in vector representations of language (Jurafsky and Martin 2000). Topic models such as latent Dirichlet allocation serve to automate the coding of texts by generating topics as sets of words that tend to cooccur (Blei, Ng, and Jordan 2003; Blei 2012). These algorithms have provided a window to the relations between documents at scale.

An active literature in computational legal studies has begun to apply these methods to legal documents. Livermore, Riddell, and Rockmore (2017) use a topic model to understand agenda formation on the US Supreme Court (see also Carlson, Livermore, and Rockmore 2016). Leibon et al. (2018) construct a geometric and network-based model to study any kind of citation-linked document corpus and use it to represent the geometric relations between US Supreme Court cases and even predict domains of heightened or diminished legal activity. Ganglmair and Wardlaw (2017) apply a topic model to debt contracts, while Ash, MacLeod, and Naidu (2018) apply one to labor union contracts.

But legal scholars are still mostly unfamiliar with a new counterpart to topic models: embedding models. The success of word embedding models, such as Google's Word2Vec algorithm, is that they "learn" the conceptual relations between words; a trained model can produce synonyms, antonyms, and even analogies for any given word in a large text corpus (Mikolov, Sutskever, et al. 2013; Levy, Goldberg, and Dagan 2015). The derived word vectors serve well as features in downstream prediction tasks by encoding a good deal of information in relatively rare word features. More recently, analogously constructed "document embeddings" have built upon the success of word embeddings to represent words and documents in a joint geometric space (Le and Mikolov 2014). Like word embeddings, these document embeddings have advantages in terms of interpretability and serve well in prediction and classification tasks.

This chapter serves to introduce document embeddings into the legal literature and illustrates the method using a corpus of US appellate court cases. Our data include the universe of US Supreme Court and US Circuit Court cases for the years 1887–2013. We construct document embeddings for each opinion in the corpus. We then construct judge vectors by taking the average of the document embeddings for the cases authored by the judge. These case vectors are used to analyze the geometry of federal appellate case law.

We ask whether the information recovered by our model provides a meaningful signal about the legal content in cases. We find that spatial clustering in these embeddings encodes differences between cases in different courts, between cases in different years, and between cases in different legal topics. The vectors can also discriminate judges based on birth cohorts but do not do well in encoding the partisan affiliation of judges or law school attended. We also demonstrate that the vectors can be used to produce a measure of similarity between the legal writings of any two judges.

In the final section, we outline a range of potential future applications for the use of embedding models in computational analysis of law. These include examining analogies and associations, experimenting with structured and categorical embeddings, and constructing embeddings for citation networks.

Vector Space Embedding Models and the Law

A first-order problem in empirical analysis of text data is the *a priori* high dimensionality of text. There is an arbitrary number of approaches for representing plain text as data. One must trade off informativeness, interpretability, and computational tractability (Ash 2018). For example, one could represent a document as a frequency distribution over words, but with a large vocabulary, say, twenty thousand words, a document is still a high-dimensional vector and interpreting such a representation is difficult.

Word embeddings came about as a dimension-reduction approach in deep learning models for prediction tasks in computational linguistics (Mikolov, Sutskever, et al. 2013). Such tasks include, for example, predicting the next word in a sequence given a set of words in a sentence. To that end, the model represents a word as a small and dense vector (say, one hundred dimensions). Initially, words are randomly distributed across the vector space. But the word locations then become features in a learning model: they move around during training to improve performance on a prediction task. In natural language settings, this process typically leads to words clustering near similar words.

Document embeddings, such as Le and Mikolov's paragraph vectors, use a separate embedding layer for both the word and the document to solve the prediction task (Le and Mikolov 2014). These models locate documents in a vector space, where documents that contain similar language tend to be located near to each other in the space. Embedding models are different from topic models (e.g., Blei 2012) because the dimensions have a spatial interpretation rather than a topic-share interpretation.

Embedding models have become popular because the spatial relations between the vectors encode useful and meaningful information (Levy, Goldberg, and Dagan 2015). To illustrate, a word embedding can identify similar words in the vocabulary. For example, the embedding of "judge" might be close to that of "jury" but far away from "flowerpot." Similarly, the proximity of document embeddings can identify similar cases in a corpus of decisions based on use of similar language. For example, in our data *Engel v. Vitale* (1962) is spatially close to *Everson v. Board of Education* (1947), presumably because they are both early US Supreme Court decisions that deal with religious freedoms in the states. Finally, judge embeddings constructed from these documents could be used to identify similar judges in the legal system. For example, we find that Antonin Scalia appears to be close to Clarence Thomas (perhaps since they are both conservative judges who tend to use originalist arguments).

Application to Federal Appellate Courts

This section illustrates the use of document embeddings in the federal appellate courts. We begin by discussing the data and how the document vectors are constructed. We then explore the visual relations between the cases. Finally, we explore similarity relations between judges.

DATA AND DOCUMENTS

The analysis utilizes a corpus of all US Supreme Court cases and all US Circuit Court cases for the years 1887–2013. We have detailed metadata for each majority opinion; we mainly use the court, date, case topic, and authoring judge. The circuit court data do not include unpublished opinions. *Per curiam* opinions and discretionary opinions (concurrences/dissents) are excluded from the analysis.

For case topic, we use the seven-category "general issue" designation coded for Donald Songer's Courts of Appeals Database (e.g., Haire, Songer, and Lindquist 2003). To make these categories, Songer's research team classified cases to a single major topic such as crime, civil rights, first amendment, due process, privacy, labor relations, economics/regulation, and others.

The cases are linked to biographical information on the judges obtained from the Federal Judicial Center. This includes a plethora of demographic and career details by judge. In the illustrative analysis, we use the birth date and political affiliation of the appointing president.

Finally, the dataset includes the full text of the authored judicial opinions. We remove HTML markup and citations. We then have each case as a list of tokens. These tokens provide the inputs for the embedding model.

CONSTRUCTION OF DOCUMENT VECTORS

The next step is to construct document vectors for each case. The model we use is Doc2Vec (Le and Mikolov 2014), implemented in the Python package *Gensim*. The objective function solved by this model is to iterate over the corpus and try to predict a given word using its context (a window of neighboring words), as well as a bag-of-words representation of the whole document. The model uses an embedding layer for the context features and the document features. Therefore, the spatial location of documents encodes predictive information for the context-specific frequencies of words in the document.

We feed the case documents in random order into Doc2Vec using standard parameter choices (Dai, Olah, and Le 2015). We used the distributed bag-of-words model over the distributed memory model, with two hundred dimensions per document vector. Other parameter choices include a context window of size ten, capping the vocabulary at one hundred thousand words (based on overall document frequency) and excluding documents shorter than forty words in length. The algorithm reads through the corpus documents in random order five times. As this chapter is an exploration and illustration, we did not explore the parameter space deeply or broadly.

VECTOR CENTERING AND AGGREGATION

We now have a set of vectors $\vec{i}$ for each case i. We normalized each vector to length 1. Each case has an authoring judge j working in court c at year t. Besides author and time, the other metadata feature is the case topic k. We use these categories for descriptive statistics, as well as to "control" for these features for more targeted analysis.

For visualization and other analysis we center and aggregate the document vectors in several ways. Let I_j be the set of cases authored by j. Let I_{jt} be the set of cases authored by j at year t. We construct a vector representation for a judge using

$$\vec{j} = \frac{1}{|I_j|} \sum_{i \in I_j} \vec{i}$$

where $|\cdot|$ gives the count of the set. Similarly, the vector for judge j at year t is given by

$$\vec{jt} = \frac{1}{|I_{jt}|} \sum_{i \in I_{jt}} \vec{i}$$

and the vector for all cases on topic k in court c during year t is given by

$$\vec{ckt} = \frac{1}{|I_{ckt}|} \sum_{i \in I_{ckt}} \vec{i}$$

Meanwhile, the same notation and corresponding aggregation formula is used to construct a vector for a year $\vec{t}$, for a court $\vec{c}$, for a topic $\vec{k}$, or for the cases in court c during a particular year t, $\vec{ct}$.

We are interested in recovering the ideological component of the judge vectors. Therefore, we explore the following steps to center the document vectors before aggregating. We represent the year-centered vector for case i as $\vec{i}_t = \vec{i} - \vec{t}_i$, where $\vec{t}_i$ corresponds to the average vector for all cases in the same year as i. Similarly, let a subscripted judge vector $\vec{j}_t$ be defined as

$$\vec{j}_t = \frac{1}{|I_j|} \sum_{i \in I_j} \vec{i}_t$$

with the average for judge j of the year-centered vectors $\vec{i}_t$.

The preferred centering specification depends on the context of the analysis. We center by interacted groups, in particular. In the results below, we variously center by topic–year $\vec{kt}$, by court–year $\vec{kt}$, and by court–topic–year $\vec{ckt}$. Only after this centering step do we aggregate by judge and perform analysis of the spatial relations between vectors. The hope is that the remaining spatial variation is purged of court-specific, topic-specific, and year-specific differences in language. The remaining variation will provide a cleaner summary of the ideological differences between judges.

Here we have used the unweighted average of the case vectors, where each case is weighted equally. Future work might explore the use of other weighting schemes. A sensible alternative would be to weight the cases by their length (in words or sentences), for example. In addition, it would be reasonable to weight the cases by the number of citations they later received (as a proxy for importance). Finally, one might normalize the vectors after centering and/or aggregating.

VISUAL STRUCTURE OF CASE VECTORS AND JUDGE VECTORS

In this section we present a variety of visualizations to understand better the spatial relationships encoded by our case vectors and judge vectors. Our visualization is a t-SNE (t-distributed stochastic neighbor) embedding plot (van der Maaten and Hinton 2008), which projects the vectors down to two dimensions for visualization purposes. We use t-SNE plots because the dimension reduction algorithm is designed to preserve local distances between points and therefore recover informative clusters (Lee and Verleysen 2007). In tests, we generated better visualizations using t-SNE than other manifold learning algorithms such as principal components, multidimensional scaling, or isomap.

We begin by exploring the institutional, temporal, and judge-level features encoded in the vectors. For figure 11.1, we centered the case vectors by topic interacted with year, as described above. We then averaged by judge and plotted the judge vectors. The vectors are labeled by court. One can see that, conditional on topic and year, the document vectors separate the courts quite well. This is consistent with systematic differences in legal language across courts, conditional on topic and year, being captured by the embedding.

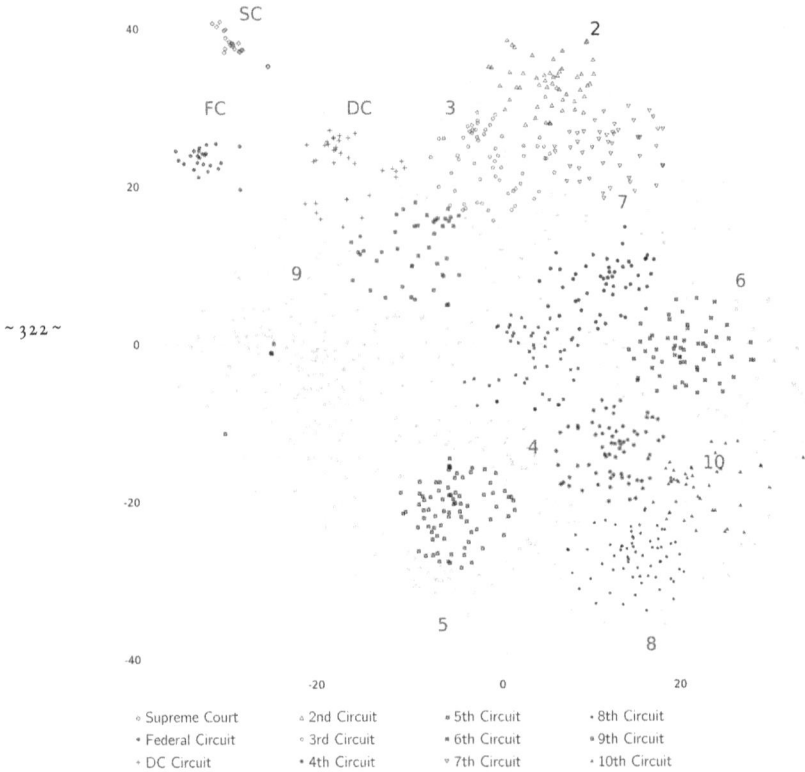

Figure 11.1. t-SNE plot centered by topic–year, averaged by judge, labeled by court. Darker data points are clustered, whereas lighter points represent outliers.

For figure 11.2, we centered on court interacted with topic. We then averaged by court–year and plotted the court–year–level averaged vectors. The dots are labeled and shaded by the decade the case was published. One can see a steady linear development of case law across the geometric space. This shows that, controlling for court and topic factors, the embedding captures systematic differences in language across time.

For figure 11.3, the cases were centered on judge interacted with year; this controls for any judge-level time-varying compo-

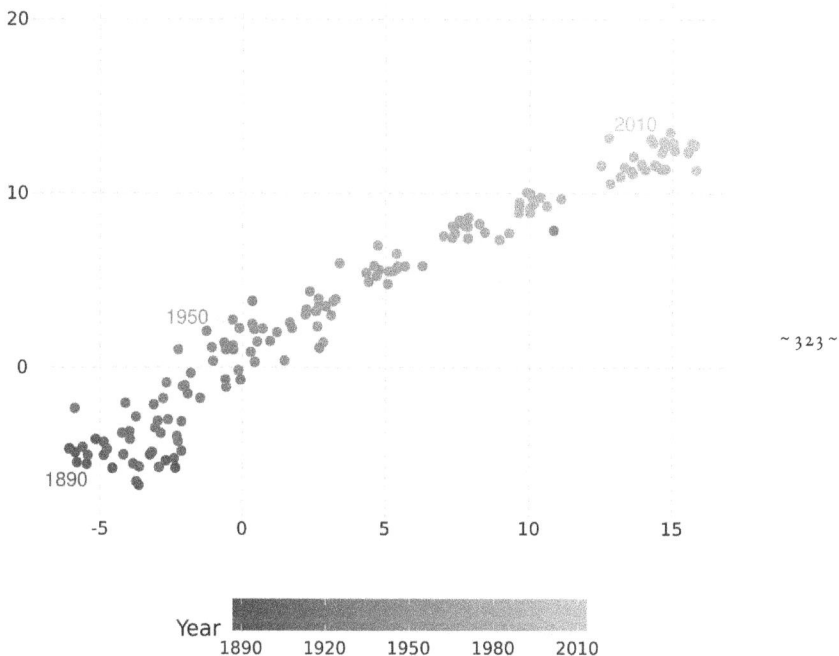

Figure 11.2. t-SNE plot centered by court–topic, averaged by court–year, labeled by decade.

nents of language. We then averaged and plotted by topic–year. The labels and colors distinguish the seven-digit general issue topic. We can see that the document embeddings discriminate topics, effectively capturing differences in language across recognized issue areas.

Next we look at whether the vectorized language in the case vectors encodes information about judge characteristics. For figure 11.4, we centered on an interacted grouping for court, topic, and year. This centering controls for any time-varying topic and court-level language variation. We then averaged by judge and plotted the judge vectors. The labels and shades are by political party—Democrat or Republican. These are randomly distributed across the

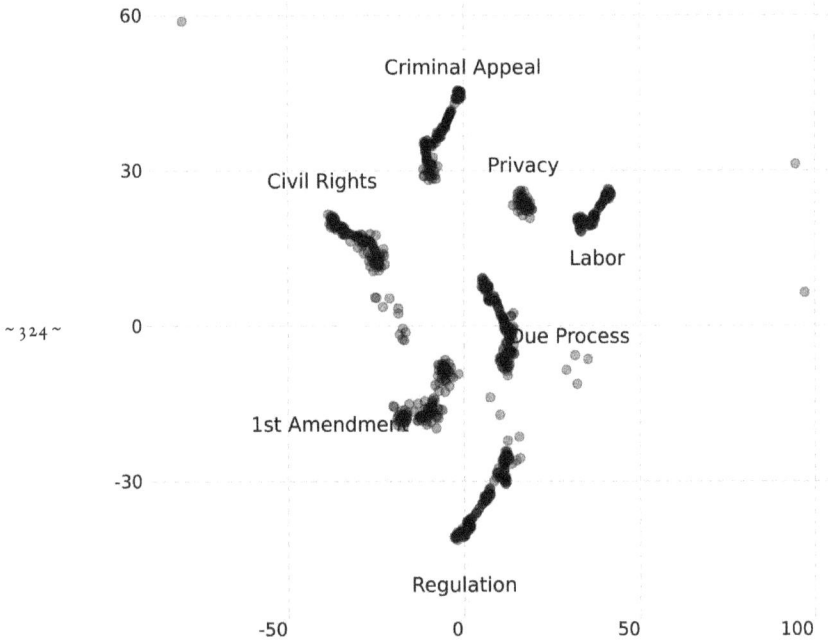

Figure 11.3. t-SNE plot centered by judge–year, averaged by topic–year, labeled by topic.

graph. It appears that the language features encoded by the document embeddings are not informative about political party.

Figure 11.5 considers another judicial biographical feature: birth cohort. As before, we centered on court–topic–year and averaged/plotted by judge. In this case, the labels and shading are by birth cohort decade (1910s through 1950s). In stark contrast to political party, there is clear segmentation across the geometric space across cohorts. Remember that this is conditioned on court–topic–year, so is not driven by time trends over the sample. The vectorized language recovers differences in the legal language used by judges from different generations.

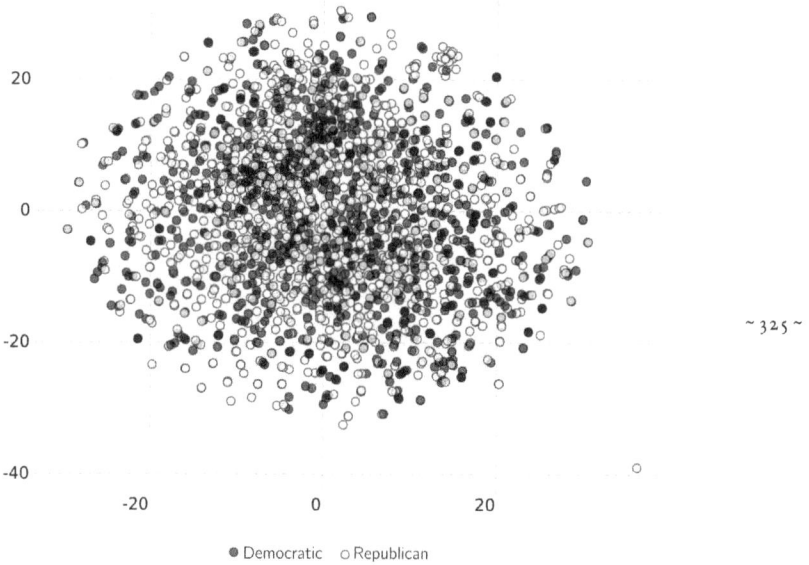

Figure 11.4. t-SNE plot centered by court–topic–year, averaged by judge, labeled by political party.

Finally, for figure 11.6, we consider law school attended as a final source of linguistic differences across judges. Conditional on court, topic, and year, we see apparent random distributions across the space in terms of law school. As with political party, it seems like language or ideological differences by school do not show up in the vectors.

ANALYSIS OF RELATIONS BETWEEN JUDGES

This section uses our vector representation of judges to produce a similarity metric between courts and judges. We adopt a measure of vector similarity that is used often for document classification—*cosine similarity* between two vectors,

$$s(\vec{v}, \vec{w}) = \frac{\vec{v} \cdot \vec{w}}{\|\vec{v}\| \, \|\vec{w}\|}$$

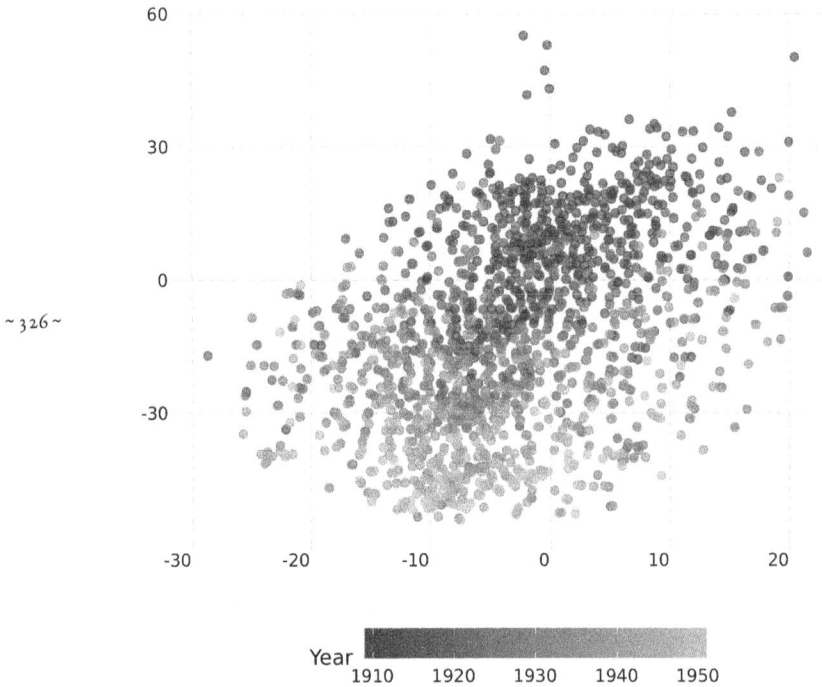

Figure 11.5. t-SNE plot centered by court–topic–year, averaged by judge, labeled by judge birth cohort.

computed from the cosine of the angle between the vectors. In the case of word embeddings, high similarity means that the words are often used in similar language contexts.

In the case of judges, we can say that similarities approaching 1 mean that the judges tend to use similar language in their opinions. Similarities approaching -1 mean the judges rarely use the same language. In between, we have a continuous metric to rank which judges are relatively more or less similar to each other.

First we look at similarity between court vectors to complement the spatial representation in figure 11.1. We centered the vectors by topic and year and then aggregated by court. We then com-

Table 11.1. Pairwise similarities between federal appellate courts.

	SCOTUS	1st	2nd	3rd	4th	5th	6th	7th	8th	9th	10th	11th	DC
1st	0.022												
2nd	-0.008	0.302											
3rd	-0.001	0.135	0.207										
4th	-0.045	-0.045	-0.081	0.126									
5th	-0.105	-0.196	-0.298	-0.269	0.038								
6th	-0.074	-0.185	-0.148	0.009	0.069	-0.107							
7th	-0.097	-0.052	-0.014	-0.055	-0.162	-0.257	0.029						
8th	-0.137	-0.215	-0.296	-0.214	-0.150	-0.184	0.050	-0.022					
9th	0.039	-0.137	-0.140	-0.182	-0.147	-0.121	-0.220	-0.265	-0.150				
10th	-0.111	-0.249	-0.361	-0.179	-0.189	0.017	0.006	-0.156	0.218	0.042			
11th	-0.086	-0.191	-0.240	-0.215	0.067	0.713	-0.039	-0.224	-0.192	-0.084	0.026		
DC	0.846	-0.085	-0.058	0.011	-0.010	-0.062	-0.097	-0.177	0.111	0.067	-0.025	0.011	
Fed.	0.178	0.200	0.132	0.116	0.124	-0.150	0.154	-0.082	-0.255	-0.116	-0.260	-0.181	0.094

-0.361   0.846

puted the pairwise similarities between the court vectors. These are reported in table 11.1.

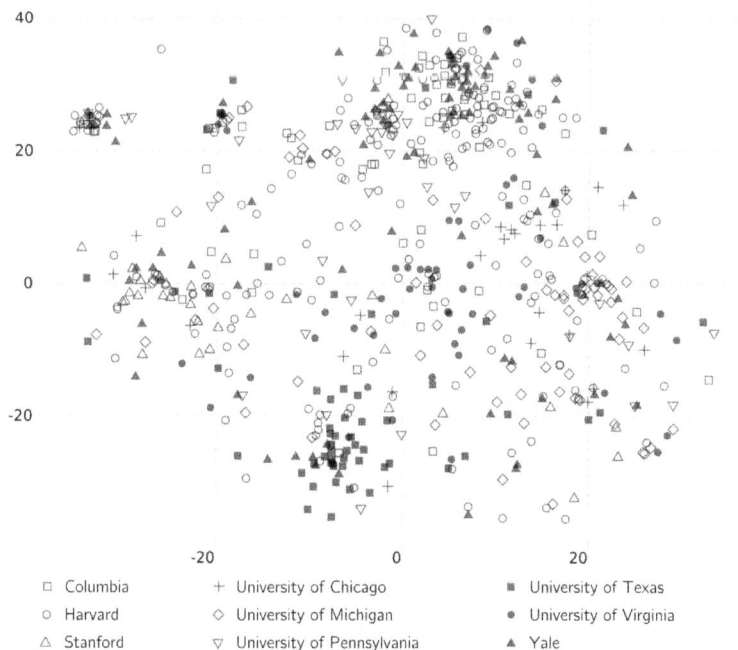

Figure 11.6. t-SNE plot centered by court–topic–year, averaged by judge, labeled by law school attended.

Legend:
- □ Columbia
- ○ Harvard
- △ Stanford
- + University of Chicago
- ◇ University of Michigan
- ▽ University of Pennsylvania
- ■ University of Texas
- ● University of Virginia
- ▲ Yale

The shading provides a gradient for similarity, with white and light gray indicating that the courts are relatively similar and dark gray meaning they are relatively dissimilar. The table has some interesting features. First, the DC circuit is most similar to the Supreme Court of the United States, which is intuitive since they both have a large portion of their document made up of cases that involve issues of federal government functioning, such as separation of powers. Second, the Eleventh Circuit is similar to the Fifth Circuit, which is intuitive since the Eleventh Circuit used to be a part of the Fifth Circuit and they share many legal precedents.

Unlike in the case of courts, the current model does not appear to capture similarity between judges. Starting with the Supreme Court, we center the document vectors on topic and year. Then we take the average of these centered vectors by judge as our representation of judge writing, reasoning, and beliefs. Unfortunately, this analysis yields few immediate insights—for example, although Justice Scalia is close in the space to Justice Thomas, he is even closer to Souter, Stevens, and O'Connor, a result that it is at odds with more conventional measures of ideological space based on their voting patterns (Epstein et al. 2007). Understanding these similarities is an area for future work. It could be that there are not enough decisions to be predictive of partisanship. Or perhaps these vectors are capturing something new about judge reasoning that is not captured by party.

Although the model does not perform well on Supreme Court justices, it is worth investigating whether appellate court judges might be better represented. For this we focus on one notable circuit court judge, Richard A. Posner, who holds well-known judicial views and has published over 3,300 opinions during his tenure. The document vectors are demeaned by court, year, and topic and aggregated by judge. Based on this information, we rank all circuit court judges by the similarity of their vector to Posner's vector.

These are reported in table 11.2. Interestingly, the most similar judge is Frank Easterbrook, who, like Posner, is known for the use of economic analysis in opinions. Stephen Breyer, of all recent justices, is most closely associated with the law and economic movement (for example, he has a published article in *The Economic Journal* on "economic reasoning and judicial review" (Breyer 2009)). In addition to his law and economics orientation, Posner has a conservative reputation, and we see other conservative judges such as Neil Gorsuch and Antonin Scalia. Henry Friendly makes an appearance—he is a well-known pragmatist, as is Posner. Finally, Michael McConnell has coauthored academic articles with Posner (McConnell and Posner 1989). These document vectors are more intuitive for connections between circuit court judges but may raise some new interesting questions about connections between Supreme Court judges.

Discussion and Future Work

To recap, we applied the document vectorization algorithm Doc2Vec to twelve decades of opinion texts from US appellate courts. While previous work has applied LDA to large corpora of legal opinions, this is the first to introduce document embeddings to the area of empirical legal studies. Our analysis of the resulting vectors serves to validate their informativeness in terms of legal jurisdiction (distinguishing courts), time (distinguishing decades), and topics (distinguishing broad legal areas). These vectors would therefore be useful for downstream prediction or inference tasks on these categories. The method has the advantage of requiring few subjective decisions by the researcher, and the resulting features are easier to work with than high-dimensional sparse representations such as n-grams.

Table 11.2. Most similar circuit court judges to Richard A. Posner.

Circuit judge name	Similarity	Rank
POSNER, RICHARD A.	1.000	1
EASTERBROOK, FRANK H.	0.663	2
SUTTON, JEFFREY S.	0.620	3
NOONAN, JOHN T.	0.596	4
NELSON, DAVID A.	0.592	5
CARNES, EDWARD E.	0.567	6
FRIENDLY, HENRY	0.566	7
KOZINSKI, ALEX	0.563	8
GORSUCH, NEIL M.	0.559	9
CHAMBERS, RICHARD H.	0.546	10
FERNANDEZ, FERDINAND F.	0.503	11
EDMONDSON, JAMES L.	0.501	12
KLEINFELD, ANDREW J.	0.491	13
WILLIAMS, STEPHEN F.	0.481	14
KETHLEDGE, RAYMOND M.	0.459	15
TONE, PHILIP W.	0.459	16
SIBLEY, SAMUEL	0.459	17
SCALIA, ANTONIN	0.456	18
COLLOTON, STEVEN M.	0.445	19
DUNIWAY, BENJAMIN	0.438	20
GIBBONS, JOHN J.	0.422	21
BOGGS, DANNY J.	0.420	22
BREYER, STEPHEN G.	0.414	23
GOODRICH, HERBERT	0.412	24
LOKEN, JAMES B.	0.410	25
WEIS, JOSEPH F.	0.408	26
SCALIA, ANTONIN (SCOTUS)	0.406	27
BOUDIN, MICHAEL	0.403	28
RANDOLPH, A. RAYMOND	0.397	29
MCCONNELL, MICHAEL W.	0.390	30

In terms of distinguishing judges, the results are more mixed. The vectors capture judge birth cohort and provide some intuitive rankings for the similarity of circuit judges to Richard Posner. But the vectors do not show a clear signal for political party or judge law school. Similarities between Supreme Court judges and the similarities of circuit judges to Supreme Court judges are not intuitive or informative.

One interpretation of these results is that judicial language is not very politicized along partisan lines. This would be consistent with the finding in Ash, Chen, and Liu (2017) that judicial language is less polarized than congressional language. In contrast, differences in policy approach, such as use of economic analysis, might be more salient for distinguishing judges. This would be consistent with the finding in Ash, Chen, and Naidu (2017) on the importance of economics (language and training) in judicial decision-making.

But there is the alternative possibility that Doc2Vec representations of language are not rich enough to encode some dimensions of judicial ideology. Richer representations, such as those constructed from grammatical relations between words (Levy and Goldberg 2014), may be needed. Another possibility is that document embeddings may require a large number of documents to form coherent ideological dimensions. Understanding the limitations of Doc2Vec and related models is important for future research.

In the rest of this section, we outline some parallel and future work in using embeddings models for empirical analysis of law. This research may serve to address the limitations identified during the research for this chapter.

ANALOGIES AND ASSOCIATIONS

The exploratory analysis taken in this chapter focused on distances and similarities. But the metric nature of these embedding models allows for richer analysis using matrix algebra. In particular, an intriguing use of word embeddings is to encode analogies. A well-known example is that word embeddings "know" that "man" is to "woman" as "king" is to "queen," through the vector algebra, *"king" – "man" + "woman" = "queen"* (Mikolov, Sutskever, et al. 2013). In a legal milieu, Ash (2016) shows that "personal income tax" – "person" + "corporation" = "corporate income tax."

Dai, Olah, and Le (2015) show that document embeddings also encode analogical relations between documents, with an application to Wikipedia articles. The document vector for the "Christina Aguilera" article, minus the vector for the "America" article, plus the vector for the "Japan" article, results in the vector for the Japanese pop star "Ayumi Hamasaki." In the case of the law, a document embedding could say something like *"Everson v. Board of Education* is to *Engel v. Vitale* as *Griswold v. Connecticut* is to *Roe v. Wade."* These cases share an analogical relation, in that the latter case is a related application of the constitutional principle articulated in the former case. In the vector math, that would be represented as *Everson – Engel + Griswold = Roe.* Finally, a judge embedding could say something like "Scalia is to Thomas as Ginsburg is to Breyer," in the sense that Scalia – Thomas + Breyer = Ginsburg.

This discussion of analogies is exemplary of the feature that directions in the embedding space encode semantic meaning, for example, those related to singular-vs.-plural, verb tense, etc. Bolukbasi et al. (2016) show how to isolate a vector direction for a semantic concept such as gender in the embedding space. Construct a list of word pairs that share the gendered analogical relation (man–woman, men–women, boy–girl, boys–girls, father–mother, etc.) and then take the average of the vectors defined by the pairwise differences. This "gender" vector defines a semantic concept rather than any particular word or pair of words. It can then be used to identify and analyze the use of gendered language.

In the law, we would be interested in isolating other types of language dimensions—notably, legal and political concepts and distinctions. For example, there might be a direction for liberal vs. conservative or procedural vs. substantive. Similarly, it would be worth investigation whether there are directions or clusters for originalists or pragmatists, or economic analysis vs. more traditional doctrinal methods.

Some of the recent work on embeddings has used these associational features to analyze partisan language (Iyyer et al. 2014) or to analyze cultural biases such as sexism and racism. Caliskan, Bryson, and Narayanan (2017) show that the results of implicit association tests are reproduced in aggregate language associations. Garg et al. (2018) and Kozlowski, Taddy, and Evans (2018) use long-run historical corpora to analyze trends in biased language over the last century.

The issue of unconscious bias is of particular significance in the legal system (e.g., Fagan and Ash 2017). Rachlinski et al. (2009) show that trial judges demonstrate the same implicit biases as the broader population on a standard psychological test, but that test was confidential, and it was not matched with the judges" actual decisions. The construction of a language-based measure of bias, available for all judges from their written opinions, would be quite useful for understanding the importance of prejudice in the judicial system.

Toward this end, Ash, Chen, and Ornaghi (2018) analyze implicit associations in judicial language. The broad descriptive results from Caliskan, Bryson, and Narayanan (2017) are replicated in the judiciary, and relations between "innocent" and "guilty" are also analyzed. Male names tend to have a stronger connotation with "guilty" (relative to "innocent") than female names. In addition, African American– and Hispanic-associated names are more closely related to "guilty" than Caucasian-associated names.

Future work should analyze whether and how these biases in language are associated to biases in decisions. One could ask, for example, whether judges with a racially based lexical bias also tend to reject discrimination complaints or to give longer criminal sentences to certain defendants. Similarly, having more traditional gender views, as detected in one's implicit gender bias, might be reflected in more conservative judicial decisions related to gender discrimination cases. We could also look for peer effects by testing whether sitting with a biased judge has an impact on a peer judge's subsequent decisions.

STRUCTURED EMBEDDINGS AND CATEGORICAL EMBEDDINGS

The document embeddings developed in the previous section were trained on the whole corpus. The embedding model did not explicitly model a time component, a court component, or other metadata categories. Differences across courts, time, and judges were encoded only through aggregating by different categories. Future work might explicitly account for differences between these categories in how the embeddings are constructed.

Along these lines, recent work in embedding models seeks to include these relations more flexibly and elegantly as a part of the data-generating process. Rudolph and Blei (2017) provide a model for learning dynamic embeddings and examine how language has changed over time in the US Congress over the last century. Rudolph et al. (2017) provide a model for structured group embeddings and allow word and document vectors to have a group component and an individual component.

In parallel work, we found difficulties in initial applications of structured embeddings to judge groups (Ash, Chen, and Ornaghi 2018). Word similarities across groups seem to be sensitive to model parameters. Systematic differences in word similarities between Republican and Democrat judges can flip based on the embedding dimension and vocabulary size, for example. While structured embeddings do not work off the shelf, we expect that there is still potential in this research area.

As discussed, the Doc2Vec embedding was not able to discriminate between judges based on assumed ideology. This may be because the language style of written decisions may not encode ideology. This information may be mostly contained in the direction of the decision (e.g., for or against plaintiffs) or in some interaction between the decision and the language. Embedding layers in the deep-learning literature provide an alternative approach for identifying spatial relations between judges in prediction of decisions.

As described, Word2Vec and Doc2Vec work by colocating words that are most similarly predictive for a deep-learning task. In that case, a word is the embedded categorical variable but embedding layers can be used for any sort of categorical variable. In future work, the judge identity could be represented with an embedding lookup layer to a relatively low-dimensional dense vector space. The location of the judge vectors, initialized randomly, would be endogenous to the model. As the model goes through further training, the locations of these vectors would be pushed around to improve predictiveness. As a by-product of the model, the judges that locate together in the vector space would be predicted to behave similarly in court, holding other factors constant. This type of model may work to analyze ideological dimensions of judging.

EMBEDDING OF CITATION NETWORKS

In this chapter, the focus has been on the language of opinions as representing legal ideas. But in a common law system, the cases cited in an opinion are another potential means to express ideological content. Ash, Chen, and Liu (2017) show that citations are more predictive of the political party of a judge than the writing style. Therefore, in the context of the geometry of law, citations could be included as features in the document embedding. This might reveal more differences, such as those between political parties.

Another approach to embedding citations is based on Rudolph et al. (2016) and Ruiz, Athey, and Blei (2017). In that paper, the model predicts occurrence of a product in a grocery shopping cart based on the cooccurrence of other products. In the legal analogue, cases could be treated as a bundle of citations to precedents in the same way that Rudolph et al. (2017) treat grocery baskets as a bundle of products. The citation embedding model would predict the presence of a particular citation using the list of cooccurring citations. As with word embeddings, cases that tend to be cited together would locate near each other in the embedding space. The model would thereby construct a "precedent space" as opposed to a language space.

An intriguing feature of the grocery cart model is that the learned parameters encode complementarity or substitutability of items. In the context of Rudolph et al. (2017), that means coffee being substitutable with tea but complementary with milk, for example. In the context of the law, we would learn which precedents are complementary (tending to be cited together) and which are substitutable (tending to appear in similar contexts but not together). By pairing substitutability metrics with ideological valence (liberal vs. conservative), we can analyze the parallel histories of liberal and conservative jurisprudence in the United States.

Conclusion

One of the fundamental challenges to using text for purposes of data analysis is the very high dimensional nature of textual artifacts. Various tools have been developed to reduce this dimensionality while preserving the kind of information that is useful to researchers, and some—such as topic modeling— have become popular as a tools for legal scholarship. Word and document embeddings have important advantages as a dimension reduction tool that researchers in law may find useful. This chapter describes a document embeddings model for a legal corpus and discusses some of the legal information, such as jurisdiction and time, that the embeddings model seems to capture well. Based on these early results, there is every reason to believe that there is substantial opportunity for future work to use these models to address significant questions of interest to scholars in empirical legal studies. ✒

Acknowledgments

We thank Brenton Arnaboldi, David Cai, Matthew Willian, and Lihan Yao for helpful research assistance. We thank Michael Livermore, Daniel Rockmore, and participants at the Santa Fe Institute Law as Data Workshop for helpful feedback on this research. Work on this project was conducted while Daniel L. Chen received financial support from the European Research Council (Grant No. 614708) and Agence Nationale de la Recherche.

36

REFERENCE NETWORKS
AND CIVIL CODES

Adam B. Badawi, University of California, Berkeley, School of Law
Giuseppe Dari-Mattiacci, University of Amsterdam

How can scholars assess the differences between legal systems? A primary way for legal comparativists to answer this question has been to sort legal systems into different "families." Some distinctions are quite clear and uncontroversial, such as the fundamental difference between common law systems (which are largely made up of former English colonies) and civil law systems. And there is general agreement that the French and German traditions are the primary influences in civil law. But things get murky quickly. While colonization played a central role in shaping the legal systems of many countries, many of these nations have sought to distinguish themselves after independence. To further confuse matters, some countries that were not subject to colonization made independent choices about their legal systems by borrowing from multiple traditions.

The diversity of legal systems has made it difficult for researchers to reach consensus on the appropriate grouping of legal families. For traditional legal comparativists, who generally use qualitative methods to categorize legal systems, there are debates about whether comparisons should be made at a macro or micro level. Moreover, the time-consuming nature of the qualitative inquiries means that it is difficult to compare a large number of legal systems. Economists have also relied on groupings of legal systems in the "legal origins" literature. This line of research seeks to attribute development to the use of certain legal systems. Doing so requires categorizing legal systems, which has generally been done using relatively blunt variables. These include whether a country uses common law or civil law and, for former colonies, the colonizing country. While these simple distinctions allow for a large number of comparisons, there have been significant challenges to this work: scholars have struggled to determine the drivers of the relevant relationships, tease out important differences, and develop meaningful and defensible classification schemes.

In this chapter, we propose a method that can capture some of the subtlety and detail that distinguishes legal systems in a way that can scale. This approach uses the digitization and machine reading of legal codes to determine the network structure of these bodies of text. The network structure depends on the use of cross-references contained within the code. We then propose to compare legal systems by using a series of metrics that analyze the similarity of network structures between codes. These comparisons can be used to validate existing categorizations, such as the proposition that the legal systems of countries that are subject to the same colonizer are more similar than those of countries subject to a different colonizer. This method can also allow more nuanced comparisons between countries. Specifically, network structure can be used to construct continuous measures to assess the similarity of legal systems. Measures of this sort should allow more refined analysis of the "legal origins" hypothesis and also hold the promise of reorienting conventional conceptions concerning the relationships between legal systems.

At the end of the chapter, we apply these methods in a pilot study of several major European countries that use the civil law system. We show substantial evidence that the network structures map on to existing understandings of legal similarity. We also uncover some additional detail in the network structures that suggests some previously unidentified ways that these civil codes differ.

This chapter proceeds as follows: We first briefly review the traditional and empirical approaches to grouping legal families and outline the limitations that each faces. Next, we provide an overview of the general structure of legal codes and constitutions, the use of cross-references, and the preservation of these cross-references as legal systems evolve and are transmitted to new environments. Finally, we describe how cross-references can be used to characterize the structure of legal codes and discuss different metrics that can be used to compare those structures between codes and constitutions.

Existing Comparative Approaches

This section provides a brief overview of the techniques used by comparative scholars and the limitations of those techniques. The two primary approaches are orthodox comparative law and empirical comparative law. The latter group includes empirical study of comparative constitutions and the methods of scholars of law and finance that focus on the association between legal origins and economic outcomes.

TRADITIONAL COMPARATIVE LAW

Traditional approaches to comparative law have a long and methodologically varied history. A significant amount of this research attempts to describe different legal systems and categorize them. The leading work in this area was performed by René David, who sought to classify legal systems into families (Gordley 2006). David divided these systems on the basis of the existence of shared philosophical and political principles and the ability of a lawyer to move between common systems (David and Brierley 1978). David saw three major classifications: Romano-Germanic, common law, and socialist. Among the most persistent classifications in this vein were the seven families identified by Arminjon, Nolde, and Wolff (1951): French, German, Scandinavian, English, Russian, Islamic, and Hindu.

~342~ Because this branch of orthodox comparative law bears most directly on this project, we discuss these efforts in some detail. Before doing so, it is worth noting that—as with many areas of comparative law—there is little agreement on the appropriate methodology for classifying legal systems. As one scholar explains, "It has been said that there are as many classifications as there are comparatists, and the number and variety of classifications is itself an indication of the failure of the enterprise" (Glenn 2006, 437). Nevertheless, there are some commonalities to classification attempts. One is the primacy of the distinction between common law and civil law, which differ substantially in their methodologies. The common law evolves through case-by-case distinctions made by judges, while civil law relies largely on existing codifications. Within the civil law tradition, the French Napoleonic code and the German civil code have had broad-ranging historical influence.

The Napoleonic code came out of the French Revolution and was drafted in a relatively short time. While the drafters were legal technocrats rather than revolutionary firebrands, a defining characteristic of the code is its clarity. Legal historians attribute this clarity to the conscious goal of making the code intelligible to lay people. Commentators describe the Napoleonic code as relatively sparse—especially in comparison to the German civil code—and also note its "clear and memorable phrases" and "the absence of cross-references and jargon" (Zweigert and Kötz 1998, 91). The German civil code (Bürgerliches Gesetzbuch, "BGB") is a stark contrast in style to the French code. Drafted at a time of relative political stability, there was no apparent attempt to make the BGB intelligible to lay people. The audience was instead trained lawyers. Stylistically, it has been described as "legal filigree work of extraordinary precision" (Isele 1949, 6), and one in which "repetitions are avoided by means of cross-references to amplifying sections" (Zweigert and Kötz 1998, 145). The use of cross-references helped to produce a code that is "highly integrated through the complex interplay of its five main books" (Reimann 2008, 880).

~ 343 ~

Both of these codes have had significant influence in Europe.[1] The Italian civil code, which was drafted prior to the BGB, took significant inspiration from the French civil code. But as time progressed, Italian lawyers began to favor the German approach to legal reasoning, and that choice expressed itself as the Italian code enacted in 1942. According to comparativists, the French imprint on Spanish law has been more lasting. Like the Italian code, the Spanish civil code was developed in the period between the enactment of the Napoleonic code and the BGB. Perhaps for that reason, Spanish law drew substantially from the French model, and many of those portions have remained in force.

Some European countries have developed legal systems that are quite distinct from the French and German traditions. Austria enacted its own civil code around the same time period as the promulgation of the Napoleonic code. Perhaps because it was not borne of a revolution, the Austrian code does not reflect Enlightenment principles as much as the Napoleonic code. Some comparativists have suggested that this

---

[1] The remaining review summarizes the analysis of Zweigert and Kötz (1998).

relative lack of modernity is one reason why the Austrian code has not had the broad influence that the German and French codes have had.

This brief review is far from comprehensive, but it does give a sense of how traditional comparativists have classified continental legal systems using qualitative methods. The subjective component of this method means that there is unlikely to be inter-rater reliability across different groupings. It also makes it difficult to provide a ranked list of how similar one legal system is to all the other European civil law systems. To the degree this is even possible, disagreements about methodology and the importance of different features would make it difficult to harmonize lists across countries.

EMPIRICAL COMPARATIVE LAW

The genesis of the recent work in empirical comparative law is a series of papers in law and finance that developed the legal origins hypothesis. These papers focused on the presence of investor and creditor protections across countries and found positive correlations between these laws and economic measures, such as per capita GDP (La Porta et al. 1998, 2000, 2002). Subsequent work in the area moved beyond the corporate sphere to hypothesize that better economic outcomes in common law countries may be a product of the common law's lack of reliance on statutes and stronger protections for an independent judiciary, relative to civil law systems (Glaeser and Shleifer 2002). Much empirical work followed, and that research sought to examine the connection between legal systems and economic measures.

Many of these studies purport to show an association between formalist and less interventionist legal systems and better economic outcomes. To pick just a few examples,

Djankov et al. (2003) use thorough survey evidence from law firms around the world to develop evidence on the speed and efficiency of evictions. The authors find that civil law countries tend to be slower with these processes than common law countries. In Djankov et al. (2008) the authors conduct a similar study that focuses on the legal restrictions on self-dealing. The study again finds that common law countries appear to have systematically better outcomes than civil law countries. There is a relatively wide-ranging literature that uses these and similar techniques. Much of the literature finds similar associations with positive outcomes and the use of common law (La Porta, Lopez-de-Silanes, and Shleifer 2008).

~ 345 ~

Studies of this sort have been subject to intense criticism. Some of these critiques focus on the difficulty of distinguishing between a country's use of a particular legal system and the identity of the colonizing country (Klerman et al. 2011). Others argue that modern political economy concerns are a more effective explanation for different economic outcomes than legal origins (Roe 2006). But the critique that is most relevant to our effort is that the measures of legal systems and families are too blunt and inaccurate to be effective.

Spamann is perhaps the most persistent and effective methodological critic of empirical comparative law. In a series of articles he has pointed out both the theoretical and applied problems posed by associating economic outcomes with legal systems. A complete review of these criticisms is beyond the scope of this chapter, but he has taken particular issue with the difficulties of coding comparative legal systems and with survey methodology. With respect to the former, he has recoded the data in prominent law and finance studies and has found quite different results (Spamann 2009b). With regard to the latter, he has argued that survey evidence can be unreliable when there are not detailed protocols for eliciting and coding information (Spamann 2015). The method that we propose would avoid these issues by using objective metrics to measure the differences between codes.

Cross-References and Networks

Both the traditional and empirical approaches to comparative law have drawbacks when it comes to the quality of the comparisons. The traditional approach lacks consensus on the appropriate way to compare legal systems, and the comparisons that exist tend to be a qualitative judgment based on the history of each country's legal system. The empirical work often uses a small number of binary measures to assess how legal systems vary. In what follows, we describe a method that can surmount some of these difficulties.

This technique—developed in Badawi et al. (2018)—uses cross-references within legal codes to characterize the network structure of these legal codes. We can then draw on techniques in the analysis of networks to describe the similarities and differences between these different legal codes. This section begins with a discussion of the role of cross-references in legal codes and shows how these references can be depicted as a network of directed links. These networks can then be summarized based on a variety of network statistics. The section concludes with an applied discussion of how to compare networks and, specifically, how to compare network representations of legal codes.

LEGAL CROSS-REFERENCES AND NETWORKS

The use of cross-references in legal codes has a very long history. The Code of Hammurabi, the oldest known written legal code, makes use of a cross-reference in Law 182, which expressly incorporates a concept from Law 181. This use of a cross-reference—to refer to a definition or idea from a separate section of the legal code—is perhaps the most common use of a cross-reference. Codes will also use cross-references to point readers to similar concepts that appear elsewhere in the code or to carve out exceptions to a rule. As the discussion on traditional comparative law implies, legal systems have a different affinity for the use of internal links. Legal systems that write codes for attorneys and others with technical legal proficiency are likely to make more use of cross-references than those that strive to make legal codes more accessible to nonlawyers. Cross-references may also be appealing where a legal system seeks to have cohesion across parts or even the entire code. For example, the definition of a particular word or concept could be used at one point and incorporated by cross-reference throughout other sections of the code.

The cross-references in legal codes can be characterized as a network. This network is composed of the cross-referenced and cross-referencing sections, called nodes, and links between these nodes, which are the cross-references themselves. These codes can be depicted as directed networks. In a legal code that uses cross-references, we characterize the direction of the link as going from the cross-referencing section to the cross-referenced section.[2] The complete representation of the code is then all the nodes that either contain a cross-reference or are cross-referenced and the links that represent the direction of the cross-references.

---

[2] In network notation, a directed network can be represented as an $n \times n$ adjacency matrix, where n is the number of nodes, a 1 in row j and column i indicates that there is a link that goes from node j to node i, and a zero indicates that there is no such link.

Network science has developed a wide-ranging suite of measures to characterize the features of networks. This section will begin by looking at some basic statistics and then explain some of the more complex measures that researchers can use to describe networks (cf. Jackson 2010, for a thorough treatment of network science in a social science context). The first basic statistic is *network density*, which is the number of actual links in a network divided by the number of potential links. For a directed graph, the network density is $d = \frac{m}{n(n-1)}$, where m is the number of actual links and n is the number of nodes. (The term $n(n-1)$ provides the number of potential links in a network.) If every section of a code provided a cross-reference to every other section of the code, the network density would be 1, while a code with no cross-references would have density zero. In practice, even legal codes that make wide-ranging use of cross-references will only use a few per section (or node). That means that even those codes that make relatively substantial use of cross-references will have quite low densities.

We characterize legal codes as directed networks with incoming and outgoing links. This structure allows us to capture the amount of influence that specific nodes have. Depending on the context of the network, nodes that have many incoming links or outgoing links may be considered particularly influential. The second measure of network complexity, *average degree*, has a close relationship with network density. Degree in a directed network is itself a directional quantity, with *indegree* counting the number of incoming links at a node and *outdegree* counting the number of outgoing links from a node. The average total degree captures the mean total of incoming and outgoing cross-references per node in the network and is equal to $\frac{2m}{n}$. For legal codes this measure provides a sense of how often cross-references tend to be used across the code.

Average indegree and outdegree are not informative because they will be equal to each other, but the variance of these measures does tell us something about the relative concentration of these types of links. If indegree (outdegree) has low variance, that means that incoming (outgoing) links are relatively evenly distributed across the network. But high variance indicates that certain nodes have substantially more incoming (outgoing) links relative to the others. In the context of legal codes, it is of particular interest when indegree and outdegree have high variance. A high indegree variance indicates that there are certain concepts and definitions in the code that are being cross-referenced at a high rate. That may mean that certain concepts are used in a relatively uniform way throughout the code. When outdegree has a high variance, that suggests that some sections of the code refer to a large number of other sections. This may mean that these sections are particularly well integrated into other sections of the code.

The size of *hubs* and *authorities* is another way to understand the degree of connectedness in a directed network. For our purposes, a hub is defined as a node that has a substantial number of outgoing links and an authority is a node that has a sizable number of incoming links.[3] We compute *Hubs* and *Authoritys* scores, which count, respectively, the number of nodes in the network that have more than five incoming links and more than five outgoing links.

The concept of a *connected component* provides an indication of how cohesive a network is. In a directed network, one tends to focus on weakly connected components, which ignore the direction of any links. A component is *weakly connected* if there is at least one undirected path from each distinct node to

---

[3] Note that this is different—although similar in spirit—to the technical definition of hubs and authorities (see Kleinberg (1999)).

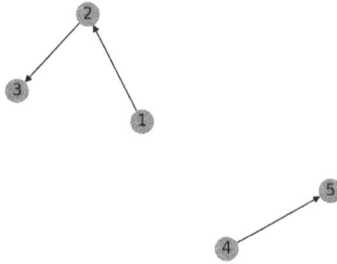

Figure 12.1. A simple network.

all other distinct nodes. Imagine, for example, that a code has five sections numbered 1–5. Further imagine that there are links from section 1 to section 2, from section 2 to section 3, and from section 4 to section 5. Figure 12.1 depicts this network. This network has two weakly connected components (the first formed by sections 1, 2, and 3 and the second by sections 4 and 5). There is a path from 1 to 2, from 2 to 3, and from 1 to 3, but there is no path from sections 1, 2, and 3 to sections 4 and 5. A network with a small number of weakly connected components will appear more cohesive than one with a large number of weakly connected components. When we report the number of weakly connected components in a code, we normalize the measure by dividing it by the number of nodes in the code.

We also compute a *modularity* statistic for each network. This modularity score, which varies between −1 and 1, takes on a negative value if there are more intercommunity links than would be predicted by chance (i.e., low modularity) and a positive value if there are more intracommunity links than would be predicted by chance (Newman and Girvan 2004). To calculate this score we must identify the communities within each civil code. To do so we use the Louvain method (Blondel et al. 2008), which proceeds in two stages. In the first stage, the algorithm assigns each node to its own community and then calculates the change in the modularity score that occurs when the node is removed from its original community and inserted into the community of its immediate neighbors. The original node is then assigned to the community that results in the largest increase in the modularity score. In the second stage, each community identified in the first stage is treated as a node. A weighted self-loop reflects the number of links between the first-stage nodes that make up that community. The first stage is then repeated on the coarse-grained community identified in the second stage. The process repeats until the application of the first stage produces no reassignments.

TRANSPLANTS, TRANSMISSION, AND NETWORK SIMILARITY

Much of the work of comparative law is assessing the origins of different legal systems. For nearly all legal systems, those origins can be traced back to one or more other legal systems. In this chapter we focus on two potential channels of transmission. The first is replication through verbatim copying. Empirical scholars have been able to track this phenomenon by identifying express citations to other legal systems as well as the exact or near exact copying of statutes from other legal systems (Spamann 2009a). A second way that influence might manifest itself is through what we call *legal style*, or, more precisely, the preferences that code drafters have for structure and technical sophistication. Some of this legal style should be reflected in how the drafters make use of cross-references in a code. As discussed above, the BGB and the Napoleonic code take quite different approaches to the use of cross-references. According to these accounts, the BGB contains many cross-references and uses them as a tool to integrate concepts across the code. The Napoleonic code, in contrast, makes less frequent use of cross-references and seeks to make each section understandable without reference to other parts of the code.

If this style is copied—as it might if lawyers from other countries receive a German or French legal education—it should be possible to observe that transmission in the text of the code. Both the verbatim and stylistic copying should also be evident in the network structure of the codes. Though it is not the focus of this chapter, understanding the potential channels of legal transmission may help validate comparisons of network structure. Beyond the obvious channel of colonization, Spamann (2009) identifies the ways that legal content and culture can continue to be transmitted. These include development aid and advice, shared commercial trade, and the migration of graduate students between countries.

With respect to the first type of copying, if a legal system were to copy another country's legal code verbatim, including the cross-references, the network structure of the two would be identical. That similarity should be evident from visual depictions of each network; those pictures should look exactly the same. That similarity will also be reflected in a comparison of the network measures discussed in the last section. Those measure should be identical for both the original code and the copied code. If one observed equality, one could be quite confident that some degree of copying occurred because the chance of two legal codes having an identical network structure is very small for a legal code of any substantial size. Similarity in those measures would at least be consistent with some copying. We do not expect to see much of this type of transmission in our analysis. One reason is that there are very few documented cases of express copying. The network structures of a copied and copying code may not be similar because some aspects could get lost in translation or the copier may only copy part of the other code. And even if high-fidelity copying occurred in the past, we face the practical limitation that we are able to analyze only the modern versions of the code.[4] Significant changes may have been made to both the copied and copying code in the time since the transmission.

Matters are more complex when it comes to detecting whether legal style has been transmitted from one country to another. Imagine, for example, two countries that are each developing a civil code. One code is being drafted by a group of lawyers who were deeply influenced by the German model, and the group of lawyers independently drafting the other code decides that using definitions and cross-references is the best way to structure the code. The visual depictions of these two codes would look relatively dense and, at least in some respects, the network measures associated with these two codes would be quite similar. But only in the former case would it be accurate to say that transmission or copying of some sort occurred. For this reason, we must be cautious in making claims that transmission has taken place when comparing codes that do not involve verbatim copying. Instead, we can only claim that the networks share similar features and, when combined with historical accounts of code development, the network analysis helps to assess the strength of the claim that transmission from one country to another took place.

With more data it may be possible to stake out stronger circumstantial cases for legal transmission. Researchers could use network statistics to perform dimensionality reduction techniques such as principal component analysis as well as clustering techniques to determine whether the results suggest the same legal families identified through qualitative methods. For an example of this type of analysis of constitutions,

---

[4]We discuss this concern more completely at the end of this section.

see Badawi et al. (2018). Even without this more extensive analysis, however, the description in quantitative terms of the differences and similarities between legal systems can facilitate interpretation and understanding.

An additional complication is the evolution of codes over time. An ideal dataset would include original versions of a civil code along with every update to that code. Unfortunately, there is very little in the way of digitized histories of civil codes. In most cases, countries have only recently digitized their codes and there are few systems in place that can accurately track amendments to a code. As a consequence, we are able only to compile network data for current or nearly current versions of the code. This feature can complicate comparisons, especially where influences occurred a long time ago. Imagine, for example, that the French code exerted a fair degree of influence on another country's code during the late nineteenth century. If we have the modern-day codes for both of these countries, it is fair to assume that both the French code and the code of the influenced country have evolved substantially in one hundred plus years since the influence was exerted. To the degree that stylistic and structural approaches persist, the influence of the French code may still be reflected in the modern codes. But if there has been a shift—especially one that relates to cross-referencing practice—it may be difficult to detect those changes. We discuss the results of our analysis with all of these limitations in mind.

A View of Civil Codes

This section will detail and apply a methodology to identify the network structures created by the cross-references in the civil codes of five European countries: Austria, France, Germany, Italy, and Spain. We first explain the data gathering process and then assess whether the data are consistent with our expectations about which codes are likely to be more networked.

METHODOLOGY

Determining the network structure of a civil code first requires a parsable digital copy of the code. Many modern nations provide these documents on government websites, although the formatting of some of them can complicate the parsing process.[5] The first step in parsing is a Python program that converts the civil code into a single XML (extensible markup language) file. This XML file preserves the hierarchical information embedded in the digital copies of the code (e.g., divisions into articles, chapters, sections, etc.). To extract this information we develop dictionaries of the words that each code uses to refer to divisions in the code.[6] In most instances, this review is conducted by speakers of the relevant language. The XML files produced by this process also preserve the names, numbering, and associated statutory content of each section.

As part of this process, we must decide the level at which the hierarchy should terminate (e.g., do we treat subsections or numbered lists within subsections as nodes?). Following previous work in this area, we identify the lowest level at which the code can terminate and use this level as the bottom of the hierarchy (Katz and Bommarito 2014). To make this concept more concrete, imagine a code that is divided into articles, sections, and subsections. Imagine further that all the articles

---

[5] As discussed above, nearly all countries supply only a modern version of their codes. We are thus unable to analyze historic versions of codes or changes to codes over time.

[6] We focus only on explicit cross-references. For a method of analyzing implicit references with an application to the BGB, see Waltl, Landthaler, and Matthes (2016).

have sections underneath them but that not all of the sections have subsections below them. In this example, the lowest level at which the hierarchy can terminate is the section, because some sections have no subsections underneath them. While taking this approach loses some amount of detail, it makes it easier to compare and interpret links within the code. This choice does so by ensuring that each node is attached to a complete unit.

The next step is to convert the XML files to JSON (JavaScript object notation) files using Python. For nearly all countries, this process is where we identify cross-references within the code.[7] As before, we develop dictionaries that contain the language used to make cross-references. The program then identifies all of the distinct nodes within the legal code and then identifies all of the links between these nodes. This information allows us to use the NetworkX package in Python to compute network statistics and to produce visualizations of each network.[8]

THE NETWORK STRUCTURES OF CIVIL CODES

This section provides a brief overview of the network structure of five major European countries that use the civil law system. The goal of this exercise is to show that a pilot application of the methodology we develop produces results that are consistent with traditional understandings of the differences between legal families. We focus on France, Germany, Italy, Spain, and Austria because there is substantial traditional scholarship on the historical development of these codes. Given the difficulty of identifying exact copying and the lack of evidence that

---

[7] The electronic version of the French code has cross-reference information embedded in the XML files, which allowed us to identify cross-references when processing the XML files.

[8] See NetworkX, https://networkx.github.io.

any direct copying occurred between these countries, we limit our focus to questions about similarities in legal style. We have two primary goals in doing this analysis. The first is to determine whether the suggestions by Zweigert and Kötz that the BGB makes more frequent use of cross-references than the Napoleonic code are correct. The second is to assess whether there is evidence that is consistent with the transmission of these legal codes. Or, to put this another way, are the codes that are purportedly influenced by the German legal tradition stylistically similar to the German civil code?

We begin with visual representations of the cross-reference networks for each of the five civil codes that we analyze. For each code we identify the largest weakly connected component. Each node in that component is depicted as a geometric shape, where the size of the shape is the degree of that node relative to the average degree in the entire network. The lines between those shapes represent a cross-reference between nodes. Figure 12.2, which shows the visualizations, provides an immediate sense of the wide variation in the use of cross-references across these different codes.

The visualizations also provide some preliminary answers to the two questions that motivate this section. For the first question—whether the BGB uses more cross-references than the Napoleonic code—the answer appears to be "yes." The largest component in the German code has more nodes and a richer network of cross-references than the largest component of the French code. As we show in our discussion of the statistics below, this is true when we compare the entirety of the networks to each other.

What about our second question: Is there evidence that would be consistent with transmission of legal style? Our discussion of the qualitative analysis of legal families suggested

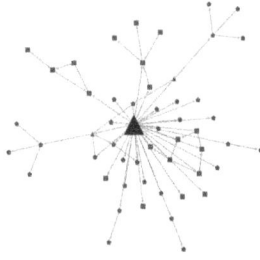

[Austria]

[France]

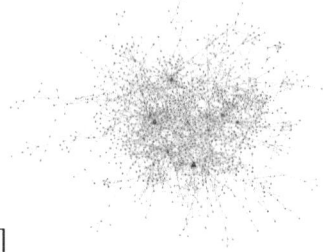

[Germany]

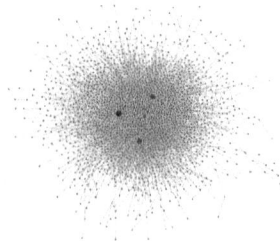

[Italy]

[Spain]

Figure 12.2. Network structure of the most connected component of each civil code.

that the French code influenced the Spanish code, the German code influenced the Italian code, and the Austrian code has had little influence. Evidence of a similar style of cross-reference use would be consistent with one legal tradition exerting influence on another. While the largest French component is substantially larger than the largest Spanish component, there appears to be some stylistic similarity. Both networks' components appear to be relatively modular; there are communities of nodes with relatively dense connections within the communities and relatively sparse connections between the communities. The German and Italian codes also share stylistic similarities. Both are quite dense—especially the Italian code—and neither code is particularly modular. To the extent there are even discernible communities in the largest weakly connected component of each code, there are dense connections between those communities. Although we should refrain from drawing strong inferences from visual depictions of part of each code, this evidence is consistent with legal style being influential in the expected ways.

We next turn to a discussion of some network statistics, which helps to confirm the impressions from the visualizations. Table 12.1 provides some basic statistics about the cross-reference networks in each of the five codes. We can use this information to provide further insight into our two motivating

Table 12.1. Basic network statistics.

Country	Nodes	Links	Density	Avg. Degree
Austria	1327	371	0.000211	0.56
France	2847	1301	0.000161	0.91
Germany	2392	2517	0.000440	2.10
Italy	3051	6697	0.000720	4.39
Spain	1992	578	0.000146	0.58

questions. With respect to the first question, whether the BGB uses more cross-references than its French equivalent, the answer is an unequivocal "yes." The BGB, which has about 450 fewer nodes than the French code, has almost double the number of cross-references. This difference is apparent in the higher density and higher average degree of the BGB relative to the Napoleonic code. These statistics show that Zweigart and Kötz are correct in their assertions that the German code makes more expansive use of the cross-reference than the French code. With regard to the second question—whether the evidence is consistent with expectations about influence—the answer is a more qualified "yes." The French and Spanish codes are the two least dense networks among the five codes and the density statistics for the two are quite close. This evidence is consistent with the Spanish code receiving influence from the French model. At the other end of the spectrum, the German and Italian codes are the two most dense codes. As suggested by the visualizations, the Italian code is substantially more dense than the German code. The average degree of the Italian code is more than double that of the German code. The relative densities of these two codes are broadly consistent with the transmission of influence, but the magnitude of the differences between the two codes warrants caution about drawing too strong an inference about the potential for influence. Austria, which we expect not to be influential, is the median code with respect to density.

Table 12.2 displays some more refined measures of network structure. The first statistic is the normalized number of weak components (the number of weakly connected components divided by the number of nodes). Recall from the discussion above that a small number of weakly connected components is indicative of a highly integrated network, while a large number of weakly connected components indicates a relative lack of

Table 12.2. Additional network statistics.

Country	Ratio	Modularity	Indeg. Variance	Outdeg. Variance	Auth5	Hub5
Austria	0.76	0.94	0.39	1.54	0	7
France	0.66	0.95	1.18	1.87	22	38
Germany	0.36	0.92	2.41	7.23	45	99
Italy	0.18	0.55	62.5	11.17	260	293
Spain	0.74	0.97	0.4	1.65	0	19

integration. The commentary from comparativists indicates that the drafters of the German legal code used cross-references to define common concepts across the entire code. To the degree this is accurate, we should expect substantial integration of this code and, thus, a relatively small number of weakly connected components in comparison to the French code, which did not seek this sort of integration. The evidence on this account is consistent with the prediction—the normalized number of weakly connected components is much lower for Germany than it is for France. The trend of substantial similarity between France and Spain also holds, as this measure is close for these two countries. The trend of Italy being extreme also holds. It has the lowest value for this measure, and the measure is substantially lower than that of Germany.

These trends are even more pronounced for modularity. This statistic measures the extent to which links in the network tend to occur within communities or between communities. As explained above, we use the Louvain method to determine the communities within each code's network. We are less confident in our expectations here because the comparativist commentary does not discuss the extent of cross-references across sections. We expect the German—and, to the degree it is related, Italian—codes to be less modular because the approach of defining general concepts that apply across the code should produce some intercommunity links. But the commentary is silent on the use of these intercommunity concepts relative to the number of intracommunity links. Looking at the statistics, we observe that Germany is less modular than France, which is consistent with the use of community-crossing links. And, again, we see similarities between France and Spain, which both have highly modular codes. But Italy is a distinct outlier. While the number is positive, indicating more intercommunity links than chance would predict, its score is much lower than the other codes. This observation is consistent with the visualization of Italy's largest component, where it is difficult to discern distinct communities. This evidence suggests that there are some meaningful differences in the use of cross-references between Germany and Italy.

The directional characteristics of the codes show some further nuances in their structures. The indegree and outdegree variances provide information about the distribution of incoming and outgoing links within the network, and the Hub5 and Authority5 statistics provide a sense of how many nodes have a substantial number of connections. We focus first on Italy, which dwarfs all the other countries in terms of the magnitude of indegree and outdegree variance. This means that Italy has a small number of nodes with extremely high levels of incoming and outgoing links. This is especially so for indegree variance. Further investigation shows that there are five nodes, which are also the first five articles of the entire code, with at least ninety incoming references. These articles lay out core principles of the code, which are then referred to throughout the rest of the code. This pattern is consistent with the German idea of certain core principles underlying different areas of law. But, curiously, we see less evidence of this pattern in the actual German code. It has a much lower indegree variance than the Italian code. And, while the node with the largest number of incoming references is the first section, the nodes with the next-highest number of incoming references are scattered throughout the code. The lack of older, digital copies of the code makes it difficult to trace whether the codes were ever similar along this dimension, but we can show that the modern codes do appear to use incoming links in a stylistically different way.

The second item of note is that, for all the countries except Italy, the outdegree variance is larger than the indegree variance. This means that, relative to the baseline for each code's network, there are more nodes that have a large number of references out to other sections than there are nodes that have incoming references. This phenomenon suggests that Italy is not only an outlier when it comes to the density of its code, it is also an outlier in the way it repeatedly refers to the core concepts that appear in its first five articles.

Conclusion

Cross-references are a widely used and longstanding feature of legal texts. They serve a variety of purposes, including directing the reader to a general-purpose definition, carving out exceptions to rules, or relating a concept to a general principle in the text. Qualitative scholars of comparative law describe the different stylistic approaches that legal traditions take to the use of cross-references. Those codes that seek to be comprehensible to nonlawyers—such as the French code—tend to minimize the use of cross-references, while those that are more technical in nature—such as the German code—make more extensive use of cross-references. To the extent that these stylistic differences are part of what is passed on when one legal tradition influences another, cross-reference networks should provide evidence of legal origin.

In this chapter, we outline a method for analyzing the cross-reference networks of legal codes and conduct a pilot study of this method. This pilot study develops evidence that is largely consistent with the accounts of qualitative comparative scholars. The present-day German code is more densely networked than the French code. We also find evidence that is consistent with qualitative accounts of transmission. The Spanish code is relatively similar to the French code when it comes to density and modularity. The differences between the Italian and German codes are more pronounced, but they are the two most dense and least modular codes.

A primary advantage of using this computational approach is the ability to scale the analysis. This allows us to uncover differences that would be difficult to do through hand coding. For example, we are able to show that the way that the Italian code uses incoming references—with a large number to the first five articles—differs from the way that the German code uses those references. An additional advantage is that we are able to analyze a much larger number of codes and constitutions than traditional qualitative methods can.

While we are able to show differences in the network structure of a handful of civil codes in our descriptive analysis, an additional goal of this chapter is to show that the technique we develop has promise in broader contexts. As discussed above, we have already begun the analysis of constitutions. The larger sample size of that study will allow for more sophisticated comparisons than are possible with the small number of codes we analyze in this chapter. Future work will be able to build on the metrics discussed here to examine changes to legal texts over time and may facilitate causal investigation into the factors—

such as revolutions, coups, or other political or social events—
that affect the channels of legal transmission. ✤

Acknowledgments

We are grateful for the support of the National Science
Foundation and the Netherlands Organisation for Scientific
Research (NWO) through the Digging into Data initiative and
for helpful comments from Mike Livermore, Dan Rockmore,
and other participants at the 2017 Computational Study of Law
working group at the Santa Fe Institute. The results presented
here are part of a larger project with Rens Bod, James Daily, Bart
Karstens, and Marijn Koolen.

ATTORNEY VOICE AND
THE US SUPREME COURT

Daniel L. Chen, Toulouse School of Economics
Yosh Halberstam, University of Toronto
Manoj Kumar, NYU Center for Data Science
Alan C. L. Yu, University of Chicago

The natural audio presentation of natural language has many sources of variance beyond simply the choice of words. Characteristics of a speech act such as pitch, diction, and intonation may be significant, even though they do not affect the semantic content of what has been spoken. There is a significant body of scholarship that examines this type of speech variation, e.g., in mate selection, leader selection, housing choices, consumer purchases, and even stock market outcomes (Nass and Lee 2001; Klofstad, Anderson, and Peters 2012; Purnell, Idsardi, and Baugh 1999; Scherer 1979; Tigue et al. 2012; Mayew and Venkatachalam 2012), but there is relatively little quantitative empirical evidence that speech variation beyond lexical choices matters for real-world behavior. Speech variation from identical utterances of "Hello" affect personality ratings (McAleer, Todorov, and Belin 2014), but linking these ratings to downstream behavior is challenging. Nevertheless, oral advocacy classes are taught at law schools, and skilled oral advocacy is a highly sought-after professional trait (Korn 2004).

In this chapter, we take up the question of the practical relevance of speech variation by examining whether specific vocal cues in the first three seconds of speech are predictive in high-stakes policymaking settings, such as the US Supreme Court.

There are many reasons to think that vocal first impressions should *not* matter very much. From the perspective of a purely rational judicial decision-maker, only the information content of a speech act should count (Posner 1973). Unless vocal characteristics carry useful information, they should be ignored. Under the attitudinal model of judging, judicial decision-making is understood as largely political, with outcomes determined by judicial attributes (Cameron 1993). Something as seemingly insignificant as vocal characteristics should not be enough to overwhelm a judge's ideological dispositions. Alternatively, under a legal model, judges would focus on the legal content of the arguments presented by litigants (Kornhauser 2012). Vocal style would again seem to be irrelevant.

More broadly, even if vocal style did influence judges' decisions, competitive pressures should work to eliminate low-cost arbitrage opportunities (Becker 2010). Specifically, if it became known that judges prefer a certain vocal style and adopting such a style was possible at relatively low cost, then all lawyers would do so or risk losing clients. By analogy, it is relatively affordable for attorneys to dress according to the norms and standards of the profession by donning business suits in court. An attorney who simply refused to conform to the norm by wearing jeans to court would risk unnecessarily biasing judges in ways that were adverse to his or her clients (and perhaps even run the risk of a malpractice lawsuit). For this reason, lawyers wear suits to court, even if that attire is not to every one of their personal tastes. At the very least, a lawyer who refuses to wear a suit would likely have to discount the cost of his or her services to reflect the lower value for clients, to the point where differences between attorney attire would be "priced into" the marketplace and, therefore, would no longer be observed.

However, if firms or clients have a preference for certain advocates beyond their performance in court, correlations between malleable characteristics and outcomes could persist. For example, if courts tended to be biased against lawyers who adopt certain behaviors but clients insisted on hiring on an equal-opportunity basis, then those lawyers might underperform compared to others at the same pay scale. Alternatively, judges may be unbiased, but law firms and clients could be biased in their hiring practices, leading to the underperformance of lawyers in the favored class. Indeed, legal theorists have suggested that discrimination, once aimed at entire groups based on "immutable characteristics," now aims at subsets that refuse to *cover*—i.e., refuse to assimilate their behavior to dominant norms (Yoshino 2006; Goffman 1963). Yoshino (2006) argues that when courts allow employment that is contingent on covering, it legitimizes second-class citizenship for the subordinated group. Subordination would conflict with values expressed in the US Constitution (Balkin 2011).

The question of covering-based discrimination is just beginning to attract attention from empirical scholars (Bertrand and Duflo 2016; Neumark 2016). When it comes to how one speaks, minorities' choice of diction has been found to be associated with long-run labor market outcomes (Grogger 2011). Female lawyers routinely pay coaches to sound more masculine[1], an indication, at the very least, of a perception of discrimination. Men with non-masculine voices have also been found to be disadvantaged in the labor market (Case 1995).

---

[1] See Starcheski, L., "Can Changing How You Sound Help You Find Your Voice?" (2014), http://www.npr.org/blogs/health/2014/10/14/354858420/can-changing-how-you-sound-help-you-find-your-voice.

The following analysis examines whether vocal characteristics have predictive power in a setting in which there are very powerful market forces that would seem well-poised to arbitrage away any low-cost competitive advantage: oral arguments before the US Supreme Court. This work reports and builds on prior analyses (Chen, Halberstam, and Yu 2016a, 2016b) and is based on the data of 1,901 US Supreme Court oral arguments between 1998 and 2012. Specifically, we examine whether voice-based snap judgments based on lawyers' identical introductory phrases, "Mr. Chief Justice" and "May it please the Court" predict court outcomes.

Data

The data come from Chen, Halberstam, and Yu (2016a). Oral arguments at the Supreme Court have been recorded since the installation of a recording system in October 1955. The recordings and the associated transcripts are made available to the public in electronically downloadable format by the Oyez Project, a multimedia archive at the Chicago–Kent College of Law devoted to the Supreme Court and its work.[2] The audio archive contains more than 110 million words in more than nine thousand hours of audio, synchronized based on the court transcripts. Oral arguments are, with rare exception, the first occasion in the processing of a case in which the court meets in person with the litigants' counsel to consider the issues. Usually, counsel representing the competing parties of a case each have thirty minutes to present their side to the justices. The justices may interrupt these presentations with comments and questions, leading to interactions involving the justices, the lawyers, and, in some cases, the *amici curiae*. All audio clips involve the

---

[2]http://www.oyez.org/

lawyers' opening statements. The first handful of words in those statements are identical: "Mr. Chief Justice" and "May it please the Court."

The labeled sample comprises almost two thousand Supreme Court advocate audio clips for fifteen years from 1998 to 2012 with ratings for confidence, masculinity, trust, intelligence, attractiveness, and aggressiveness. Each audio clip was rated by approximately twenty Mechanical Turk workers, and a total of 20,888 ratings are available in this database. These data serve as a training set for 14,932 unrated audio clips of Supreme Court advocates from 1946 to 1997 and 2013 to 2014, spanning roughly seventy years.

The raters were asked to use headphones and to rate on a Likert scale from 1 (low) to 7 (high) the characteristics of masculinity, attractiveness, confidence, intelligence, trustworthiness, and aggressiveness. These six traits were selected based on previous research on listeners' perceptual evaluations of linguistic variables (Eckert 2008; Campbell-Kibler 2010; McAleer, Todorov, and Belin 2014). They are also similar to the ones used in Todorov et al. (2005), which presented subjects with pictures of electoral candidates' faces and asked them to rate their perceived attributes. That study found that perceptions of competence predicted election outcomes. Male and female lawyers were rated in separate blocks, such that participants either rated male advocates or female advocates but not both, so raters would not be comparing females and males on the degree of masculinity. Female lawyers were rated in terms of femininity instead of masculinity.

The elicitation algorithm randomized the order of the questions and whether "masculine" or "feminine" occupied the left or right portion of the scale (i.e., the *polarity* of the scale). The order and polarity of questions were held fixed for any particular rater to minimize cognitive fatigue. For additional nudges across experimental designs and to ensure attention by the rater, listening attention checks were employed. If raters failed, they would be dropped from the sample. There were six alertness trials, three with beeps and three without. The beep comes at the beginning of the lawyer's voice. For these questions subjects were asked if they heard a beep but not to rate the lawyer's voice.

Raters were also asked to rate the quality of the recording. While there was no time limit on how long a subject could spend on each trial, they were given a minimum of five seconds to respond; they were not allowed to proceed to the next trial until the five seconds was up (and all the questions completed) in order to ensure that subjects were given enough time to complete the ratings and to discourage them from speeding through the trials. No information regarding the identity of the lawyer or the nature of the case was given to the participants.

To control for the possibility of within-voice modeling by raters, instead of the basic design (in which the listener was presented with one voice sample and rated the sample on all scales), Chen, Halberstam, and Yu (2016b) also employed a design with only one question, randomly selected for each voice sample with only sixty clips and fewer subjects. Each voice clip was played aloud only once in order to capture the respondents' first impressions and prevent overthinking of responses (Ballew and Todorov 2007). There was a high degree of correlation of individual perceptions across experimental designs and stimulus presentation methods (see fig. 13.1).

Figure 13.1 plots the mean rating for each of the advocates using the two approaches discussed above. The x-axis reflects the mean ratings obtained from raters who were asked to rate each advocate on the full set of attributes, and the y-axis reflects the mean ratings obtained from raters who were randomly

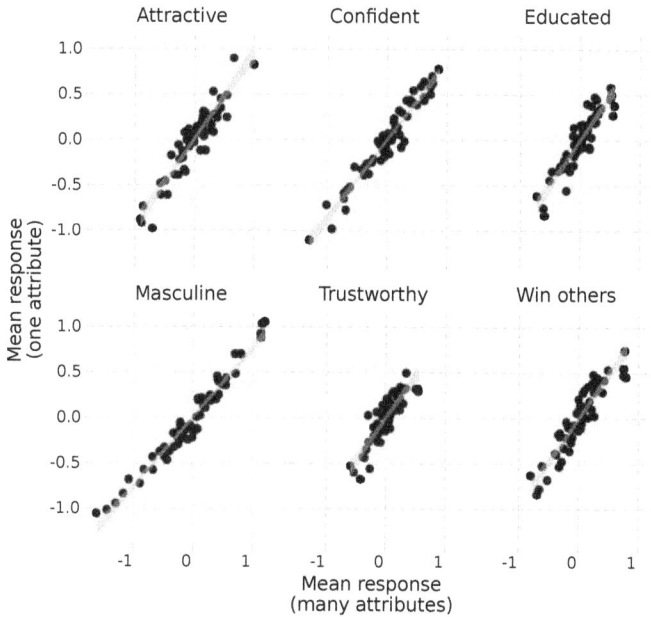

Figure 13.1. Correlation in average voice perceptions across experiments (many vs. one attribute). Note: This figure plots the mean rating of sixty voice samples in the pilot, where the x-axis reflects mean ratings obtained from raters who were asked to rate each advocate on the full set of attributes, whereas the y-axis reflects the mean ratings obtained from raters who were randomly assigned to rate each advocate on only one attribute. Standard error is estimated with a linear model.

assigned to rate each advocate on only one attribute. The ratings are highly correlated across these experimental designs, suggesting that trait judgments obtained from listening to a voice are quite stable.

Additional data for predicting Supreme Court outcomes were drawn from Katz, Bommarito, and Blackman (2017), which sought to predict Supreme Court decisions using pretrial characteristics (which collectively received roughly 25% of the importance weight) as well as court and judge historical trends specific to issue, parties, and lower courts (which collectively received roughly 75% of the importance weight). These features are divided into seven categories: (a) Justice and Court Background Information (e.g., justice year of birth), (b) Case Information (e.g., legal issue), (c) Overall Historic Supreme Court Trends (e.g., ideological direction), (d) Lower Court Trends (e.g., circuit court ideological trends), (e) Current Supreme Court Trends (e.g., mean agreement level of the current court), (f) Individual Supreme Court Justice Trends (e.g., mean justice ideological direction), and (g) Differences in Trends (e.g., difference between justice and circuit court directions). In Katz, Bommarito, and Blackman (2017), random forest—a weighted nonparametric model that forms weighted predictions based on nearest neighbors—was found to perform well.

Baseline Model and Performance Evaluation

Chen, Halberstam, and Yu (2016a) used the Katz, Bommarito, and Blackmun prediction of justices' votes as a control in a linear regression. Chen, Halberstam, and Yu asked whether vocal characteristics had an explanatory effect above and beyond the predictors generated by the Katz, Bommarito, and Blackmun model, and if so, how much. That paper argued that the reason vocal characteristics performed well relative to the best prediction model is that, as noted in Katz, Bommarito, and Blackman (2017), that model performs best on cases on which the court was in agreement (9–0) and performs worst on cases with high levels of disagreement among members of the court (5-4). In fact, in close cases affirming the lower court, the model predicts the outcome with only 25% accuracy.[3] Chen, Halberstam, and Yu (2016a) showed that vocal characteristics

---

[3] Figure 6 in Katz, Bommarito, and Blackman (2017).

are predictive of outcomes with the swing voter, which is where the random forest model may do poorly. In brief, the vast majority of judge votes are in easy cases, where extra-legal and extra-ideological factors may play a smaller role. In hard cases, where judges are closer to indifference (i.e., for close calls), human bias could tip the swing vote, the importance of which is magnified when case outcomes are examined.

This chapter extends the analysis to the available universe of Supreme Court oral arguments. Again, the best existing predictive model is used as a baseline. Features are then added to determine whether they increase accuracy. Chen, Halberstam, and Yu (2016a) showed that vocal characteristics are predictive of Supreme Court votes depending on the political party of the judge, and the correlation persists after controlling for available characteristics of the lawyer and the case, as well as the best prediction of Supreme Court votes. Here, we add the predicted voice trait ratings over a longer time frame in the prediction model of Supreme Court decisions. A binary outcome is constructed based on whether the justice reversed or affirmed the lower court opinion. Katz, Bommarito, and Blackmun's model uses a large number of judge and case characteristic features, as well as court trends and lower court trend features. However, their model does not include advocate audio features.

Establishing causality is beyond the scope of the current work. The ideal experiment would be to randomize the voice of the lawyer unbeknownst to the justices or the lawyer, and then test for effects and outcome. Although such an experiment is impossible, future clever research designs may find a way to ground strongly supported causal claims. It is worth noting, however, that prior results assuage some of the concern of omitted-variables bias: for example, if the vocal cues are correlated with case weakness, then all judges should respond to vocal cues in the same way.

Features and Feature Engineering

Given the availability of the raw audio data, a choice must be made as to whether to use the raw data or the predicted features in the machine learning prediction. For example, in macroeconomic forecasting that relies on principal components or factor analysis, the predicted trait is commonly used. In this case the underlying factor driving multiple economic indicators (eigenvectors) is believed to have a continuous distribution. Moreover, since the eigenvectors underlying common trait characteristics are likely to be highly correlated, a sparse model like LASSO is less appropriate. Both principal components analysis and regularization approaches aim to reduce dimensionality. However, regularization is a type of supervised learning (built on the assumed relationship between the outcome and the predictors), whereas principal components analysis is a type of unsupervised learning (considering only the predictors). Given that the goal of our task is to test for the predictive information in audio cues, we opt to use the (predicted) trait features rather than some version of the raw data or unsupervised representations thereof.

The ratings were on a 1–7 scale. Each rater's rating was normalized by subtracting their average rating and dividing by the standard deviation of their ratings (i.e., z-score). The aggregated z-scores corresponding to every lawyer yields a continuous voice trait rating for every lawyer. Then the z-scores were made binary: if a z-score was positive, it was replaced with 1; if it was negative it was replaced with -1.

Next, every audio clip of a lawyer's opening statements from 1946 to 2014 was processed into a fixed number of frames, and each frame was vectorized into thirteen dimensions following the standard approach used in voice analysis (MFCC) (Ganchev, Fakotakis, and Kokkinakis 2005). A trained random forest classifier model—a weighted nonparametric model that forms weighted predictions based on nearest neighbors—was used to generate the predicted traits. It was most accurate in predicting perceived masculinity (65.79%) and least accurate in predicting perceived trustworthiness (56.02%). The greater predictability in perceived masculinity is consistent with some results reported in Chen, Halberstam, and Yu (2016a). That study also played the voice clips backward and asked raters to rate the backward clips. Among the perceptual questions, ratings for perceived masculinity were most strongly correlated for the forward and backward clips.

These predicted voice trait ratings were appended to the original dataset. More specifically, for the audio clips of 1998–2012, the binarized version of the originally obtained continuous z-score ratings was appended, and for audio clips from 1946 to 1997 and 2013 to 2014, the binary voice traits predicted from the above-mentioned model were appended.

Results

The model is evaluated with the binary voice features, which improve casewise accuracy by 1.1 percentage points, from 0.634 to 0.645, and decrease justicewise accuracy by 0.1 percentage points, from 0.649 to 0.648.

The following charts show the feature weights. To present the relative scale, the intercept at the bottom of the figure corresponds to the most important feature present in the

Audio features
feature importance (jittered by petitioner and respondent)

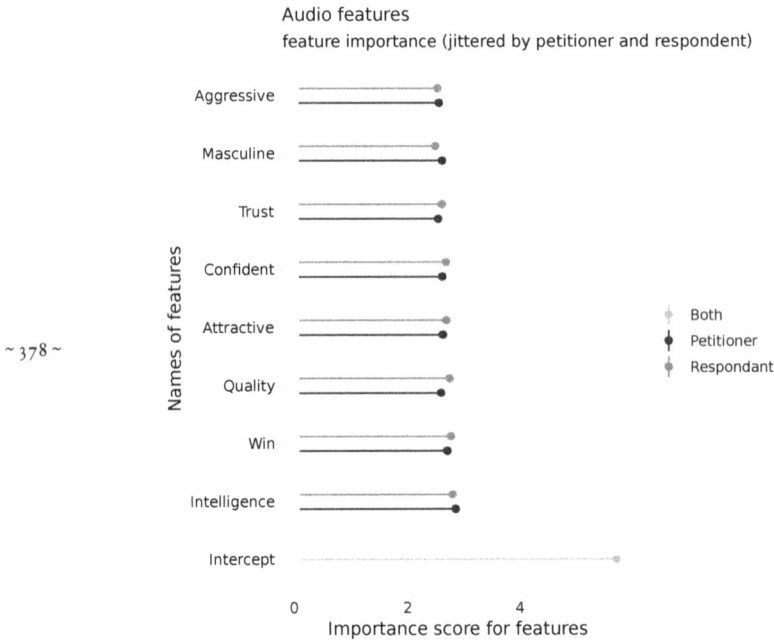

Figure 13.2. Feature weights relative to top feature weight.

model.[4] Since this is a random forest model, the feature charts do not speak to the directionality of the features' effects.[5]

An extension of this study can examine the predictive power of other audio features, such as the number of

---

[4] The most important feature is "justice_cumulative_lc_direction_diff" (the difference between the lower court disposition direction and the justice's cumulative direction). "Disposition direction" is a measure of whether the decision of the court whose decision the Supreme Court reviewed was itself liberal or conservative. "Previous" refers to previous Supreme Court terms, and "cumulative" refers to all prior terms. As such, these two indicators are measurements related to ideology, and, in particular, the ideological differences between the justice and the lower court opinion.

[5] Additionally, to address the question of whether the audio features were really detecting lawyer gender, including the gender variables in subsequent analysis did not increase the accuracy of the model, and the average gender scores for petitioner and respondent sides were not in the top thirty most important features.

interruptions, the political dialect (Kadiri et al. 2018), vocal implicit bias (Jaiswal et al., 2019 (forthcoming)), or phonetic accommodation (Chen and Yu 2016). A richer high-dimensional characterization or hand-labeling of the audio clips may also prove fruitful.

Discussion

This chapter examines the possibility that vocal features can influence judicial decisions. In Chen, Halberstam, and Yu (2016a, 2016b) an econometric analysis reveals that vocal features impact court outcomes, even based on a speech sample of less than three seconds. The connection between vocal characteristics and court outcomes was specific only to perceptions of masculinity and not other characteristics. Consistent with employers mistakenly favoring lawyers with masculine voices, perceived masculinity was negatively correlated with winning and the negative correlation was larger in more masculine-sounding industries. The first lawyer to speak was the main driver. Among these petitioners, males below median in masculinity were seven percentage points more likely to win in the Supreme Court. Republicans, more than Democrats, vote for more feminine-sounding females, while Democrats, but not Republicans, vote for less masculine-sounding men. Perceived masculinity explains additional variance relative to and is orthogonal to the best random forest prediction model of Supreme Court votes.

In this chapter, we extend the initial work and show that the best predictive model of Supreme Court votes improves with the addition of voice characteristics of Supreme Court advocates for almost seventy years of data, rather than fifteen. The improvement appears robust for predicting Supreme Court case outcomes and appears limited for predicting Supreme Court justice votes, similar to the finding of Chen, Halberstam, and Yu (2016a). Chen and Kumar (2018) find that the facial features of lawyers also predict Supreme Court votes. A surprising finding across these papers is that the characteristics studied were shown to be half as important as the most important feature typically attributed to legal advocates—political ideology. ✒

Acknowledgments

We thank Vishal Motwani and Phil Yeres for early assistance. Work on this project was conducted while the first author received financial support from the European Research Council (Grant No. 614708) and Agence Nationale de la Recherche.

oJle

DETECTING IDEOLOGY IN
JUDICIAL LANGUAGE

Marion Dumas, London School of Economics and Political Science

~383~

To what extent do the political affiliations of US judges influence the way they interpret law and craft their opinions? The literature on the political economy of law, which sought to challenge an idealized view of judging as apolitical and impartial (Rodriguez, McCubbins, and Weingast 2009), advances the argument that judges, like each one of us, carry ideology and policy preferences and that these preferences shape their legal practice (Segal and Spaeth 1996). In the canonical tests of this hypothesis, ideology and policy preferences are coarse grained into a binary distinction between conservative and liberal (or Republican and Democrat). Many studies then correlate these ideologies with the voting patterns of judges—again, frequently reducing the dimensionality of outcomes in judicial decision-making to a single liberal–conservative axis (Sunstein, Schkade, and Ellman 2004; Revesz 1997). For example, Democratic judges and justices are found to be more likely to rule in favor of the defendant in criminal cases, for the government in regulatory questions, and for plaintiffs against corporations in civil cases, each of which is interpreted as a "liberal" outcome (Segal and Spaeth 2002). Substantively, this legal realist school represents the decision-making of judges exactly as that of other political strategic actors, with the law itself playing no special part (Rodriguez and McCubbins 2006).

Recently developed methods in computational linguistics give us an opportunity to revisit the question of the relationship of law to politics. Indeed, the studies mentioned above focus on judges' votes concerning case outcomes and largely ignore the legal texts written by judges (other than to extract outcome information). Votes on the disposition of a case might require weighing various elements of the case or the law to decide on the outcomes for the parties, but outcomes alone do not directly influence the direction of future legal interpretation. Rather, the law is modified through the written opinions that are issued by the court. Thus, although partisan affiliation of judges may be correlated with the relative weight judges put on elements of a case and its final outcome (Kornhauser 1992), the discussion of precedent and legal texts found in judicial opinions may be unaffected by partisan influences.

Computational linguistic models are an interesting approach to the study of judicial politics. In other contexts, the statistical analysis of features of written text has been used for purposes of author attribution (Mosteller and Wallace 1963), to examine trends in writing style over time (Hughes et al. 2012), and to estimate the influence of judicial clerks on the law (Carlson, Livermore, and Rockmore 2016). In all of these applications, quantitative information that is extracted from a written text—the frequency of use of certain words—is used to draw conclusions either about the texts themselves or about the characteristics of the authors. In the context of judicial politics, the question asked is whether it is possible to predict a judge's partisanship and other attributes on the basis of quantitatively identifiable characteristics found in the judge's writing. Stated another way, these models ask, "Is there *enough* divergence in what people of different groups say or write to predict a person's group membership from listening to or reading them?"

When researchers have asked this question in the context of explicitly partisan communication, the answer has been a resounding "yes." According to Gentzkow, Shapiro, and Taddy (2016), the two parties now speak in different languages: computational analysis of congressional speeches robustly identifies partisan differences in the vocabularies used by the two parties. Moreover, computational analysis reveals statistically robust splits in political language in elite discourse more generally, whether in the media (Gentzkow and Shapiro 2010) or elite discourse (Jensen et al. 2012).

These studies are based on identifying linguistic patterns that are predictive of ideology or partisanship, where the latter are measured by some ground-truth, such as voting, official party affiliation, or explicit branding. Regularity in linguistic patterns can arise from using different frames to discuss the same issue (e.g., "illegal aliens" versus "undocumented immigrants"), or by tending to focus on different subjects and issues (e.g., defense versus health). One of the first papers to apply these methods to the study of ideological divisions in society was Gentzkow and Shapiro (2010), extending the earlier work of Groseclose and Milyo (2005). In Gentzkow and Shapiro (2010), the authors identify a subset of "partisan phrases" from the *Congressional Record* by taking the one thousand phrases that are most highly associated with a particular party (using the χ^2 statistic for the null hypothesis that the propensity to use a given phrase is equal for Democrats and Republicans). Using these phrases, they then build an estimator for the ideology of a congressperson or media outlet based on the frequency of use of these phrases weighted by the strength of their correlation with ideology. The resulting estimator was found to have correlation of over 0.6 with true ideology. Jensen et al. (2012) follow the same method to measure slant in elite discourse over 140 years using the Google corpus of digitized books. To validate the linguistic estimator of ideology, they show that they can make predictions about partisanship for members of Congress whose speech was not used to estimate the partisan phrases.

It is perhaps not surprising that political actors should express ideological differences in their speech. Interestingly, Jelveh, Kogut, and Naidu (2015) found that these results and methods generalize to texts that are not explicitly political, namely, the papers of professional economists. They found that linguistic patterns can clearly predict their membership to two ideological groups, liberal or conservative. They measure ideological affiliation by linking economists to their political campaign contributions to the Democratic or Republican party and the number of left-leaning and right-leaning petitions they signed. As Gentzkow and Shapiro (2010) did with the *Congressional Record*, the authors identify partisan phrases in the very large corpus of peer-reviewed economics research papers produced between 1991 and 2008, controlling as well for topic of the papers. The correlation between predicted and ground-truth ideology for economists in the test set was found to be 0.42, and importantly, the receiver operating characteristic curve (a standard method to evaluate the quality of a binary classifier, described in detail further in the paper) showed clearly that their linguistic ideology estimator performs much better than a random model and is therefore clearly not overfitted and picking up noise.

Peer-reviewed articles, unlike political speeches, are not typically thought to intentionally signal ideological group membership via strategic use of language. They tend, on the contrary, to be technical and, seemingly, to signal subject expertise and scientific objectivity. Thus, the result of Jelveh, Kogut, and Naidu (2015) is surprising and suggests that ideology shines through despite professional norms that would seek to subdue it. Hence, we might expect that judicial language, once passed through the filter of these models, may also betray ideological membership despite its technicality.

This paper applies computational linguistics models to the corpus of all judicial decisions of the district and appellate federal courts on matters of environmental regulation from 1970 to the present—a policy area that has seen growing partisan polarization over the study period (Shipan and Lowry 2001)—to determine whether these decisions bear a clear imprint of the partisan affiliation of judges. The partisan affiliation of judges can be measured either by their direct (self-reported or official) affiliation with a party or by the party of the president who appointed them (Segal and Spaeth 2002). These variables are very closely correlated.

Data collection focused on a specific area of law in order to have an exhaustive and well-circumscribed corpus that spans the judicial hierarchy. This narrows the semantic field, which seems important since ideological variation, if it exists, may express itself in issue-specific ways (e.g., variation in the adjectives used to qualify liability regimes in hazardous waste). Since these methods have not in the past been applied to judicial texts, this paper tests the joint hypothesis that the partisan affiliation of judges affects the way they interpret the meaning of environmental laws and that these differences can be detected by computational linguistic models. Testing this hypothesis is a first step toward developing a set of tools that can detect the interaction of politics with the use and development of precedent. Rejecting this hypothesis should spur the search for more fine-grained models of language and more subtle models of the ideological content of law. Indeed, political scientists often reduce ideology to partisanship (or a one-dimensional left or right, liberal or conservative spectrum). Yet, one definition of ideology is a system of political beliefs or ideas that motivate action (Sypnowich 2014). A system of beliefs that successfully helps legal actors adjudicate between complex trade-offs may not be reducible to a single dimension.

Having a reliable tool to uncover the influence of political power and political ideology on the evolution of law would be very useful. Indeed, important debates on the role of the judiciary in a democratic system revolve around the question of the judiciary's responsiveness to the political environment. Commentators have long argued that a core function of courts in rule-of-law societies is to protect minority rights against the "tyranny of the majority" (Dahl 1957). To successfully fulfill this function, the judiciary must maintain a level of independence relative to Congress. The judiciary can also serve as a check on the exercise of executive power and can help stabilize democratic polities in the face of authoritarian challenges (Huq 2018). If judges are beholden to or fearful of the executive branch, they are unlikely to resist an executive with tyrannical impulses. Furthermore, courts, lacking either "the sword or the purse," rely on sociological legitimacy for authority rather than using direct coercion (Bickel 1986). Political scientists who have studied sources of support for the courts in the United States have found that they are grounded, at least in part, in the perception that courts make decisions for distinctly "legal" rather than "political" reasons (Gibson and Nelson 2014). More broadly, a "social equilibrium" characterized by the rule of law requires that the law be stable, predictable, and transparent (Hadfield and Weingast 2012; Waldron 2008). This would seem difficult to achieve if individual judges didn't share a certain level of consensus on what the law is at a given point in time, whatever their partisan affiliations.

These questions carry particular importance in contemporary American politics. Roll-call votes almost perfectly sort along party lines, and party competition has intensified. The everyday tactics of governing are geared toward scoring partisan points and undermining the opposing party rather than resolving public policy problems (Lee 2015). Also, the policy positions of politicians have drifted apart on a number of issues relative to the post–World War II period (Bateman, Clinton, and Lapinski 2017). However, the degree of polarization manifest in Congress and other elected bodies seems to be not as widely experienced in the public at large. Indeed, the vast majority of the public has, by and large, remained moderate.[1] This fact does not stand in contradiction to findings showing that partisan affect and identity have intensified, especially since the public believes itself to be more polarized on issues of policy than it really is. Note that the extent of polarization in the public is debated by scholars, some seeing stronger divisions (Bafumi and Shapiro 2009). The intensity of polarization among politicians may contribute to the growing view of the public that the government is dysfunctional and untrustworthy.[2] Since the 2016 presidential election, the level of conflict between the parties has intensified to the point that it may threaten core constitutional norms. In this climate, the courts play a salient role as the arbiter of partisan disputes and serve as a forum where stakeholders can challenge actions taken by policymakers that they regard as illegitimate. In this context, it is important to understand whether courts are themselves

---

[1] See M. P. Fiorina, "Has the American Public Polarized," Hoover Institution *Essays on Contemporary American Politics* (2016), https://www.hoover.org/research/has-american-public-polarized

[2] See Pew Research Center, "Public Trust in Government Remains Near Historic Lows as Partisan Attitudes Shift" (2017), https://www.people-press.org/2017/05/03/public-trust-in-government-remains-near-historic-lows-as-partisan-attitudes-shift/.

gripped by the divides that characterize the rest of the elite and political class, or if they maintain a level of neutrality that allows a resolution of these conflicts.

Given the political context of the last forty years in the United States, we can expect that if the judicial branch displayed a growing partisan polarization, as have other branches of government, then the particular context of environmental rulings would be a setting where it would be evident. Indeed, elite partisan polarization regarding environmental regulation has grown over the last decades. Estimated ideal points of policymakers have diverged over time after enjoying bipartisan support in the 1970s (Shipan and Lowry 2001). Meanwhile, public support for existing environmental laws remains high. For example, a majority of the public is in favor of regulating carbon dioxide as a pollutant (Marlon, Fine, and Leiserowitz 2017). Although, in general, public opinion has polarized to some extent, an increase in pro-environmental preferences among Democrat voters mostly accounts for this (Eun Kim and Urpelainen 2017). Environmental regulation has been one of the targets of the conservative movement (Jacques, Dunlap, and Freeman 2008; Skocpol and Williamson 2016), possibly exacerbating the divide between elite and public attitudes.

Data

The dataset used for the following analysis is derived from the *Federal Reporter* and *Federal Supplement*, published by Westlaw and consisting of the full record of every case for which an opinion was published in the appellate courts and district courts. Westlaw clerks identify each legal issue discussed in the opinion and classify this issue according to the Westlaw nomenclature of legal issues. In this nomenclature, there are 589 legal issues in environmental law. The dataset comprises all court cases between 1970 and 2014 that include at least one issue classified as an environmental issue.

The raw text of the opinions was parsed to extract the following information: the year, the court, the Westlaw headnote issues,[3] the associated text of the opinion, and finally, the name of the judge authoring the opinion. These names were then matched with the database of federal judge attributes developed by Gryski and Zuk. (2008), which includes the party affiliation of judges and the party of the president who appointed them. It was not always possible to parse the name of the authoring judge and match it to the database. Some cases are not actual opinions but merely summary judgments, others are ruled by magistrate judges who do not feature in the database, and others are missing from the database because they were appointed after 2000 (the database covers appointments up to 2000). In other cases, the formatting prevented automatic parsing. Table 14.1 shows the number of opinions at the district, appellate, and Supreme Court levels, as well as those unmatched. The table also shows the number of Republican and Democrat judges at the district and appellate levels who have authored at least one opinion in this corpus (among those that could be matched to the Gryski and Zuk. (2008) database). The analysis that follows uses the corpus of 6,541 cases at the district and appellate levels in which at least 30% of headnotes are classified as relating to an environmental issue and the corresponding dataset of 1,458 judges. Supreme Court decisions are not included because the sample is too small.

---

[3] A headnote is a brief summary of a particular point of law that editors at Westlaw add to the beginning of an opinion. The headnotes provide an overview of all the points raised by the opinion and help the reader find the discussion of these points in the opinion.

Table 14.1. Number of judges and opinions in the corpus. The second row reflects the application of a filter that keeps only opinions where 30% of the headnotes are on an environmental issue.

Data	District	Appeals	SC	Not matched	Total
Whole corpus	4441	2889	113	1358	7506
Environment	3200	2165	63	1113	6541
Rep. judges	595	203			798
Dem. judges	502	158			660

Language Modeling

The analysis in this paper starts by following the algorithm developed by Jelveh, Kogut, and Naidu (2015), already shown to be successful at detecting partisan affiliation in another corpus of professional writing (specifically, academic economics publications). Briefly, this is a bag-of-words algorithm that first finds phrases correlated with the partisan affiliation of the authoring judge; these are called "slanted phrases." It then uses the counts of the occurrences of these slanted phrases to train a predictive model (called here "the slanted phrases model"). The model is then used to predict the party affiliation of judges whose opinions were not used to train the model (a test set). The performance of this model relative to a random model is then assessed. As a preview of the results, the language model is unable to detect partisanship in judicial writing. The model is then extended to include other textual features, called "judge embeddings," in an attempt to tap into more subtle textual features (we call this model "the model with embeddings"). Here, again, the model is unable to detect partisanship. We now describe these methods in detail.

The first step of language modeling consists of preprocessing the texts. The standard procedure in bag-of-words modeling is to tokenize, lemmatize, and remove extremely common words ("stop-words"), punctuation, and other nonalphabetical characters. Numbers are also removed. Because this is a legal corpus, any references to a law or opinion were transformed to a standardized format (e.g., "$151F3d23$" becomes a single token 151_f3d_23) before removing any remaining numbers so that these references could be conserved as terms. All n-grams of length 1—7[4] were counted for each opinion. These n-grams (from here on called "phrases") were then filtered to include only those that occurred at least ten times in the whole corpus and in at least ten different opinions.

The second step is to extract phrases that seem correlated with ideology. The set of judges was split into twelve equal groups (with the same distribution of Democrat and Republican appointees) for the district and appeals levels separately. Two of these groups were retained as a test set. Using the ten remaining groups (i.e., the training set), the following process was performed tenfold: in each fold, a different group of the ten in the training set was left out, and the phrase counts were computed for all judges who were appointed by a Republican president on the one hand and for all judges who were appointed by a Democrat president on the other. The correlation of each phrase with this party grouping was computed (Pearson's χ^2 statistic and the associated p-values) and phrases with $p \leq 0.05$ were kept. The ten folds were then combined, keeping only the phrases that were retained in six out of ten folds ("permissive filter") and ten out of ten folds ("restrictive filter"). We thus obtained twenty thousand to fifty thousand phrases seemingly correlated with ideology (from here on "slanted phrases") at the district level with the permissive filter and about

---

[4]As pointed out by Jelveh, Kogut, and Naidu (2015), Margolin, Lin, and Lazer (2013) found that longer phrases can capture ideology. This may be particularly important for legal language, in which long and qualified noun phrases may carry more ideological meaning.

four thousand to five thousand slanted phrases with the restrictive filter, and at the appeals level, about thirteen thousand to 330,000 slanted phrases with the permissive filter and about two thousand to three thousand with the restrictive filter.[5]

Given this set of (seemingly) slanted phrases, can we predict the party of the president who appointed judges in our test set? Here we apply the ensemble learning approach described in Jelveh, Kogut, and Naidu (2015), which has a partial least-squares (PLS) model at its core. PLS is a dimension-reduction modeling technique appropriate when the number of predictors is very large compared to the number of observations. PLS tries to find underlying dimensions that balance the twin goals of maximizing both the correlation with the outcome variance and the proportion of the variance in the predictors (as in principal component analysis). The ensemble learning procedure works as follows:

1. Set K the number of models and resulting predictions.

2. For each $k = 1, ..., K$, sample with replacement 80% of the judges in the training set, and sample without replacement five times the square root of the number of slanted phrases to obtain a smaller set of slanted phrases.

3. Build the frequency matrix $\mathbf{F}$ in which row j gives the number of times judge j used slanted phrase p in the selected set of slanted phrases across all of the opinions he or she authored. Note that some judges may have a count of zero for all phrases in that set, in which case that judge is dropped from this particular model iteration. Similarly, some phrases are not used by that set of authors, in which case they,

---

too, are dropped from this particular model iteration. The resulting counts are divided by the norm of each row and then by the standard deviation of each column (so that the vector of counts for each phrase has unit variance).

4. A PLS model with $n = 3$ latent dimensions is fit to the resulting frequency matrix **F** and the vector of party appointments for the same set of judges. The estimation itself is done with the *scikit-learn* package in Python, which uses the original algorithm proposed by Wold et al. (1984) (the so-called NIPALS algorithm).

5. A threshold f_k is determined by maximizing both the rate of judges correctly classified as appointed by a Republican and by a Democrat president.

6. This model iteration k (a fitted PLS model for the subset of slanted phrases, the threshold, and the standardization used for each phrase) is then applied to the test set of judges. In other words, a frequency matrix $\mathbf{F}_{test}$ is built for this set of judges to predict a response variable $\hat{y}$. If the judges in the test set have not used any of the slanted phrases, they are dropped (no prediction is generated). If $\hat{y}_j < f_k$, judge j is classified as appointed by a Republican president and vice versa. This constitutes a "vote" for this particular ideological classification.

7. The procedure is repeated K times, each yielding a vote for each judge in the test set (except those who are dropped because they did not use a slanted phrase).

The votes are then aggregated to yield a probability that a given judge in the test set was appointed by a Republican versus Democrat president.

Results

The analysis was carried out separately for the district and appeals levels (recognizing the difference in decision-making in these different types of courts). The set of 1,097 district judges was split into twelve stratified samples (i.e., with a constant ratio of Democrat and Republican judges), ten of which were used for building the model and two of which were used to test it. The set of 361 appeals judges was also split into twelve stratified samples, ten for training and two for testing.

Recall that each prediction is probabilistic. Given a threshold t, if the probability that the judge is Democrat is below that threshold, we predict the judge to be Republican, and if the probability is above, we predict that he or she is Democrat. From these predictions, we compute the true positive rate (the number of judges correctly predicted to be Democrat over the number of Democrats in the sample) and the true negative rate (the number of judges correctly predicted to be Republican over the number of Republicans in the sample). Mechanically, as the threshold increases, the true negative rate should increase and the true positive rate decrease. In a random model (that simply predicts on the basis of the baseline probabilities of being Democrat or Republican), these rates would both be equal to 0.5 when the threshold is 0.5, and when plotting them against each other, they would vary linearly along the 45° line. If the model has predictive power, there should be a range of thresholds where the true positive and true negative rates are both higher than this 45° line. To check this, we thus plot the true negative and true positive rates against each other as the threshold varies from 0 to 1, which is called the "receiver-operating curve" (ROC).

Figure 14.1 shows the resulting ROC plots for the district and appeals levels for a model developed with the phrases selected by the permissive filter and a model developed with the phrases

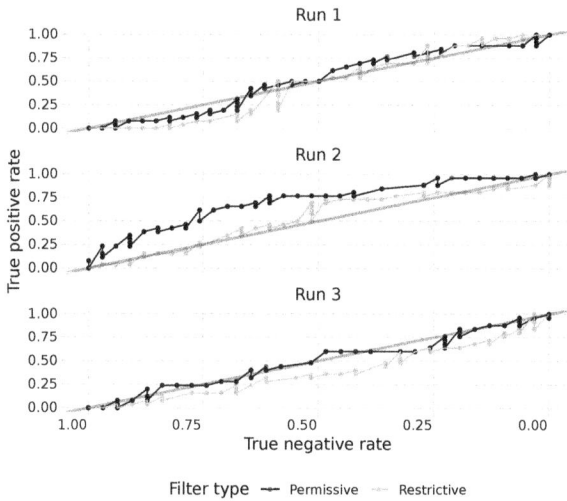

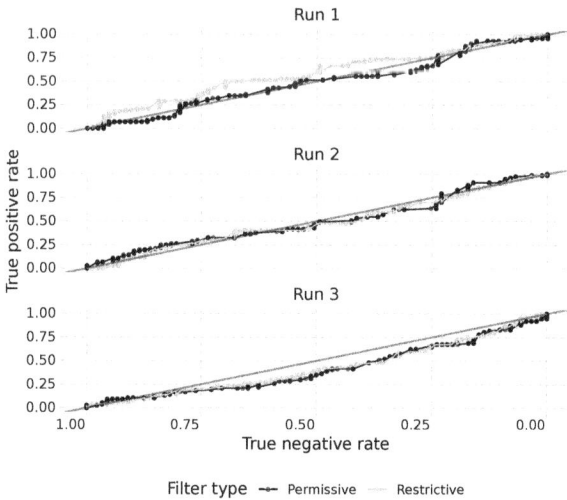

Figure 14.1. Appeals level (top) and district level (bottom). The ROC curves of the language model. The plot shows the results for both the permissive and restrictive sets of slanted phrases. It also shows that the performance is variable (especially at the appeals level) but remains low when resampling the dataset to obtain different splits between training and the test set (Runs 2 and 3).

selected by the restrictive filter. We see that the resulting curves are indistinguishable from a random model. This means that the phrases detected as being correlated with ideology are not used by judges in a sufficiently consistent manner to allow us to predict the ideology of judges in a new sample.

Beyond Bag-of-Words: Adding Embeddings

In bag-of-word models, texts are reduced to collections of n-grams (which are formed by adjacent or closely situated words) and their respective counts. In many spheres of discourse, different ideological meanings are expressed by different concepts (e.g., "death tax" versus "estate tax"). In the law, opinions attempt to identify under which conditions a rule applies. Hence, different ideologies may not be expressed by different concepts but by different ways of classifying context-specific conditions, in which case the full meaning conveyed by grammar and text structure may be needed to perceive ideological differences in meaning. At the very least, the context in which a word appears should matter.

For these reasons, we might expect that models that find latent dimensions of discourse through word embeddings, such as Word2Vec (Mikolov, Chen, et al. 2013), would stand a better chance of detecting judicial ideology. The logic behind Word2Vec is that "you know a word by the company it keeps." The idea is to embed words in an N-dimensional space that represents the context of the word's occurrence. Nearby words in this space have similar uses and meanings. Doc2Vec (Le and Mikolov 2014) is similar but contains an additional layer that represents a meaningful unit of text (a document, the corpus of an author, a sentence). Doc2Vec has been shown to perform better than other state-of-the-art models in detecting duplicate questions on online forums, sentiment analysis, and semantic textual similarity (Lau and Baldwin 2016). An additional reason embeddings are worth experimenting with is that they suffer less from sampling variability. Indeed, in bag-of-words models, each word is a variable, so the data are highly multidimensional and highly sparse. In contrast, embedding models learn a small set of underlying dimensions, reducing sparsity and dimensionality. We now explore whether vectorizing each judge's corpus of writing using the Doc2Vec embedding model (Le and Mikolov 2014) and adding these vectors as extra features to the model changes our ability to detect ideology.

The procedure is as follows: A Doc2Vec model is trained (implemented in the Python package *Gensim*) using all opinions written by judges in the training set, aggregated by judge. The model was estimated for several parameter choices: for fifty, one hundred, and two hundred embedding dimensions, a sliding window of lengths eight and twenty, and a minimum word count of two. The resulting model is then used to obtain a vector representation for each judge in the training set and each judge in the test set. These vectors are then added as features in the partial least-squares model and ensemble learning model described in the previous section.

Figure 14.2 shows the results of this experiment. The results are presented for the permissive filter combined with the embeddings for one hundred embedding dimensions and the sliding window of size eight for three different runs (i.e., three different stratifications of the data by random sampling). We also estimated the model with a sliding window of length twenty, with either fifty or two hundred embedding dimensions, with no change in results. The area under the ROC curve, called AUC, a quantitative measure of model quality, is given in table 14.2 for each of the six models in figure 14.2. We see that the vector representations slightly improve performance at the appeals level in some of the runs (especially Run 3) but not in a reliable

Table 14.2. AUC values at the appeals level with and without embeddings for three different splits of the data (into training and test set).

	Slanted phrases model	Model with embeddings
Run 1	0.5	0.48
Run 2	0.7	0.72
Run 3	0.49	0.6

way, since in Run 1, they provide no improvement at all. At the district level, embeddings make no difference and the model with embeddings performs no better than a random model.

Discussion

While it has been shown that the language model used here can detect a conservative versus liberal bent in the peer-reviewed economics literature (Jelveh, Kogut, and Naidu 2015), we find here that it fails to identify partisan affiliation of judges via their writings. When we added embeddings, we saw a slight improvement for appellate courts but nowhere near the extremely high levels of accuracy that these embedding models achieve in cases like sentiment analysis classification.[6] How should we interpret the inability of the language model to predict judicial ideology in this corpus?

First, it is worth noting that other modeling strategies may lead to a different conclusion. The Doc2Vec model could be optimized using cross-validation (to find an optimal number of embedding dimensions and sliding window size). Finally, the analysis here aggregated opinions for the entire period of 1970 to the present, whereas polarization has increased over time in the political branches of government. It may be that breaking down the analysis into successive time periods would reveal partisan legal

---

[6]Le and Mikolov 2014 obtain a 12% error rate for classification of movie reviews as positive or negative.

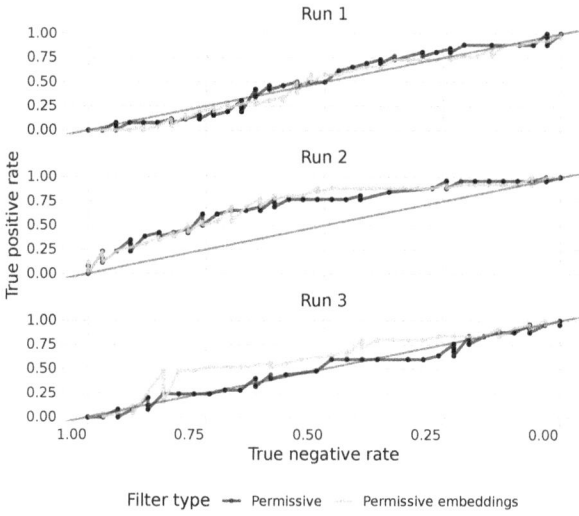

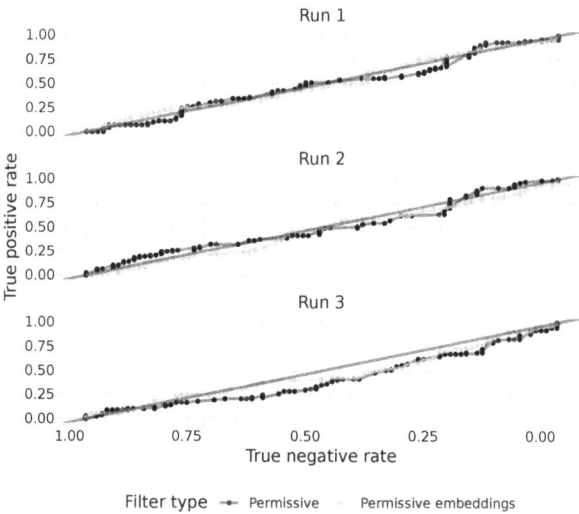

Figure 14.2. Appeals level (top) and district level (bottom). Comparison between the ROC curves of the models with and without embeddings. The plot shows the results for the permissive set of slanted phrases.

writing in the more recent record. (This wasn't attempted because the corpus would be too small.)

Putting aside a possible technical explanation, let us focus on the substantive possibility that, indeed, judicial partisanship does not lead to easily detectable differences in legal language, at least at the district and appellate levels. As mentioned earlier, in other areas of discourse, partisan affiliation leads to clear semantic differences readily identified by algorithms, even in professional writing. In the law, it does not. This suggests that any partisan influences in judicial opinions would also be hard to perceive by the lay reader, which means that they would have limited cultural influence.

Note that partisan ideology could still influence law itself, i.e., the rules that map facts to case outcomes (the doctrinal space, Kornhauser 1992). The models built here are not subtle enough to test whether this is the case. However, what we can provisionally conclude here is that legal discourse, i.e., the concepts, ideas, and tone of writing that create public meaning around an issue, is relatively homogeneous across the partisan divide among judges. This finding gives credence to the claim in Sunstein (1996) that the legitimacy of law can more easily be sustained in a pluralistic society if legal reasoning is incompletely theorized. He contrasts analogical reasoning to a reasoning based on deep normative principles concerning concepts such as freedom, merit, and equality. The argument is that society is confused and divided on deep theories, so it would be dangerous to use them as a basis to justify legal rulings. Instead, people who disagree on these broad principles can achieve "incompletely theorized agreement" on specific rules and the resolution of particular conflicts. This view of legal practice suggests that it does not and should not bring to the fore the fundamental differences in basic assumptions that distinguish conservative and liberal political thought. In turn, this neutrality regarding ultimate principles should lead to a more homogeneous form of discourse when justifying decisions.

The puzzle is that science is also an institution whose legitimacy depends in part on maintaining a standard of objectivity unbiased by politics (Cash et al. 2003). Why are we able to detect a clear conservative versus liberal bent in the peer-reviewed economics literature (even after controlling for topic) and not in legal opinions? First, it is harder for social science to live up to a standard of political neutrality, especially if social scientists attempt to deliver research that is timely and relevant to address current issues. Second, economics may be a special case, as it is more tightly engaged with policy (and decisions with distributional impacts on people) and more directly addresses questions such as the proper reach of markets versus governments that are at the core of ideological debates. Indeed, a recent study finds that the American public's trust in economist experts is tepid (Johnston and Ballard 2016). It is also possible that social scientists have not developed norms of neutrality that are as strong as judges'. This opens up an interesting set of questions concerning the relationship between social scientists' use of language, i.e., their ability to engage in public debate, promote new ideas, and take stances, and the overall legitimacy of social sciences as professions, especially in contrast to that of party leaders and courts.

Conclusion

This paper applied natural language modeling techniques to determine whether judicial opinions bear the imprint of partisanship. In order to start from a known benchmark, we used methods that had already been used in several other studies of elite political language. We find that these methods do not identify a strong partisan ideological influence on judicial opinions in one area of the law (environmental regulation), unlike in other spheres of policy discourse (Congress, the media, academic economic writing). The analysis therefore rejects the joint hypothesis that existing language models can detect ideology in judicial language and that partisan affiliation is a salient divide in the way judges justify decisions.

Rejection of the working hypothesis raises a number of questions. First, what is the relationship between legal language and the ability of courts to perform their functions in society? According to some theories of law, the judiciary's authority rests on its ability to maintain characteristics of the rule of law. The rule of law can emerge if citizens perceive court decisions to be relatively stable and, on average, congruent with the values they hold (Hadfield and Weingast 2012). In *The Will of the People*, Barry Friedman[7] retraces the history of constitutional law and argues that the Supreme Court has maintained a tight dialogue with the public, each leading the other in turn, the court never diverging too far without a backlash. Furthermore, the authority of the court is greater if the public sees it as a useful tool to patrol the government and protect democracy from its excesses. For this reason, we can expect legal opinions to not follow the partisan polarization of the elite. In the context of this theory, the technical language of law would not be a way of hiding underlying ideological differences but rather part and parcel of the system of norms that keeps the judiciary cohesive and able to fulfill its role in the constitutional system.

Second, both courts and academic institutions are meant to aggregate information and search for objective judgments (or at least aspire to this standard). Why is the linguistic signature

---

[7]Friedman (2009)

of partisan affiliation so much stronger in the economics academic literature than in legal opinions? One possibility is that social scientists engage in the market of ideas and attempt to build coherent systems of social thought. In doing so, they build differentiated schools of thought, which then influence elite and political discourse more generally (Jensen et al. 2012). In contrast, common law judges resolve one conflict at a time, without aspiring to develop full theories (Sunstein 1996). More humble in their aims, they may achieve more cohesion and arrive closer to a standard of neutrality than social scientists can (or should) achieve. ❧

৩৭

OPINION CLARITY IN STATE AND FEDERAL TRIAL COURTS

Adam Feldman, SCOTUSBlog

The Importance of Clarity

In the Anglo-American tradition, courts and the judicial interpretation of the law have a central place, with judicial opinions and the case law that is created by those opinions forming a core component of the legal system. As a *genre* of lawmaking (Livermore, Riddell, and Rockmore 2017), judicial opinions have many defining features: they take a narrative structure, often recounting in detail the circumstances that led to a dispute; and they focus on interpreting preexisting legal norms and then applying those norms to adjudicate that particular, concrete dispute. Once an opinion has been published, it takes on precedential status, either binding on other courts within the same jurisdiction, or at the very least as a persuasive authority that may be relied on in subsequent cases. Beyond the directly binding character of judicial opinions, private parties and their legal counsel look to the opinions issued by courts for guidance on the content of the law.

As with any other legal text, judicial opinions range in the clarity with which they communicate their legal content. Just as statutes or regulations may involve vague language, confusing definitions, or a Byzantine structure of cross-references and interlocking requirements, judicial opinions can be opaque, confusing, meandering, or incoherent and contradictory—characteristics that make it difficult for those opinions to serve their public purpose of communicating the content of the law to the public and future courts. Although commentators disagree on the extent to which judicial opinions should be judged on their literary character (Schauer 1995), the desirability of clarity in judicial writing is broadly recognized.

For both normative and professional reasons, judges have many reasons to care about the clarity of their writing (Baum 2009, 109). From a normative perspective, judges who care about fulfilling their duty to faithfully announce the law and provide useful guidance to future litigants and courts will, to the best of their abilities, endeavor to write clearly. From the perspective of professional incentives, judges have many reasons to communicate clearly (although their goals may also, sometimes, be served by writing in ways that obscure rather than clarify).

Most generally, clarity in writing can help promote the policies that are implicated by a judicial decision. For example, a statute may include a vague definition that could be interpreted in ways that are either more or less protective of some social interest, say, antidiscrimination norms. In making a decision in a particular case, some judges may be inclined to read the statute in ways that promote that social interest. For those judges, maximizing the policy goal would involve issuing a judicial opinion that announces a very expansive reading of the statutory language in question. The more clearly that opinion is written, the easier it will be for private parties to understand the import of the decision and the more difficult it will be for future courts to retreat to a less protective reading.

In a judicial hierarchy, judges whose decisions face review may be well served by more clearly written opinions if clarity increases the chances of convincing the reviewing court of the merits of a decision. Appellate courts concern themselves with courts of final review, i.e., the federal or state supreme courts. These final arbiters are ultimately reviewed by future high courts, which can overturn prior decisions, as well as legislatures that can overturn statutory interpretations through legislation. Clearly setting out the grounds for a decision may help protect even high courts from subsequent reversal.

Judges also have institutional reasons to promote clear writing. According to Judge Patricia Wald, "Modern judges write opinions mainly for two reasons,"

> [F]irst, to reinforce our oft-challenged and arguably shaky authority to tell others—including our duly elected political leaders—what to do. . . . One of the few ways [unelected judges] have to justify our power to decide matters important to our fellow citizens is to explain why we decide as we do.
>
> [S]econd. . . is to demonstrate our recognition that under a government of laws, ordinary people have a right to expect that the law will apply to all citizens alike. . . . Even to approach a goal of consistency, litigants, lawyers, reviewing judges, the press, and ordinary citizens need to know why a particular judge came to a particular decision in a particular set of circumstances. (Wald 1995, 1372)

Both the explanatory (or reason-giving) function and the consistency function of opinions are promoted by clear writing. There is little legitimizing value to an explanation that cannot be readily understood. Indeed, unclear writing may undermine judicial legitimacy by creating the impression that a court is shirking its obligation to explain the reasons for its decisions. Likewise, the goal of consistency is undermined by unclear writing when subsequent courts cannot easily extract the underlying principles or considerations that motivated an earlier decision.

Although judges collectively may have institutional reasons to promote written clarity, any individual judge may be tempted to free-ride on the efforts of others, if clear writing is difficult or otherwise costly. There are, however, mechanisms to enforce group norms concerning writing quality. The simplest is based on the natural human desire for respect: judges who are attuned to their colleagues' thoughts and perceptions of their writings will be inclined to tailor their opinions to meet their colleagues' expectations. More concretely, future prospects of promotion may depend on favorable evaluations of their opinions from other judges and institutional higher-ups (Posner 2004, 1265–73). To the extent that written clarity is a factor in these evaluations, promotion-oriented judges will be inclined to invest in more clearly written opinions. In jurisdictions where judges are selected through elections, the public may also evaluate judges based on the clarity of their writing.

It is worth noting that the incentives faced by judges may not always point in the direction of greater clarity. Several commentators have emphasized the fact that judges frequently write for a variety of audiences (Baum 2009; Mikva 1987; Posner 2010). Judge Posner provides the example of how an opinion dense with citations may be intended to inspire the "confidence" of a professional audience but may have the unintended consequence of leaving the law unclear (Posner 1995, 1442-43). Texts that are steeped in legalese and jargon may improve the standing of an opinion with one audience while leaving another confused and concerned. In addition, in controversial cases, judges may obscure the grounds of a decision with the goal of avoiding criticism.

Finally, the clarity of judicial writings also has broader consequences for equality and the rule of law. In particular, unclear legal dispositions may disproportionately affect certain sectors of the population (Yamamoto 1990). Those with higher socioeconomic status, who can parse complex legal language by virtue of educational advantages or (more to the point) who can hire legal professionals, are less exposed to risks from unclear legal texts. For those with less education and who lack access to lawyers, unclear legal writings leave them alienated from the system of laws that is meant to structure their behavior and that can be used to justify serious sanctions for noncompliance. In this way, unclear legal writing is more than simply an inconvenience or a problem for courts but generates unjust costs and undermines the fairness and legitimacy of the law.

Analyzing Clarity

Traditionally, judicial writing style has been largely discussed in a qualitative fashion, both by commentators and judges themselves. Judge Posner (Posner 1995, 1429–30) is among the most well-known arbiters of judicial writing style. Posner separates judicial writing into the "pure" and "impure" styles and favors the latter over the former. The pure style is the conventional approach that is associated with legalese, convoluted structure, and "technical legal terms without translation into everyday English." The impure style, on the other hand, addresses a "hypothetical audience of laypersons" and attempts to explain "why the case is being decided in the way that it is." This dichotomous understanding of judicial writing provides a useful starting point for examining judicial style but fails to concretely define features that could be used in replicable empirical analysis, instead leaving much to the subjective judgment of the reader.

Qualitative analyses often proceed through the use of exemplars. In this vein, Richard Wydick (1994, 5) cites the clarity of Cardozo's writings, describing characteristics such as an "economy of words," "no archaic phrases," "no misty abstractions," verbs in the "active voice," and "no ambiguities to leave us wondering who did what to whom." These descriptors all relate to the concept of clear writing and can be used as a starting place to begin to establish the features that relate to this concept with more specificity.

While often interesting, qualitative accounts suffer from a lack of standard parameters of evaluation. With advances in natural language processing and computational text analysis, it is possible to quantitatively measure at least certain aspects of legal writing. Although there is no generally agreed upon gold standard for judicial writing, certain conventional measures of writing quality exist that are widely accepted as important. "Clarity" is among the aspects of style that are, at least arguably, well suited to quantitative measurement. Even if clarity is not easily measured itself, there are well-accepted proxy measures that are widely understood to track the clarity of a written text.

Recent scholarship has begun to apply these quantitative tools to judicial writing. Much of this work focuses on Supreme Court briefs and opinions. Carlson, Livermore, and Rockmore (2016) used function word frequencies and sentiment analysis as measures of the writing style of justices on the Supreme Court and estimated several temporal trends, finding, for example, that sentiment has become more negative over time, that there is a "style of the time" that characterizes the justices' writings, and that intrajustice variation increased as the justices took on a greater number of clerks over time. Corley and

Wedeking (2014) used LIWC's dictionary-based software to find that, as the court uses language with greater certainty, lower courts are more likely to treat their decisions favorably. In recent works, Black et al. (2016b) and Owens and Wedeking (2011) examine how Supreme Court justices vary their writing clarity depending on their intended audience(s). Other works expand their loci beyond the United States Supreme Court to apply similar tools to forums such as state supreme courts (Goelzhauser and Cann 2014).

Several works also examine how responses to Supreme Court opinions are affected by writing clarity. Hansford and Coe (2019) conduct a survey experiment that manipulates the complexity of language in Supreme Court decisions to measure the effects of clarity on the political support for the decision. These authors find a complicated set of effects, with complex language decreasing acceptance of a decision but legalistic terminology increasing acceptance. Black et al. (2016a) build an estimate of clarity based on nineteen quantitative readability measures that are used as the basis for a principal component analysis. Using the top principal component (which accounts for 77% of variability), the authors compare the clarity of opinions to various measures of public acceptance, finding that when the court issues opinions that are contrary to public opinion, it tends to write more clearly, presumably in an effort to increase acceptance of its decisions.

Feldman (2016) used dictionary-based software to assess the writing quality of Supreme Court briefs based on features such as wordiness, passivity, sentence complexity, and sentiment. That analysis found that Supreme Court briefs with higher quality writing were more likely to affect opinion content. Spencer and Feldman (2018) recently developed a metric akin to the one used by Black et al. (2016a) based on a factor analysis of readability measures that showed that motions for summary judgment with clearer writing were more likely to succeed in both state and federal trial courts. Prior to this, Long and Christensen (2011) used the Flesch Reading Ease scale and Flesch–Kincaid Grade-Level Scale (discussed in more detail below) to examine whether the readability of Supreme Court briefs affected litigants' likelihood of success, failing to find a meaningful relationship between more readable briefs and successful outcomes.

This chapter builds on prior work on writing quality in the courts as well as earlier comparative analyses of state and federal courts, which include Eisenberg et al. (1995)—comparing award sizes and trial times in federal and state jury trials— and Galanter (2004)—comparing the frequency of trials in federal and state courts. The following section focuses on data, methodology, and basic hypotheses. With these in view, the chapter describes the implementation of various metrics to compare writing clarity in the opinions from state and federal courts. The dataset used is a stratified random sample of over four thousand state and federal court opinions. This exploratory analysis provides a basic comparison of opinions in both forums that can serve as a starting place to analyze distinctions in opinion clarity. The chapter concludes by discussing how this work can be applied in future studies interested in quantitatively comparing judicial opinions and other types of legal writing.

Methodology and Analysis

The data for this analysis consist of a sample of over four thousand federal and state court opinions. All opinions were obtained through Westlaw. To control for the issue content of decisions, the sample was restricted to product-liability decisions. The dates of the opinions range from 2000 to 2015. The search parameters were first set to include only state trial court cases and then only federal district court cases. Opinions were saved as plain-text files. The Quanteda package in R Markdown[1] was then used to generate readability metrics for each opinion.

Clarity Scores

Quanteda in R provides multiple text statistics metrics including several different readability scores. The readability scores that will be used as the basis for the following analysis are the Flesh–Kincaid Grade Level, the Gunning Fog score, the Coleman Liau Index, the Smog Index, the Automated Readability Index, and Spache scores. These algorithms differ in how they use and weigh certain linguistic factors, such as word and sentence count and length, and so they measure clarity and readability from slightly different angles. Nevertheless, although each of the underlying readability measures was designed for a different purpose and application, when applied to naturally produced texts, they tend to move in the same direction, as the complexity of documents tends to increase along several dimensions. Stated another way, documents with many polysyllabic words also tend to include longer sentences.

Flesch–Kincaid scores were developed for use by the US Navy in the mid-1970s (Kincaid et al. 1975). They have since been used in applications ranging from evaluating the ease of reading naval training manuals (Kincaid, Aagard, and O'Hara 1980) to hospitals' patient information leaflets (Williamson and Martin

---

[1] https://quanteda.io/

2010). The algorithm for Flesch–Kincaid Grade Level is:

$$11.8\frac{\text{syllables}}{\text{words}} + 0.39\frac{\text{words}}{\text{sentences}} - 15.59.$$

The *Automated Readability Index (ARI)* was also created to gauge the readability of training manuals (in the case of ARI these were US Air Force manuals) and has been employed within a variety of industries (Senter and Smith 1967). The ARI has also been used to measure the readability of online product reviews (Korfiatis, Garcia-Bariocanal, and Sanchez-Alonso 2012). The algorithm for the ARI is:

$$4.71\frac{\text{characters}}{\text{words}} + 0.5\frac{\text{words}}{\text{sentences}} - 21.43.$$

The *Simple Measure of Gobbledygook (SMOG) Index*, designed for applications in the field of psychology, estimates the age necessary to read a given passage of writing (McLaughlin 1969). It has been used in a diverse set of applications, including measuring the readability of job analysis questionnaires (Ash and Edgell 1975). The SMOG algorithm is:

$$1.043\sqrt{\text{polysyllables} \times \frac{30}{\text{total sentences}}} + 3.1291.$$

The *Gunning Fog Index* was created in order to estimate the appropriate grade level necessary to read given writings. Its applications have ranged from measuring the readability of newspaper excerpts to business materials (Gunning 1969). The Gunning Fog Index categorizes words with three or more syllables as complex. The algorithm for the Gunning Fog Index is:

$$100\frac{\text{complex words}}{\text{words}} + 0.4\frac{\text{words}}{\text{sentences}}.$$

The *Coleman Liau Index* was one of the first readability measures designed for use on a computer and focuses on characters in a word rather than syllables. It also is based on reading grade level and was designed to help the United States Department of Education tailor textbooks to appropriate reading levels for a given grade (Coleman and Liau 1975). The algorithm for the Coleman Liau Index is based on the average number of letters per one thousand words (ANL) and average number of sentences per one hundred words (ANS):

$$0.058ANL - 0.296ANS - 15.8.$$

Finally, the *Spache Index* was created to gauge the age appropriateness of school reading materials (Spache 1953). Spache scores have an associated set of words that are applied in the algorithms. These include over nine hundred common words, such as "about," "child," and "made." Difficult words are coded as those that are not covered by the list of common words. The algorithm for the Spache Index is based on a sample of text one hundred to 150 words in length:

$$0.086 \frac{words}{difficult\ words} + 0.141 \frac{words}{sentences} + 0.83.$$

This chapter uses these six readability measures to create a single index variable for opinion clarity. As is readily apparent, the clarity metrics track each other in fairly obvious ways—for example, the number of words per sentence appears several times, as do various measures of the sophistication of the words in a text. Factor analysis is a statistical technique to reduce the number of dimensions in data with multiple (possibly overlapping) features. In broad strokes, factor analysis constructs new dimensions that best account for the variability within a dataset, allowing for compact descriptions of observations in lower-dimensional space that nonetheless capture relevant differences (Cudeck 2000; Brown 2014; Fabrigar and Wegener 2011, 268). Factor analysis thus provides information about the factors that underpin a set of correlated measures. The estimates of influence of the various measures calculated through factor analysis are known as *factor loadings*.

Factor analysis may lead to the generation of multiple factors if the scores cannot be properly captured with just one factor. This generally is the case when individual variables are particularly unique from one another and so the correlation between them is low. While factor analysis may yield multiple factors if the underlying values present more than one dimension of results, factor analysis in this instance generated only a single factor—this result is not surprising given the facial similarity between the measures. The following section provides a check on the validity of this procedure by applying this method to several nonjudicial and publicly accessible texts.

External Validity Check

To first assess the validity of these readability measures and their combination through factor analysis, this procedure was applied to several nonlegal texts. Three sets of texts were used for this assessment. The first is a collection of three children's stories.[2] The second set is comprised of three of the Federalist Papers, a famous collection of essays written (individually) by Alexander Hamilton, James Madison, and John Jay, and published anonymously in 1787–1788 with the goal of convincing residents of the newly formed State of New York to ratify the US Constitution. The Federalist Papers are of particular interest in the history of quantitative style determination, as their authorship was largely determined through one of the most important "stylometric studies" ever

---

[2] The three stories, *The Three Little Pigs*, *Jack and the Beanstalk*, and *Little Red Riding Hood*, can be found at https://americanliterature.com/.

accomplished by statisticians (Mosteller and Wallace 1963). The third set includes introductory chapters from machine learning and organic chemistry textbooks (Shalev-Shwartz and Ben-David 2014; Neuman Jr. 1999). Intuitively, the children's books should be expected to be most clearly written, with the textbooks and the Federalist Papers being more complex.

Isolating the text from these eight documents and applying factor analysis led to the creation of six factors. The first factor was the only significant one, accounting for over 96% of the variance between the underlying variables. The factor loadings for the six variables show that they are all strongly associated with the latent variable. The loadings for five of the variables was over 0.99, and the sixth measure, Coleman Liau, had a loading of 0.95. The underlying variables' uniqueness measures provide more evidence that the variables measure a similar characteristic of writing clarity: the variables' uniqueness scores varied from 0.001 to 0.06. Finally, the scoring coefficients for the six variables (which is a measure of the influence of the underlying variable on the latent measure) further confirm their mutual information for this dataset: the scoring coefficients for the six variables was within the range of 0.164 to 0.172.

The clarity scores derived through factor analysis ran from −1.4 to 1.06 with lower scores equating to clearer writing. They lined up in the expected manner. The three documents that scored easiest to read were the children's stories. The scores for these stories were −1.4, −1.2, and −0.76. Next were the textbook chapters. Their scores were 0.24 and 0.29. Finally, the scores for the Federalist Papers were the highest at 0.77, 1.04, and 1.06.

The relative reading ease of these texts is also apparent when examining them qualitatively. The children's stories tend to have short sentences with simple vocabulary. One example from *The*

Three Little Pigs reads,

> The first little pig was very lazy. He didn't want to work at all and he built his house out of straw. The second little pig worked a little bit harder but he was somewhat lazy too and he built his house out of sticks.

The sentences in the textbooks are also relatively short and to the point. The machine learning text opens, for example,

> The subject of this book is automated learning, or, as we will more often call it, Machine Learning (ML). That is, we wish to program computers so that they can "learn" from input available to them. Roughly speaking, learning is the process of converting experience into expertise or knowledge.

There is no excessive jargon in these introductory chapters and they are written for a nontechnical audience.

The sections from the Federalist Papers are written in the most complex manner of the three sets of texts and are rated as the least clear. Here is a snippet of text from Federalist Paper 1 as an example:

> And yet, however just these sentiments will be allowed to be, we have already sufficient indications that it will happen in this as in all former cases of great national discussion. A torrent of angry and malignant passions will be let loose. To judge from the conduct of the opposite parties, we shall be led to conclude that they will mutually hope to evince the justness of their opinions, and to increase the number of their converts by the loudness of their declamations and the bitterness of their invectives.

The text has multiple clauses and complicated language, all of which contribute to its measured level of clarity.

Clarity of Judicial Opinions

What do legal writings at both extremes of clarity (high and low) look like? Here is an example of a sentence from an opinion in the dataset that scored high (i.e., lower in clarity) according to this index:

> The aforementioned Defendants moved the Court to enter an order precluding Plaintiff's counsel from referring to, mentioning, engaging in commentary and/or otherwise eliciting testimony, whether direct or indirect, concerning the amount of profits or royalties derived from any of the operations that occurred on the Defendant's property and the amount of consideration paid by the Defendants for the properties at issue.[3]

On average, sentences from opinions with higher scores are lengthier, with more clauses and longer words, while sentences from opinions with lower scores are more direct, less loquacious, and more to the point.

The following excerpt provides several sentences from a written opinion with one of the highest clarity scores:

> 1. The accident, which forms the basis of Plaintiff's Amended Complaint, occurred on December 12, 1995, at the construction site of Solar Sources Inc.'s (hereinafter 'Solar') coal preparation plant in Cannelburg, Indiana.
>
> 2. At the time of the accident, the construction of the coal preparation plant was approximately halfway completed. (Deposition of Steven L. Vaughn taken on February 26, 1999, 191, hereinafter 'Vaughn Dep.').

~421~

---

[3] *In re Flood Litigation*, 2006 WL 6460516 (W.Va. Cir. 2006).

3. Plaintiff was employed as a pipe-fitter by Trimble Engineers and Constructors.[4]

This passage lays out the case's fact pattern with short, declarative statements. The sentences are easy to follow, and they flow from one to the next. This is in sharp contrast to the excerpt from the previous opinion that scored higher according to the clarity measure.

To derive clarity scores, a similar factor analysis was performed on the opinions in the dataset based on the six underlying variables for writing clarity. The overlap between the six variables and how well they were captured by the dominant factor was much less pronounced for the opinion data than in the validity check. Again, there was one significant factor, but it accounted for only 74% of the variance between the underlying variables. Two of the underlying variables, Coleman Liau and Gunning Fog, have negative factor loadings of −0.86 and −0.65, respectively. The other factors have positive loadings: 0.75 for SMOG, 0.93 for ARI, and 0.96 for both Flesch-Kincaid and Spache scores. These loadings lead the latent variable to take on more disparate scoring coefficients for the six measures. The scoring coefficients in this instance are −0.19 for Coleman Liau, −0.15 for Gunning Fog, 0.17 for SMOG, 0.21 for ARI, 0.22 for Flesch-Kincaid, and 0.22 for Spache.

The values for the clarity scores were derived from the first factor, with higher scores associated with texts that are less clear. Table 15.1 provides some basic descriptive statistics. Clarity scores ranged from −1.40 to 15.0, with all but three scores under 5.79, again with lower scores equating to more readable texts. From these data, there are readily apparent differences between the statistics for state and federal opinions. The state scores, for

---

[4] *Vaughn v. Daniels Company*, 2001 WL 36039333 (Ind. Cir. 2001).

instance, tend to fall below those for the mixed federal and state (overall) values, while the federal scores are generally higher. The spread of the data is slightly greater in the federal sample with a standard deviation more than twice as large. The two histograms in figure 15.1 show the distribution of values for the two sets.

Table 15.1. Summary statistics for clarity scores.

Federal		State		Overall	
N	2100	N	1946	N	4046
25th percentile	0.49	25th percentile	-1.03	25th percentile	-0.92
75th percentile	0.95	75th percentile	-0.81	75th percentile	0.69
99th percentile	2.81	99th percentile	-0.29	99th percentile	2.45
Mean	0.81	Mean	-0.88	Mean	0.00
Median	0.68	Median	-0.93	Median	0.21
Std. dev.	0.69	Std. dev.	0.30	Std. dev.	1.00

The clarity scores for the state opinions are clearly lower than those of the federal courts. This suggests that federal court opinions in the sample tend to be less clear than state court opinions. The federal court's scores also have more variation.

Additional analyses confirm that the differences between the state and federal distributions are significant. A two-sample Kolmogorov–Smirnov test for the equality of distribution showed that state court opinion clarity scores tended to be smaller than those for federal courts at $p < 0.01$.

Word Counts

The clarity variables discussed so far focus on what might be referred to as "microclarity," in that they apply to documents of varying lengths. An additional feature that can contribute to the complexity of a document is its length—other things being equal, longer documents are more difficult and time consuming to read and understand, and may contain extraneous information

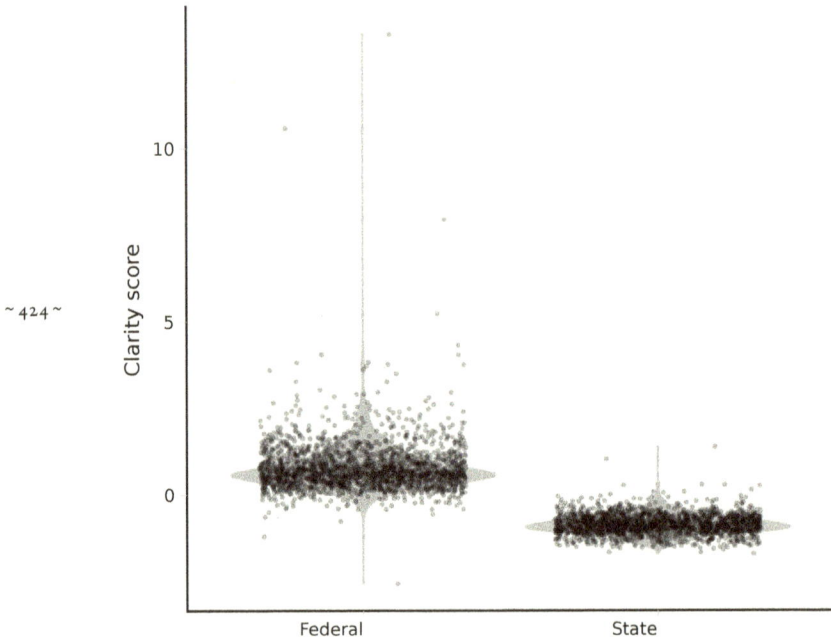

Figure 15.1. Federal and state clarity score histograms. The violin distribution plot represents a mirrored density. Data points randomly "jittered" on the x-axis.

that distracts from the core message, a phenomenon related to the notion of "information overload" that has been discussed in the context of mandatory disclosure requirements (Persson 2018). Overly lengthy judicial opinions have been a point of concern of jurists and commentators for decades (McComb 1949). The virtues of shorter opinions are discussed by Lebovits (2004, 64): "Brief opinions hold the reader's attention, allow readers to move on to other things, and distill the opinion's essence."

In addition to their lower clarity, federal court opinions are also longer. A density plot in figure 15.2 contrasts the length of opinions in the state and federal courts, and the difference between

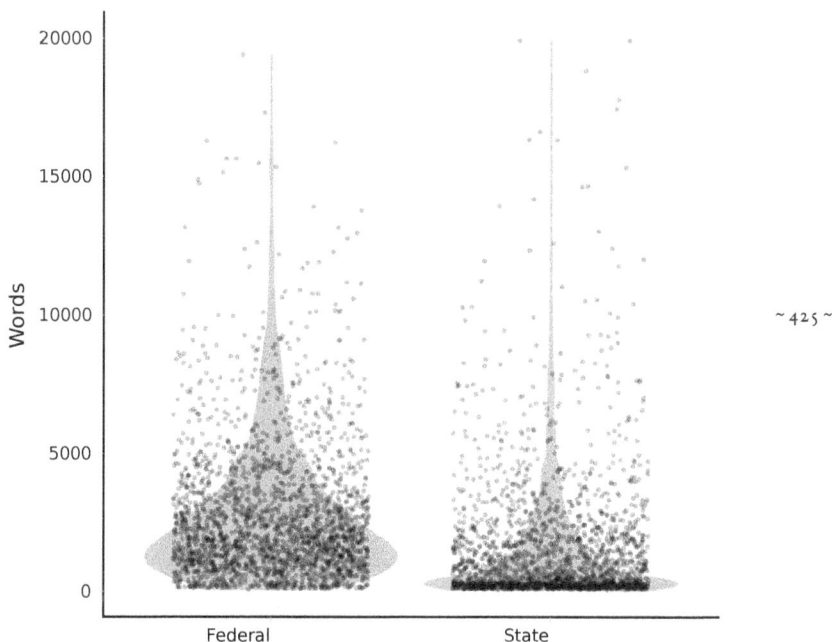

Figure 15.2. Kernel density plot of federal and state court opinion length. The violin distribution plot represents a mirrored density. Data points randomly "jittered" on the x-axis. *Note*: Twenty-five opinions over twenty thousand words removed from plot.

the two is visually apparent. To further verify this graphical finding on opinion length, a two-sample Kolmogorov–Smirnov test for whether state opinion lengths were shorter than those for federal courts was run as a robustness check, showing that the difference was significant at $p < 0.01$. The median opinion length for the entire set of federal and state court opinions is $1,532$ words. The median length for state opinions is 790 words, while the median length for federal court opinions is 2119.5 words.

Discussion

The preceding analysis provides a quantitative description of an important phenomenon that raises a range of empirical questions concerning the *causes* and *consequences* of differences in written clarity and opinion length in the state and federal courts (Livermore 2018).

In terms of causes, there are a number of potential explanations. One class of explanations concerns differences in the underlying cases. Although the sample for the preceding analysis was restricted to opinions that pertain to product liability, there are, nonetheless, many potential sources of variation in case complexity between the federal and the state level. Perhaps most obviously, federal product-liability cases are predicated on underlying state law claims and are in federal court based on diversity or federal question jurisdiction. This procedural posture creates several avenues for potential differences. Most obviously, there is at least one additional legal issue present in the federal cases: the question of jurisdiction itself is one that can be (and often is) litigated. Opinions that deal with a larger number of legal subjects are naturally longer. These types of procedural issues may also involve more complicated legal issues that are naturally associated with writing that performs poorly on clarity measures. To test for this possibility, language from a small sample of federal and state opinions identified as not relating to jurisdictional questions was extracted and compared. While the median clarity score for state opinions was still lower, the difference in median values was considerably less than it was for the entire sample.

In addition to the added question of diversity jurisdiction, cases that are removed to federal court may share other characteristics. A motion to remove a case to federal court may be a tactic that is more likely to be deployed in cases involving high stakes. Cases that involve certain federal questions may be inherently more complex—for example, if the civil provisions of the Racketeer Influenced and Corrupt Organizations Act are invoked. The type of defendants that find themselves in federal court may also have certain shared characteristics; for example, they may, in general, be able to able to afford more sophisticated lawyers. Any of these unobserved differences between the cases heard by the federal and state courts may account for differences in the clarity and length of opinions.

A second class of explanations involves differences in judicial characteristics. All federal judges share the same process of nomination by the president and confirmation by the Senate, and hold life-tenured positions. State judges, by contrast, are selected through a range of different appointment or electoral processes and with many different employment conditions. There are differences in the workloads of federal and state judges, the types of cases they are likely to see, the norms concerning collegiality they follow, and the prestige they enjoy. There may also be a variety of differences in their backgrounds, including their prior work experience, their education, and their connection to party politics. Any of these differences in the characteristics of judges or the courts in which they work might affect how they write.

A final category of potential explanations involves the law and law-related practices that influence the opinions that are published. State and federal law are different, and even when federal courts apply state law in diversity cases, several federal doctrines still govern those contexts. Differences in the law being applied to a given case could be expected to be reflected in the text of judicial opinions and may affect writing clarity or opinion length. In addition, the law can affect the types of cases that lead to published opinions—for example, by encouraging (or discouraging) settlement. Even when a court reaches a final resolution, not every disposition results in a formally published opinion, at least in the federal courts. Indeed, in the data for this analysis, many of the state court opinions appear shorter because they were summarily decided in a few paragraphs— it may be that similar federal dispositions are simply not published and therefore are not reflected in the data.

The consequences of differences in the clarity and length of opinions in the federal and state courts may turn, in part, on their underlying causes. For example, if the differences arise because of different publication practices, it may be that the corpus of federal opinions is *easier* to navigate, even if the opinions are less clearly written and longer, because there is less extraneous clutter interfering with the ability of legal researchers to identify relevant law. On the other hand, if the educational background of federal judges results in opinions filled with with legalese and jargon, then the law will be unnecessarily inaccessible and obscure.

Conclusion

This chapter presents one of the first attempts to quantitatively compare writing clarity in federal and state court trial court opinions. The writings of both federal and state court judges must be read and understood by multiple audiences, including lawyers, other judges, academics, and the lay public. Well-crafted opinions explain the reasoning behind a decision and, in that way, legitimate the exercise of authority by the judiciary. Opinions also provide a means of coordinating across judges by stating the legal principles and rules that undergird decisions in particular cases. A lack of clarity interferes with both of these functions. Although much prior work has focused on opinion writing at the appellate level, the opinions of trial court judges are often important: they explain outcomes to the parties of a case, coordinate the actions of external institutions such as enforcement bodies or agencies, inform the appeals process, and frequently serve as the final word on many legal points.

The initial steps taken in this chapter are descriptive in nature, and the preceding analysis is not intended to examine the relevant causal processes. The value of this analysis is in calling attention to a potentially important phenomenon and developing methods for subjecting that phenomenon to quantitative measurement, which is a necessary precursor to additional empirical study. The basic finding—that federal court opinions are longer and less clearly written than state court opinions—has a range of potential social consequences and is worthy of future research.

~429~

There are many possible explanations for differences in the opinions of federal and state judges, having to do with the characteristics of the cases being heard, the attributes of the judges, or the law. Although it might be tempting to blame differences in writing style on the judges authoring the opinions, there may be a variety of other causal influences. At the very least, it is quite possible that the added complexity of federal cases, especially where jurisdictional questions are in play, is a relevant factor. Publication decisions and, in particular, the greater inclination for state courts to publish short summary orders, may also account for some of the difference.

There are many possible avenues for future work to build on the preliminary analysis provided here. Off-the-shelf approaches like those used in this chapter provide a first step in the development and application of quantitative methods for the study of clarity of legal writing, but future developments in computational text analysis are likely to generate even more accurate techniques for gauging the clarity of writing in general, as well as legal writing specifically. These techniques can continue to supplement and be supplemented by increased causal understanding of the processes that affect the clarity of legal documents as well as the ground-truth qualitative assessments made by lawyers, academics, and the public concerning their ability to interact fluently with the law as written. ✒

Acknowledgments

I would like to thank Michael Livermore, Dan Rockmore, Shaun Spencer, Elliott Ash, and the various participants of the Online Workshop on the Computational Study of Law (OWCAL) for all of their helpful comments and feedback.

CONJECTURE

MACHINE LEARNING AND THE RULE OF LAW

Daniel L. Chen, Toulouse School of Economics

Predictive judicial analytics holds the promise of increasing the fairness of law. Much empirical work observes inconsistencies in judicial behavior. By predicting judicial decisions—with more or less accuracy depending on judicial attributes or case characteristics—machine learning offers an approach to detect when judges are most likely to allow extralegal biases to influence their decision-making. In particular, low predictive accuracy may identify cases of judicial "indifference," where case characteristics (interacting with judicial attributes) do not strongly dispose a judge in favor of one or another outcome. In such cases, biases may hold greater sway, implicating the fairness of the legal system.

Arbitrary Decision-Making

There is ample social scientific evidence documenting arbitrariness, unfairness, and discrimination in the US legal system. To give just a flavor of the relevant research:

- United States federal appeals court judges become more politicized before elections and more unified during war (Berdejo and Chen 2016; Chen 2016b).

- Refugee asylum judges are two percentage points more likely to deny asylum to refugees if their previous decision granted asylum (Chen, Moskowitz, and Shue 2016).

- Politics and race also appear to influence judicial outcomes (Schanzenbach 2005; Bushway and Piehl 2001; Mustard 2001; Steffensmeier and Demuth 2000; Albonetti 1997; Thomson and Zingraff 1981; Abrams, Bertrand, and Mullainathan 2012; Boyd, Epstein, and Martin 2010; Shayo and Zussman 2011), as does masculinity (Chen, Halberstam, and Yu 2016b, 2016a), birthdays (Chen and Philippe 2018), football game outcomes (Chen 2017; Eren and Mocan 2016), time of day (Chen and Eagel 2016; Danziger, Levav, and Avnaim-Pesso 2011b), weather (Barry et al. 2016), name (Chen 2016a), and shared biographies (Chen et al., 2019 (forthcoming)) or dialects (Chen and Yu 2016).

- There are also various papers showing clear judicial biases in laboratory environments, such as the influence of anchoring, framing, hindsight bias, egocentric bias, snap judgments, representative heuristics, and inattention (Guthrie, Rachlinski, and Wistrich 2000, 2007; Rachlinski et al. 2009; Rachlinski, Wistrich, and Guthrie 2013; Simon 2012).

Thus, the primary question is not whether these problematic features of the legal system exist. Rather, the dilemma facing policymakers is what, if anything, can be done. This chapter will argue that predictive judicial analytics in the form of applied statistical/machine learning (from causal inference to deep learning) holds at least some promise on this front.

Prior empirical work has focused on evaluating judges to observe the influences on their behavior, helping to diagnose the problem of bias but offering little in terms of remedy. The advent of machine learning tools and their integration with legal data offers a mechanism to detect in real time and thereby remedy judicial behavior that undermines the rule of law. This commentary presents a conceptual framework for understanding a large set of behavioral findings on judicial decision-making and then taking steps to ensure more fair treatment of legal subjects by the legal system.

The theoretical basis for the following argument is the observation that behavioral biases are most likely to manifest in situations where judges are closer to indifference between options. Such contexts are also those where there are likely to be the highest levels of disparities in interjudge accuracy of algorithms predicting judicial decisions—essentially conditions where judges are unmoved by legally relevant factors. If algorithms can identify the contexts that are likely to give rise to bias, they can also reduce those biases through behavioral nudges and other mechanisms, such as through judicial education. The following discussion fleshes out these claims.

The Problem of Indifference

Imagine a legal outcome of interest, such as asylum designations by immigration judges. Let's denote that outcome Y. Imagine further that there is some set of covariates (or "features" in the language of machine learning) X such that these features can be used to generate predictions of Y via some function $Y = f(X) + \epsilon$, where ϵ denotes some small "error" or variation. The covariates X are *legally relevant* inasmuch as prevailing legal norms require or at least permit their use by legal decision-makers for the relevant decision. In the case of an asylum adjudication, the political circumstances of an applicant's home country would be legally relevant.

There might also be a set of covariates W that are legally irrelevant and that *should not* predict a legal outcome: $y \perp W, var(\varepsilon) \perp W$. The set W might include litigant characteristics that decision-makers are not permitted to take into account—such as race—or they may include irrelevant features in the environment, such as the weather. Since judges are randomly assigned, judicial characteristics fall into W, because the judge that a litigant randomly draws is not legally relevant to the outcome of a decision. Of course, as mentioned briefly above, there is substantial literature showing that features in W are in fact predictive of legal outcomes in a variety of settings (Berdejo and Chen 2016; Chen 2016b; Chen, Moskowitz, and Shue 2016; Schanzenbach 2005; Bushway and Piehl 2001; Mustard 2001; Steffensmeier and Demuth 2000; Albonetti 1997; Thomson and Zingraff 1981; Abrams, Bertrand, and Mullainathan 2012; Boyd, Epstein, and Martin 2010; Shayo and Zussman 2011; Chen and Philippe 2018; Eren and Mocan 2016; Chen 2017; Chen and Eagel 2016; Danziger, Levav, and Avnaim-Pesso 2011b; Barry et al. 2016; Chen et al., 2019 (forthcoming); Chen and Yu 2016).

The preferences of decision-makers (e.g., judges) over X may also affect the influence of W over outcomes. A judge could be said to have *strong* preferences over X when it is costly to depart from the legally optimal outcome, defined as the outcome that would be generated through consideration of X alone. Judges might have such strong preferences based on ideology, personal psychology, or some set of institutional characteristics. But a judge may also have *weak* preferences over X, meaning that there was a relatively low cost in departing from the legally optimal outcome. In such cases of *legal indifference*, the factors within W can be expected to have greater influence. Stated another way, when the predictive power of X wanes, the potential scope of influence for W waxes.

A Role for Machine Learning

Chen et al. (2017) conceptualize the notion of early predictability. The basic idea is that machine learning could be used to automatically detect judicial indifference—i.e., instances where the judges appear to ignore the circumstances of the case when making decisions. This information could then be used to trigger debiasing information or other interventions to prevent decisions that would undermine the fair and nonarbitrary operation of the justice system.

How would this work? Continuing our example, let's consider asylum courts. In this important context it turns out that, using only the information available at the time that a case opens, judges with the highest and lowest grant rates are much more predictable than others (Chen et al. 2017). These judges seem to have strong prior preferences concerning outcomes, and the legally irrelevant fact that an applicant is assigned to a low– or high–grant rate judge largely determines outcomes.

There is, however, a category of less predictable judges, and these judges tend to have middling grant rates. Given their unpredictability, one possibility is that they lack strong preferences and are therefore guided by random factors when making a decision—essentially flipping a coin. Another possibility is that they are more sensitive to the circumstances of the cases. There is some evidence pointing to this second alternative: the less predictable judges also tend to have substantially more hearing sessions than the judges who rarely grant asylum.[1]

At this level of granularity—identifying judges whose behavior is predictable at relatively early procedural stages—some interventions might be possible. For example, training programs could be targeted toward these judges, either with the goal of debiasing or to help them learn how to use the hearing process to better advantage. Simply alerting judges

---

[1] Interestingly, judges who grant at a high rate also hold relatively more hearing sessions, perhaps to collect more information to justify their likely more controversial decision.

to the fact that their behavior is highly predictable in ways that may indicate unfairness may be sufficient to change their behavior.

Higher levels of granularity in the analysis may free up even more targeted interventions. It may be possible, for example, to not only identify early deciding judges, but also to examine how case characteristics interact with this judicial attribute to learn the types of case–judge pairs that are most predictable at early stages. When such pairs are found, judges can be given a "red flag" that they should be particularly attuned to subsequent information, essentially as a counterweight to confirmation bias or other nonlegal sources of influence.

Just as machine learning can be used to identify judges who tend to be unmoved by legally relevant factors, these techniques can also be used to detect instances where judicial decisions can be predicted by legally irrelevant factors. There is a substantial social science literature establishing this possibility and, in the asylum example, Chen and Eagel (2016) find these influences are highly prevalent. They include whether a hearing was before lunch or toward the end of the day, the size of the applicant's family, the weather, the number of recent grants by the court, whether genocide has been in the news, and the date of the decision. While the literature typically studies one behavioral feature at a time, Chen and Eagel (2016) demonstrate the possibility that machine learning could automate the detection of inconsistencies between judges due to legally irrelevant factors.

The asylum example is just one of many where machine learning techniques can be used to detect bias. Amaranto et al. (2017) use a very large dataset concerning prosecutorial decisions in New Orleans over a twelve-year period (430,000 charges and 145,000 defendants) to test for racial bias and the effectiveness of prosecutors in pursuing the riskiest defendants. The authors construct a predictive model of recidivism and then test whether prosecutors who are relatively strict (i.e., drop relatively few cases) screen based on risk of recidivism. In fact, more stringent prosecutors do not appear to target riskier defendants, and this phenomenon has differential effects across races, with less-risky African American defendants actually receiving relatively harsh treatment.

~439~

More generally, machine learning techniques can be used on data of any sort and, in the context of a legal decision, a wide range of data, from the weather conditions to judge characteristics, has proven informative. Given the textual nature of the law and the importance of argumentation and reason-giving to legal decision-making, there are substantial textual data that can be used to examine how legally relevant and legally irrelevant factors affect legal outcomes. For example, Ash and Chen (2017) use judges' writings to predict the average harshness and racial and sex disparities in sentencing decisions. That work finds that the information contained in written opinions can improve significantly on naïve predictions of punitiveness and disparity.

Again, this information could be used to aid decision-makers in ways that reduce bias in the system. Informing judges about the predictions made by a model decision-maker could help reduce judge-level variation and arbitrariness. Potential biases that have been identified in prior decisions or writing could be brought to a judge's attention, where they could be subjected to higher-order cognitive scrutiny. Such efforts would build on the already significant push to integrate risk assessment into the criminal justice process and help inform judges of the objective risks posed by defendants.

Judicial Education

An additional pathway for machine learning to improve the quality of legal decision-making is by informing, and to some extent comprising, efforts at judicial education. The first goal would be to expose judges to findings concerning the effects of legally relevant and legally irrelevant factors on decisions, with the goal of *general* rather than *specific* debiasing. For example, Pope, Price, and Wolfers (2013) found that awareness of racial bias among NBA referees subsequently reduced that bias. The second goal would be to educate legal decision-makers in the tools of data analysis so that they can become better consumers of this information when it is present during legal proceedings, and more generally to provide a set of thinking tools for understanding inference, prediction, and the conscious and unconscious factors that may influence their decision-making.

Efforts at judicial education have had considerable success in the past. By 1990, 40% of federal judges had attended an economics training program. This law and economics program was founded in 1976 as a two-week training course with lectures by Nobel Prize–winning economists Milton Friedman, Paul Samuelson, and other luminaries. Ash, Chen, and Naidu (2017) test for effects from this training, finding dramatic results. Economics language used in academic articles becomes rapidly prevalent in judicial opinions. Economics training affects both the trained judges and their peers as economic language travels from one judge to another and across legal areas. Perhaps most tangibly, economics training changed how judges perceived the consequence of their decisions. Judges in economic regulation cases shifted their votes in an antiregulatory direction by 10%. In the district courts, when judges were given discretion in sentencing, economics-trained judges immediately rendered 20% longer sentences relative to their non-economics-trained counterparts.

Part of what made the economics training program successful is likely because theory provided structure for judges to understand the patterns they saw. The question for theorists and researchers now is whether machine learning, text-as-data analysis, and other similar developments allow for a further step. If judges are shown the behavioral findings, will they become less prone to behavioral biases? If judges are taught the theoretical structures that drive behavioral bias, will they become better judges? Perhaps a new generation of theory and evidence from behavioral and social sciences could not only enhance understanding of the law, but also provide better justice and increase cooperation, trust, recognition, and respect. ✒

Acknowledgments

Work on this project was conducted while Chen received financial support from the European Research Council (Grant No. 614708), Swiss National Science Foundation (Grant Nos. 100018-152678 and 106014-150820), and Agence Nationale de la Recherche.

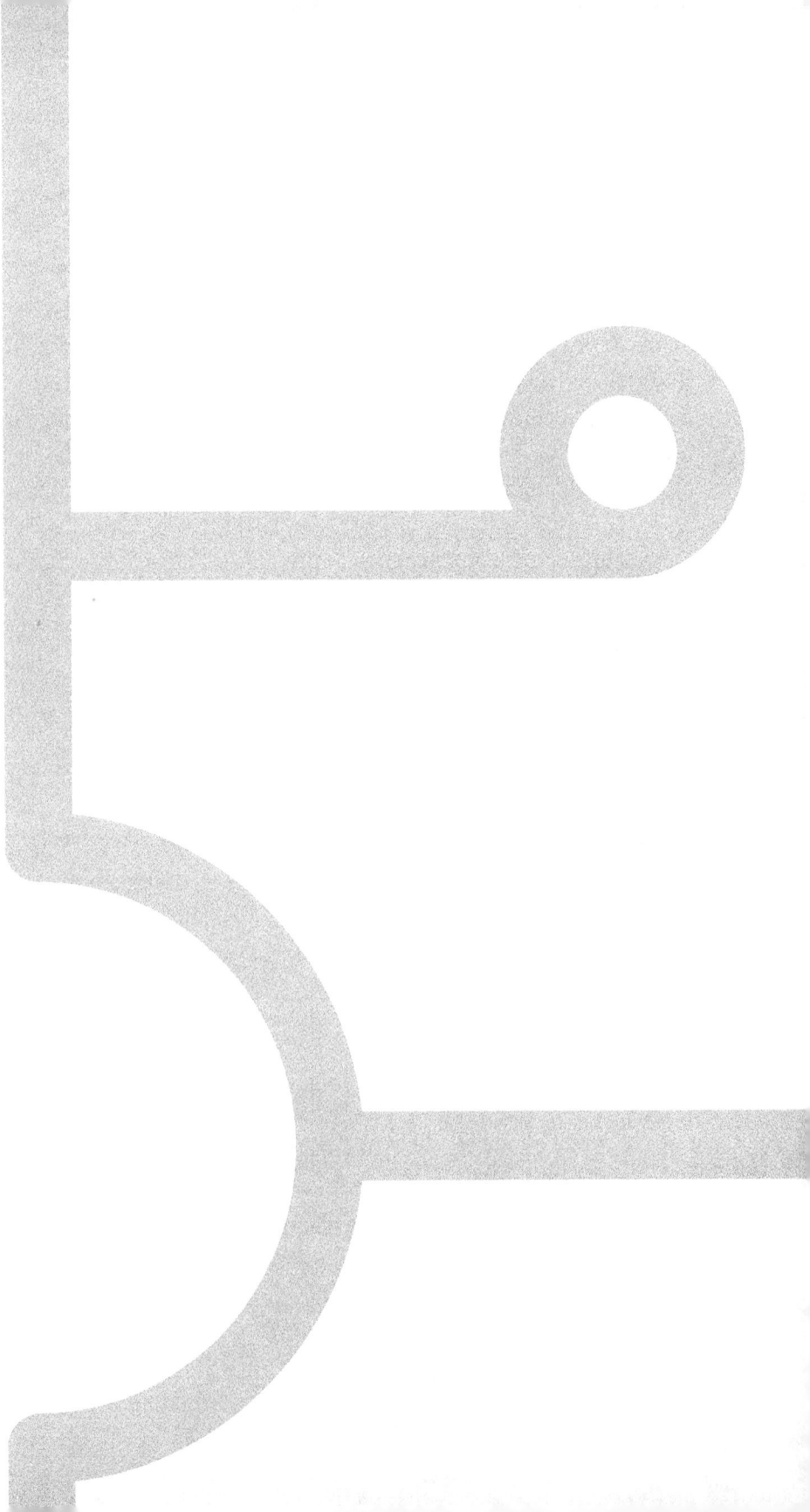

۰ٟٔ

THE LAW SEARCH TURING TEST

Michael A. Livermore, University of Virginia
Daniel N. Rockmore, Dartmouth College

One of the defining features of AI has been the field's desire and ability to articulate and then surmount increasingly sophisticated challenges toward an overarching goal of human-level (or more than human-level) general intelligence. The most famous of these challenges is the so-called Turing test, suggested by Alan Turing in one of the foundational papers in artificial intelligence (Turing 1950; A. P. Saygin 2000). Turing suggested that an honest benchmark of machine intelligence would be a machine capable of engaging in natural language conversation with a person in such a way that the person can't tell the difference between the computer and another human. With recent advances, Google's new assistant technology appears to have pulled off the feat, at least within the limited domain of setting up appointments over the phone.[1]

Turing's test was an adaptation of a party pastime of the day called the Imitation Game, in which the challenge was for an interlocutor to carry on a conversation by passing notes with a hidden player with the goal of determining the gender of the challenger. This framing of a *test* or *competition* soon gave rise to other kinds of Turing tests. Recently, these challenges have become less about appearing human than about vanquishing the human. Such "Terminator tests" (as one of the authors here

---

[1] See https://www.theverge.com/2018/5/8/17332070/google-assistant-makes-phone-call-demo-duplex-io-2018.

has called them[2]) have included mastering board games: first checkers, then chess, then *Jeopardy!*, and, most recently, Go and poker.

Each of these challenges offers its own obstacles. The game of poker, for example, requires fairly sophisticated models of human psychology and not just mastery of a set of formal rules. When the program DeepStack showed that it could beat professional-level poker players at heads-up no-limit Texas Hold'em, it showed that machine code could represent models of fairly complex human behavior, such as bluffing (Moravčík et al. 2017). When IBM's Watson beat *Jeopardy!* champion Ken Jennings, it demonstrated sophisticated natural language processing in addition to encyclopedic knowledge of a variety of fields of human knowledge and culture (pop and otherwise). The complexity of the game Go and its long history of human play offered another set of challenges that were surmounted by DeepMind's AlphaGo program, which beat Go world champion Ke Jie in 2017. Moving even further, DeepMind developed a new algorithm, referred to as AlphaZero, that learns the game "solely through self-play reinforcement learning, starting from random play, without any supervision or use of human data" (Silver et al. 2017). This new algorithm has exhibited a level of *creativity* in its play that is so sophisticated that expert players have characterized its gameplay as "games from far in the future," "Go from an alternate dimension," "alien," and "inhuman."[3]

---

[2] Krakauer, D., and D. N. Rockmore, "Never Mind Turing Tests. What About Terminator Tests?," *Chronicle of Higher Education*, Chronicle Review, August 10, 2015, https://www.chronicle.com/article/Never-Mind-Turing-Tests-What/232183.

[3] Dawn Chan, "The AI That Has Nothing to Learn From Humans," The Atlantic (Oct. 20, 2017 issue), https://www.theatlantic.com/technology/archive/2017/10/alphago-zero-the-ai-that-taught-itself-go/543450/.

A central advantage of many recent challenges is clear predefined outcomes that are attached to very large datasets that have either already been generated by humans (e.g., translation and images) or that can be machine generated (as in the case of AlphaZero, which essentially played itself for millions of games). It turns out that law includes many possible training sets like this, either where there are hand-labeled features (such as Westlaw headnotes) or other training data that can be extracted from the law. We believe that the domain of law has the potential to generate many potential AI challenges. In this chapter we articulate one law-related AI challenge that we have worked on together for several years.

"Attorney-matons"

As the law increasingly becomes a domain for the use of AI, it is natural to ask if there are associated Turing test–like challenges. Legal argumentation might be one. Imagine replacing the focus on casual conversation in Turing's imitation game with a legal setting, with the goal being for a human client to guess whether the counselor is human or machine. Or, for a Terminator test version, one might imagine an adversarial setting—such as litigation proceedings—where the goal for the designers is not only to write code that mimics a human, but rather to *win the game*.[4] Along similar lines, machines working as assistants could outperform humans at tasks such as analyzing contracts for problematic terms, or classifying documents as privileged in the course of discovery.

One of the defining features of *legal* advice and *legal* argumentation is that they have, at their base, legal materials, such as constitutions, statutes, regulations, and judicial opinions

---

[4]We are tempted to dub this imagined "law bot" Leibniz, in honor of the philosopher Gottfried Leibniz's conjecture of an era when disputation could be "reduced" to computation.

(Schauer and Wise 1996). Reliance on the law is what distinguishes legal advice and argument from, say, business advice or ethical argument. Because of the centrality of law in legal reasoning, a necessary function of an operational artificial lawyer is identifying the relevant legal materials, a process we refer to as "law search"(Livermore et al. 2018). The law search task that all lawyers undertake can also give rise to a related AI challenge that admits a Turing test framing.

One of the first things that a law student learns is how to navigate the morass of legal materials that he or she will be called on to master in the coming years.[5] In the analogue days, law students and lawyers relied on books and indexes. As the law has migrated into a digital format, law search has become digitized as well, with most materials available in one form or another online. Some of this material is publicly available, such as the judicial opinions collected and published through the Free Law Project at CourtListener.Org, but much of the law continues to exist behind the paywall, either private, as is the case with services such as LexisNexis, or public, as is the case for the information collected by the US courts and published on its PACER service.

The ability to execute a law search well in either digital or physical domains is closely related to the process of legal reasoning more generally. In the digital domain, law search is not simply a matter of typing a search term and collecting results. Rather, there is an iterative process at work. A starting question leads to initial search results, which then shape the framing of a legal argument or legal question, which in turn influences subsequent searches, and so on. In addition, some algorithms may take into account search history

---

[5] This material is usually integrated into the first-year curriculum as a course in "legal research." See, e.g., First-Year Legal Research and Writing Program, Harvard Law School, http://hls.harvard.edu/dept/lrw/.

when generating results (Noble 2018; Hannak et al. 2013). If a person's prior behavior affects search results, then the questions and answers they generate in dynamic interaction with the digital world may be idiosyncratic. All of these effects occur against a backdrop of research indicating that even minor variation in how search results are presented can have meaningful effects on subsequent beliefs and decisions (Epstein and Robertson 2015).

By the time that Google capitalized on the commercial potential of search online, private companies had already recognized the economic opportunities presented by the digitization of law search. Commercial law search software is ubiquitous, and digital databases with search interfaces are now fully integrated into the standard workbench of the twenty-first-century lawyer. In the early days, the value proposition of these firms was largely in their access to digital versions of legal documents, but as a considerable portion of this material has migrated out from behind paywalls, more of the primary value added by these firms is in providing law search services *per se*— both in the form of more sophisticated algorithms as well as via high-quality curated metadata (such as Westlaw headnotes) that can aid searches. With the deluge of digital information, the problem is not finding information but finding and separating the truly relevant from the superficially related.

A law search AI challenge would thus focus on the ability for a search algorithm to find relevant legal documents given some kind of well-framed inquiry. Following the Turing test approach, relevance can be defined sociologically, relative to a baseline of human judgment. In the common law tradition, judicial opinions provide a treasure trove of data on such judgments. Under the norm of *stare decisis*, judges are obligated to apply the law as interpreted in prior decisions.

This norm sets the stage for advocacy and deliberation over the appropriate set of authoritative earlier decisions (i.e., *precedent*) that will be applied in an individual case. After arriving at a decision about the relevant authorities, judges then construct opinions that cite and discuss that precedent and its applicability to the current decision. These lists of citations are hand constructed by legal experts (judges) with the help of law clerks, parties' attorneys, and other sources and represent a judgment about the legal documents that are relevant to a given decision.

Thus we arrive at the following explicit challenge: to predict, based purely on the semantic content of a judicial opinion, the citation information in that document.[6] This law search Turing test matches the underlying structure of the original Turing test by seeking to emulate human beings in a natural setting—a law search task that lawyers engage in every day.

From a research perspective, perhaps the most exciting feature of this test is that there is a large amount of labeled data that has been certified by judges who are both experts and lawmakers. The citation data encode judgments about what other documents in a corpus are relevant to the source document (i.e., the opinion being constructed). The words in the source document are natural language representations of the legal issues that are present in a case. Successfully predicting citations from the semantic content of opinions would be a significant step in demonstrating machine intelligence, as it would mirror fairly sophisticated judgments made by highly trained professionals in a domain that is dominated by natural

---

[6]One could imagine performing a similar operation with other legal documents, such as statutes. These documents, however, are much sparser in their textual and citation information.

language as well as cultural, political, and social norms about legal relevance.

From a practical perspective, the challenge is highly relevant to the goal of creating a law-bot that is a truly high-end digital workplace assistant, a technology with the potential to transform a multi-billion-dollar industry.

A First Law Search Turing Test

In recent work (Livermore et al. 2018), we took a small step forward in constructing a law search Turing test by considering a restricted version of the problem, narrowing down the world of judicial opinions and restricting the information used in search solely to the data that are found in those documents. We worked with opinions issued by the US Supreme Court in the period 1946–2005 that contained at least one citation, for a total of 9,575 files. Given this corpus, we then considered three algorithmic strategies for predicting the citations within the corpus (i.e., citations to other opinions in the corpus) from a randomly chosen source opinion.

Our search strategies take advantage of two obvious ways of connecting documents: mutual citation and semantic similarity. Based on these connections, we create a notion of distance for documents in the corpus (Leibon et al. 2018). Note that our positing of this structure on the dataset should already be viewed as a part of the algorithm. Any algorithm would necessarily need a representation of the data—effectively a preprocessing step—and this is the one our algorithms use. We refer to this as the construction of a *search space*.

For mutual citation, we assume a directed network of citation information such that opinion A points to opinion B if opinion A cites opinion B and vice versa. This structure is grounded in the qualitative observation that searchers often use citations as one way to identify documents of interest.

A second set of edges between documents is constructed based on their semantic content (i.e., the words contained in those documents). Semantic content can be understood as a proxy for several different actual search mechanisms used by law searchers, including keyword searches, curated categorizations (i.e., Westlaw headnotes), and other sources (such as treatises). The operating assumption is that documents with similar words will show up together in keyword searches, be grouped under similar headnotes, and appear in the same treatises. We rely on topic modeling to represent semantic similarity in a low-dimension space (Blei 2012).

The network representation used here is actually a "multinetwork," as the nodes (opinions) have multiple forms of connections. Given this multinetwork there is a natural notion of distance. The formalization behind the search space is discussed in detail in Leibon et al. (2018). In short, we use a generalization of the famous Google PageRank algorithm (Page et al. 1999) that views the corpus as a state space for a Markov chain (Grinstead and Snell 1997). The distance between two nodes in this state space is effectively a measure of the difficulty of random walking from one to the other.

Next we need to operationalize the challenge. For that, we take the randomly chosen source opinion (and create the multinetwork anew with the chosen opinion removed), strip out all of its citations, and ask how well an algorithm can predict the citations within the judicial opinion based exclusively on the topic proportions within that citation-stripped source opinion. The representation of the source opinion as topics loosely reflects the weighted constellation of ideas that the law researcher has in mind as she approaches the task of law search. The more accurate the reconstruction of the citation list from the source data, the better the algorithm is understood to perform.

We experiment with three different search mechanisms based on the source document as represented by a topic distribution. In each case, we ask for a specific number (N) of documents, which we call recommendations. Correct recommendations are those that predict the actual citations in the source document. We assess the performance of the three approaches based on recall and precision.[7]

The *proximity* algorithm finds the document in the multinetwork whose topic distribution is closest to the source document and returns the N closest documents. The *covering* algorithm is a parametrized approach to search that allows a searcher to investigate in many directions in the space. Finally, the third approach is an *adaptive* algorithm that takes advantage of reinforcement learning to use the search space as a training set. The learner seeks to maximize correct recommendations by optimizing parameters concerning how to trade off depth of search with breadth (essentially by balancing the proximity and covering approaches) (Livermore et al. 2018).

We ran the challenge for $N = 10$ and $N = 20$. All algorithms worked orders of magnitude better than random (i.e., choosing a document from the corpus at random). The adaptive algorithm was the best performer, covering generally better than proximity. The parameter values learned by the adaptive algorithm seemed to indicate a need for a balance between exploring multiple issues and pursuing some depth. Table 17.1 provides a summary of the results.

---

[7]Recall is the number of correct recommendations divided by the total number of citations in the source document. Precision is the percentage of correct recommendations divided by the total number of recommendations (N).

Table 17.1. Results of three search approaches.

	precision@10	precision@20	recall@10	recall@20
Proximity	13%	10%	3%	7%
Covering	16%	18%	5%	13%
Adaptive	19%	18%	7%	15%

Conclusion

As the tools of artificial intelligence, especially machine learning, intersect more and more deeply with the law, it is interesting to create useful challenges to spur algorithmic development with real-world applications. The problem of search—and specifically law search—is one such challenge. In this brief essay we outline a simple framing of the law search problem, discuss potential training data and means of testing the success of approaches, and summarize recent work of our implementing this approach on a sample of Supreme Court opinions. Our early results indicated that the methods of machine learning hold substantial promise in meeting the challenge of law search. Future researchers who turn to this challenge are likely to find fertile ground. ☙

Acknowledgments

This chapter draws from Livermore et al. (2018) and Leibon et al. (2018). We thank Greg Leibon, Reed Harder, Allen Riddell, Faraz Dadgostari, Mauricio Guim, and Peter Beling for their contributions to those two articles. We also thank the editors and peer reviewers of the journal *Artificial Intelligence and Law* for many helpful comments and suggestions.

REFERENCE

BIBLIOGRAPHY

A. P. Saygin, V. Akman, I. Cicekli. 2000. "Turing Test: 50 Years Later." *Minds & Machines* 10 (4): 463–518.

Abbott, A. 2001. *Time Matters: On Theory and Method.* Chicago, IL: University of Chicago Press.

Aberbach, J. D., and B. A. Rockman. 2006. "The Past and Future of Political–Administrative Relations: Research from Bureaucrats and Politicians to 'In the Web of Politics'—And Beyond." *International Journal of Public Administration* 29 (12): 977–995.

Abraham, K. S., and G. E. White. 2013. "Prosser and His Influence." *Journal of Tort Law* 6 (1): 27–74.

Abrams, D., M. Bertrand, and S. Mullainathan. 2012. "Do Judges Vary in Their Treatment of Race?" *Journal of Legal Studies* 41 (2): 347–383.

Albonetti, C. A. 1997. "Sentencing Under the Federal Sentencing Guidelines: Effects of Defendant Characteristics, Guilty Pleas, and Departures on Sentence Outcomes for Drug Offenses, 1991-1992." *Law & Society Review* 31 (4): 789–822.

Alexander, C. S., Z. J. Eigen, and C. G. Rich. 2016. "Post-Racial Hydraulics: The Hidden Dangers of the Universal Turn." *New York University Law Review* 91 (1): 1–58.

Alexander, C. S., and M. J. Feizollahi. 2019 (forthcoming). "Dragons, Claws, Teeth, and Caves: Legal Analytics and the Problem of Court Data Access." In *Computational Legal Studies: The Promise and Challenge of Data-Driven Legal Research,* edited by R. Whalen. Cheltenham, UK: Edward Elgar.

Amaranto, D., E. Ash, D. L. Chen, L. Ren, and C. Roper. 2017. "Algorithms as Prosecutors: Lowering Rearrest Rates Without Disparate Impacts and Identifying Defendant Characteristics 'Noisy' to Human Decision-Makers." Working paper, Social Science Research Network (SSRN). https://ssrn.com/abstract=2993003.

Angrist, J. D., G. W. Imbens, and D. B. Rubin. 1996. "Identification of Causal Effects Using Instrumental Variables." *Journal of the American Statistical Association* 91 (434): 444–455.

Angrist, J. D., and J.-S. Pischke. 2008. *Mostly Harmless Econometrics: An Empiricist's Companion.* Princeton, NJ: Princeton University Press.

———. 2010. "The Credibility Revolution in Empirical Economics: How Better Research Design is Taking the Con out of Econometrics." *Journal of Economic Perspectives* 24 (2): 3–30.

Arminjon, P., B. Nolde, and M. Wolff. 1951. *Traité de Droit Comparé.* Paris, France: Librairie Générale.

Arnull, A. 2006. *The European Union and its Court of Justice.* Oxford, UK: Oxford University Press.

Ash, E. 2016. "The Political Economy of Tax Laws in the US States." Working paper, Columbia University, New York, NY. https://pdfs.semanticscholar.org/1a4d/ 365571252a70c9db0e94b0a5d98f128e7c78.pdf.

———. 2018. "Emerging Tools for a 'Driverless' Legal System: Comment." *Journal of Institutional & Theoretical Economics* 174 (1): 206–213.

Ash, E., and D. L. Chen. 2017. "Predicting Punitiveness from Judicial Corpora." Working paper. http : / / users . nber . org / ~dlchen / papers / Predicting _ Punitiveness_and_Sentencing_Disparities_from_Judicial_Corpora.pdf.

Ash, E., D. L. Chen, and S. Naidu. 2017. "Ideas have Consequences: The Impact of Law and Economics on American Justice." Working paper. https://users.nber. org/~dlchen/papers/Ideas_Have_Consequences.pdf.

Ash, E., D. L. Chen, and A. Ornaghi. 2018. "Implicit Bias in the Judiciary: Evidence from Judicial Language Associations." Working paper. https://users.nber.org/ ~dlchen/papers/Implicit_Bias_in_the_Judiciary.pdf.

Ash, E., D. Chen, and W. Liu. 2017. "The (Non-)Polarization of US Circuit Court Judges, 1930–2013." Working paper. https://users.nber.org/~dlchen/papers/ Polarization_of_US_Circuit_Court_Judges_slides.pdf.

Ash, E., W. B. MacLeod, and S. Naidu. 2018. "The Language of Contract: Promises and Power in Union Collective Bargaining Agreements." Working paper. http: //elliottash.com/wp-content/uploads/2019/03/paper-ash-macleod-naidu- 2019-03-30.pdf.

Ash, R. A., and S. L. Edgell. 1975. "A Note on the Readability on the Position Analysis Questionnaire (PAQ)." *Journal of Applied Psychology* 60 (6): 765–766.

Ashley, K. D. 2017. *Artificial Intelligence and Legal Analytics.* Cambridge, UK: Cambridge University Press.

Athey, S., and G. W. Imbens. 2017. "The State of Applied Econometrics: Causality and Policy Evaluation." *Journal of Economic Perspectives* 2 (31): 3–32.

Axelrod, R. 2015. *Structure of Decision: The Cognitive Maps of Political Elites.* Princeton, NJ: Princeton University Press.

Badawi, A. B., R. Bod, J. Daily, G. Dari-Mattiacci, T. Ginsburg, B. Karstens, and M. Koolen. 2018. "Internal Reference Networks and Legal Origins." Working paper.

Baekgaard, M., J. Blom-Hansen, and S. Serritzlew. 2015. "When Politics Matters: The Impact of Politicians' and Bureaucrats' Preferences on Salient and Nonsalient Policy Areas." *Governance* 28 (4): 459–474.

Bafumi, J., and R. Y. Shapiro. 2009. "A New Partisan Voter." *The Journal of Politics* 71 (1): 1–24.

Balkin, J. M. 2011. *Living Originalism.* Cambridge, MA: Harvard University Press.

Ball, W. D. 2011. "Tough on Crime (on the State's Dime): How Violent Crime Does Not Drive California Counties' Incarceration Rates—And Why It Should." *Georgia State University Law Review* 28 (4): 987–1084.

Balla, S. J. 1998. "Administrative Procedures and Political Control of the Bureaucracy." *American Political Science Review* 92 (3): 663–673.

Ballew, C. C., and A. Todorov. 2007. "Predicting Political Elections from Rapid and Unreflective Face Judgments." *Proceedings of the National Academy of Sciences* 104 (46): 17948–17953.

Bang, H., and J. M. Robins. 2005. "Doubly Robust Estimation in Missing Data and Causal Inference Models." *Biometrics* 61 (4): 962–973.

Barry, N., L. Buchanan, E. Bakhturina, and D. L. Chen. 2016. "Events Unrelated to Crime Predict Criminal Sentence Length." Working paper. https://users.nber.org/~dlchen/papers/Events_Unrelated_to_Crime_Predict_Criminal_Sentence_Length.pdf.

Barzun, C. L. 2015. "Inside-Out: Beyond the Internal/External Distinction in Legal Scholarship." *Virginia Law Review* 101 (5): 1203–1288.

Bateman, D. A., J. D. Clinton, and J. S. Lapinski. 2017. "A House Divided? Roll Calls, Polarization, and Policy Differences in the US House, 1877–2011." *American Journal of Political Science* 61 (3): 698–714.

Bateman, D. A., and J. Lapinski. 2016. "Ideal Points and American Political Development: Beyond DW-NOMINATE." *Studies in American Political Development* 30 (2): 147–171.

Baude, W., A. S. Chilton, and A. Malani. 2017. "Making Doctrinal Work More Rigorous: Lessons from Systematic Reviews." *Chicago Law Review* 84 (1): 37–58.

Baum, L. 2009. *Judges and Their Audiences: A Perspective on Judicial Behavior.* Princeton, NJ: Princeton University Press.

Becker, G. S. 2010. *The Economics of Discrimination.* Chicago, IL: University of Chicago Press.

Bendor, J., and A. Meirowitz. 2004. "Spatial Models of Delegation." *American Political Science Review* 98 (2): 293–310.

Bennett, A. 2008. "Process Tracing: A Bayesian Perspective." In *The Oxford Handbook of Political Methodology,* edited by J. M. Box-Steensmeier, H. E. Brady, and D. Collier. Oxford, UK: Oxford University Press.

Berdejo, C., and D. L. Chen. 2016. "Electoral Cycles Among US Courts of Appeals Judges." *Journal of Law and Economics* 60 (3): 479–496.

Bergstra, J. S., R. Bardenet, Y. Bengio, and K. Balázs. 2011. "Algorithms for Hyper-Parameter Optimization." In *Proceedings of the 24th International Conference on Neural Information Processing Systems,* 2546–2554. Red Hook, NY: Curran Associates.

Bergstra, J., D. Yamins, and D. D. Cox. 2013. "Hyperopt: A Python Library for Optimizing the Hyperparameters of Machine Learning Algorithms." In *Proceedings of the 12th Python in Science Conference,* 13–20. Austin, TX: SciPy.

Berk, R. A., S. B. Sorenson, and G. Barnes. 2016. "Forecasting Domestic Violence: A Machine Learning Approach to Help Inform Arraignment Decisions." *Journal of Empirical Legal Studies* 13 (1): 94–115.

Berry, M. W., ed. 2004. *Survey of Text Mining.* New York, NY: Springer.

Berry, M. W., and M. Castellanos, eds. 2008. *Survey of Text Mining II.* New York, NY: Springer.

Bertrand, M., M. Bombardini, R. Fisman, and F. Trebbi. 2018. "Tax-Exempt Lobbying: Corporate Philanthropy as a Tool for Political Influence." NBER Working Paper No. 24451, National Bureau of Economic Research, Cambridge, MA. http://www.nber.org/papers/w24451.

Bertrand, M., and E. Duflo. 2016. "Field Experiments on Discrimination." NBER Working Paper No. 22014, National Bureau of Economic Research, Cambridge, MA. https://www.nber.org/papers/w22014.

Bickel, A. M. 1986. *The Least Dangerous Branch: The Supreme Court at the Bar of Politics.* New Haven, CT: Yale University Press.

Binongo, J. N. G. 2003. "Who Wrote the 15th Book of Oz? An Application of Multivariate Analysis to Authorship Attribution." *Chance* 16 (2): 9–17.

Black, R. C., R. J. Owens, J. Wedeking, and P. C. Wohlfarth. 2016a. "The Influence of Public Sentiment on Supreme Court Opinion Clarity." *Law & Society Review* 50 (3): 703–732.

———. 2016b. *US Supreme Court Opinions and Their Audiences.* Cambridge, UK: Cambridge University Press.

Black, Ryan C., and James F. Spriggs II. 2008. "An Empirical Analysis of the Length of US Supreme Court Opinions." *Houston Law Review* 45 (3): 622–682.

Blei, D. M. 2012. "Probabilistic Topic Models." *Communications of the ACM* 55 (4): 77–84.

Blei, D. M., and J. Lafferty. 2007. "A Correlated Topic Model of Science." *Annals of Applied Statistics* 1 (1): 17–35.

Blei, D. M., A. Y. Ng, and M. I. Jordan. 2003. "Latent Dirichlet Allocation." *Journal of Machine Learning Research* 3:993–1022.

Blondel, V. D., J.-L. Guillaume, R. Lambiotte, and E. Lefebvre. 2008. "Fast Unfolding of Communities in Large Networks." *Journal of Statistical Mechanics: Theory & Experiment* 2008 (10): P10008.

Bobek, M. 2015a. "The Changing Nature of Selection Procedures to the European Courts." In *Selecting Europe's Judges,* edited by M. Bobek, 1–23. Oxford, UK: Oxford University Press.

———. 2015b. "The Court of Justice of the European Union." In *The Oxford Handbook of European Union Law,* edited by A. Arnull and D. Chalmers, 153–177. Oxford, UK: Oxford University Press.

Bolukbasi, T., K.-W. Chang, J. Y. Zou, V. Saligrama, and A. T. Kalai. 2016. "Man is to Computer Programmer as Woman is to Homemaker? Debiasing Word Embeddings." In *Advances in Neural Information Processing Systems 29,* 4349–4357. Red Hook, NY: Curran Associates.

Bonica, A., J. Chen, and T. Johnson. 2012. *Automated Methods for Estimating the Political Ideology of Individual Public Bureaucrats Across Time and in a Common Ideological Space.* Paper presented at the Annual Summer Meeting of the Society for Political Methodology, July 2003, Chapel Hill, NC.

Boustead, A. E., and K. D. Stanley. 2015. "The Legal and Policy Road Ahead: An Analysis of Public Comments in NHTSA's Vehicle-to-Vehicle Advance Notice of Proposed Rulemaking." *Minnesota Journal of Law, Science & Technology* 16 (2): 693–733.

Bowie, J. B., D. R. Songer, and J. Szmer. 2014. *The View from the Bench and Chambers: Examining Judicial Process and Decision Making on the US Courts of Appeals.* Charlottesville, VA: University of Virginia Press.

Boyd, C. L., L. Epstein, and A. D. Martin. 2010. "Untangling the Causal Effects of Sex on Judging." *American Journal of Political Science* 54 (2): 389–411.

Boyd, C. L., and D. A. Hoffman. 2017. "The Use and Reliability of Federal Nature of Suit Codes." *Michigan State Law Review* 2017 (5): 997–1032.

Bradley, A. P. 1997. "The Use of the Area Under the ROC Curve in the Evaluation of Machine Learning Algorithms." *Pattern Recognition* 30 (7): 1145–1159.

Brady, H. E., and D. Collier. 2010. *Rethinking Social Inquiry: Diverse Tools, Shared Standards.* Lanham, MD: Rowman & Littlefield.

Brehm, J. O., and S. Gates. 1999. *Working, Shirking, and Sabotage: Bureaucratic Response to a Democratic Public.* Ann Arbor, MI: University of Michigan Press.

Breiman, L. 1996. "Stacked Regressions." *Machine Learning* 24 (1): 49–64.

———. 2001. "Statistical Modeling: The Two Cultures." *Statistical Science* 16 (3): 199–215.

Breyer, S. 2009. "Economic Reasoning and Judicial Review." *Economic Journal* 119 (535): F215–F135.

Brown, L. N., and T. Kennedy. 2000. *The Court of Justice of the European Communities.* London, UK: Sweet & Maxwell.

Brown, T. A. 2014. *Confirmatory Factor Analysis for Applied Research.* New York, NY: Guilford Publications.

Buntine, W. 2009. "Estimating Likelihoods for Topic Models." In *Advances in Machine Learning: First Asian Conference on Machine Learning,* 51–64. Berlin, Germany: Springer.

Bushway, S. D., and A. M. Piehl. 2001. "Judging Judicial Discretion: Legal Factors and Racial Discrimination in Sentencing." *Law & Society Review* 35 (4): 733–764.

Calabresi, G., and A. D. Melamed. 1972. "Property Rules, Liability Rules, and Inalienability: One View of the Cathedral." *Harvard Law Review* 85 (6): 1089–1128.

Caldeira, G. A., and J. L. Gibson. 1992. "The Etiology of Public Support for the Supreme Court." *American Journal of Political Science* 36 (3): 635–664.

Caldeira, G. A., J. R. Wright, and C. J. W. Zorn. 1999. "Sophisticated Voting and Gate-Keeping in the Supreme Court." *Journal of Law, Economics & Organization* 15 (3): 549–572.

Caliskan, A., J. J. Bryson, and A. Narayanan. 2017. "Semantics Derived Automatically from Language Corpora Contain Human-Like Biases." *Science* 356 (6334): 183–186.

Camerer, C. F., A. Dreber, E. Forsell, T.-H. Ho, J. Huber, M. Johannesson, M. Kirchler, J. Almenberg, A. Altmejd, and T. Chan. 2016. "Evaluating Replicability of Laboratory Experiments in Economics." *Science* 351 (6280): 1433–1436.

Camerer, C. F., A. Dreber, F. Holzmeister, T.-H. Ho, J. Huber, M. Johannesson, M. Kirchler, et al. 2018. "Evaluating the Replicability of Social Science Experiments in Nature and Science between 2010 and 2015." *Nature Human Behaviour* 2 (9): 637–644.

Cameron, C. M. 1993. "New Avenues for Modeling Judicial Politics." Prepared for delivery at *Conference on the Political Economy of Public Law*, October 1993, W. Allen Wallis Institute of Political Economy, University of Rochester, NY.

Cameron, C., and L. Kornhauser. 2017. "What Courts Do . . . And How to Model It." Draft chapter of a book-in-progress on the positive political theory of courts.

Campbell-Kibler, K. 2010. "Sociolinguistics and Perception." *Language & Linguistics Compass* 4 (6): 377–389.

Campbell, D. T. 1979. "Assessing the Impact of Planned Social Change." *Evaluation & Program Planning* 2 (1): 67–90.

Canfield-Davis, K., S. Jain, D. Wattam, J. McMurtry, and M. Johnson. 2010. "Factors of Influence on Legislative Decision Making: A Descriptive Study." *Journal of Legal, Ethical & Regulatory Issues* 13 (2): 55–68.

Carlson, K., M. A. Livermore, and D. Rockmore. 2016. "A Quantitative Analysis of Writing Style on the US Supreme Court." *Washington University Law Review* 93 (6): 1461–1510.

Carpenter, D. P. 2001. *The Forging of Bureaucratic Autonomy: Reputations, Networks, and Policy Innovation in Executive Agencies, 1862–1928.* Princeton, NJ: Princeton University Press.

———. 2014. *Reputation and Power: Organizational Image and Pharmaceutical Regulation at the FDA.* Princeton, NJ: Princeton University Press.

Case, M. A. C. 1995. "Disaggregating Gender from Sex and Sexual Orientation: The Effeminate Man in the Law and Feminist Jurisprudence." *Yale Law Journal* 105 (1): 1–105.

Cash, D. W., W. C. Clark, F. Alcock, N. M. Dickson, N. Eckley, David H. Guston, J. Jäger, and R. B. Mitchell. 2003. "Knowledge Systems for Sustainable Development." *Proceedings of the National Academy of Sciences* 100 (14): 8086–8091.

Chalfin, A., A. M. Haviland, and S. Raphael. 2013. "What do Panel Studies Tell Us About a Deterrent Effect of Capital Punishment? A Critique of the Literature." *Journal of Quantitative Criminology* 29 (1): 5–43.

Chen, D. L. 2016a. "Implicit Egoism in Courtrooms: First Initial Name Effects with Randomly Assigned Defendants." Working paper.

———. 2016b. "Priming Ideology: Why Presidential Elections Affect US Judges." Working paper, SSRN (Social Science Research Network). https://ssrn.com/abstract=2816245.

———. 2016c. "This Morning's Breakfast, Last Night's Game: Detecting Extraneous Factors in Judging." IAST Working Paper No. 16-49, Institute for Advanced Study in Toulouse, Tolouse, France. http://iast.fr/pub/31020.

———. 2017. "Mood and the Malleability of Moral Reasoning." TSE Working Paper n. 16-707, Institute for Advanced Study in Toulouse, France. https://www.tse-fr.eu/publications/mood-and-malleability-moral-reasoning.

Chen, D. L., X. Cui, L. Shang, and J. Zhang. 2019 (forthcoming). "What Matters: Agreement Between US Courts of Appeals Judges." *Journal of Machine Learning Research*.

Chen, D. L., M. Dunn, L. Sagun, and H. Sirin. 2017. "Early Predictability of Asylum Court Decisions." In *Proceedings of the 16th Edition of the International Conference on Artificial Intelligence and Law*, 233–236. New York, NY: Association for Computing Machinery.

Chen, D. L., and J. Eagel. 2016. "Can Machine Learning Help Predict the Outcome of Asylum Adjudications?" In *Proceedings of the 16th Edition of the International Conference on Artificial Intelligence and Law*, 237–240. New York, NY: Association for Computing Machinery.

Chen, D. L., Y. Halberstam, and A. C. L. Yu. 2016a. "Covering: Mutable Characteristics and Perceptions of (Masculine) Voice in the US Supreme Court." TSE Working Paper 16-680, Toulouse School of Economics, Toulouse, France. https://www.tse-fr.eu/publications/covering-mutable-characteristics-and-perceptions-voice-us-supreme-court.

———. 2016b. "Perceived Masculinity Predicts United States Supreme Court Outcomes." *PLoS ONE* 11 (10): e0164324.

Chen, D. L., and M. Kumar. 2018. "Is Justice Really Blind? And Is It Also Deaf?" Working paper. http://users.nber.org/~dlchen/papers/Is_Justice_Really_Blind_slides.pdf.

Chen, D. L., T. J. Moskowitz, and K. Shue. 2016. "Decision-Making Under the Gambler's Fallacy: Evidence from Asylum Judges, Loan Officers, and Baseball Umpires." *Quarterly Journal of Economics* 131 (3): 1181–1242.

Chen, D. L., and A. Philippe. 2018. "Clash of Norms: Judicial Leniency on Defendant Birthdays." TSE Working Paper n. 18-76, Institute for Advanced Study in Toulouse, France. http://www.iast.fr/publications/clash-norms-judicial-leniency-defendant-birthdays.

Chen, D. L., and A. Yu. 2016. "Mimicry: Phonetic Accommodation Predicts US Supreme Court Votes." Working paper. http://users.nber.org/~dlchen/papers/Mimicry.pdf.

Chen, Y., R. M. Rabbani, A. Gupta, and M. J. Zaki. 2018. "Comparative Text Analytics via Topic Modeling in Banking." Piscataway, NJ: IEEE.

Chilton, A. S., and M. K. Levy. 2015. "Challenging the Randomness of Panel Assignment in the Federal Courts of Appeals." *Cornell Law Review* 101 (1): 1–56.

Choi, S. J., and G. M. Gulati. 2005. "Which Judges Write Their Opinions (And Should We Care?)" *Florida State University Law Review* 32 (4): 1077–1122.

Chow, G. C. 1960. "Tests of Equality Between Sets of Coefficients in Two Linear Regressions." *Econometrica* 28 (3): 591–605.

Clermont, K. M., and S. J. Schwab. 2009. "Employment Discrimination Plaintiffs in Federal Court: From Bad to Worse?" *Harvard Law & Policy Review* 3 (1): 103–132.

Clinton, J. D., A. Bertelli, C. R. Grose, D. E. Lewis, and D. C. Nixon. 2012. "Separated Powers in the United States: The Ideology of Agencies, Presidents, and Congress." *American Journal of Political Science* 56 (2): 341–354.

Clinton, J. D., and D. E. Lewis. 2008. "Expert Opinion, Agency Characteristics, and Agency Preferences." *Political Analysis* 16 (1): 3–20.

Clinton, J., S. Jackman, and D. Rivers. 2004. "The Statistical Analysis of Roll Call Data." *American Political Science Review* 98 (2): 355–370.

Cohen, L., K. B. Diether, and C. Malloy. 2012. "Legislating Stock Prices." NBER Working Paper No. 18291, National Bureau of Economic Research, Cambridge, MA. https://www.nber.org/papers/w18291.

Coleman, M., and T. L. Liau. 1975. "A Computer Readability Formula Designed for Machine Scoring." *Journal of Applied Psychology* 60 (2): 283.

Copus, R., and R. Hübert. 2017. "Detecting Inconsistency in Governance." Working paper, Social Science Research Network (SSRN). https://ssrn.com/abstract= 2812914.

Corley, P. C. 2008. "The Supreme Court and Opinion Content: The Influence of Parties' Briefs." *Political Research Quarterly* 61 (3): 468–78.

Corley, P. C., and J. Wedeking. 2014. "The (Dis)advantage of Certainty: The Importance of Certainty in Language." *Law & Society Review* 48 (1): 35–62.

Craver, C., and J. Tabery. 2017. "Mechanisms in Science." In *The Stanford Encyclopedia of Philosophy,* Spring 2017, edited by E. N. Zalta. Palo Alto, CA: Metaphysics Research Lab, Stanford University.

Cudeck, R. 2000. "Exploratory Factor Analysis." In *Handbook of Applied Multivariate Statistics & Mathematical Modeling,* edited by H. E. A. Tinsley and S. D. Brown, 265–296.

Dahl, R. A. 1957. "Decision-Making in a Democracy: The Supreme Court As a National Policy-Maker." *Journal of Public Law* 6 (2): 279–295.

Dai, A. M., C. Olah, and Q. V. Le. 2015. "Document Embedding with Paragraph Vectors." Working paper.

Danziger, S., J. Levav, and L. Avnaim-Pesso. 2011a. "Extraneous Factors in Judicial Decisions." *Proceedings of the National Academy of Sciences* 108 (17): 6889–6892.

———. 2011b. "Reply to Weinshall-Margel and Shapard: Extraneous Factors in Judicial Decisions Persist." *Proceedings of the National Academy of Sciences* 108 (42): E834.

David, R., and J. E. C. Brierley. 1978. *Major Legal Systems in the World Today: An Introduction to the Comparative Study of Law.* New York, NY: Simon & Schuster.

de Waele, H. 2015. "Not Quite the Bed That Procrustes Built: Dissecting the System for Selecting Judges at the Court of Justice of the European Union." In *Selecting Europe's Judges,* edited by M. Bobek, 24–50. New York, NY: Oxford University Press.

DeMuth, C. 2016. "Can the Administrative State be Tamed?" *Journal of Legal Analysis* 8 (1): 121–190.

Devins, N., and D. E. Lewis. 2008. "Not-So-Independent Agencies: Party Polarization and the Limits of Institutional Design." *Boston University Law Review* 88 (2): 459–498.

Diamond, A., and J. S. Sekhon. 2013. "Genetic Matching for Estimating Causal Effects: A General Multivariate Matching Method for Achieving Balance in Observational Studies." *Review of Economics & Statistics* 95 (3): 932–945.

Djankov, S., R. La Porta, F. Lopez-de-Silanes, and A. Shleifer. 2003. "Courts." *Quarterly Journal of Economics* 118 (2): 453–517.

———. 2008. "The Law and Economics of Self-Dealing." *Journal of Financial Economics* 88 (3): 430–465.

Drake, C. 1993. "Effects of Misspecification of the Propensity Score on Estimators of Treatment Effect." *Biometrics* 49 (4): 1231–1236.

Dworkin, R. 1978. *Taking Rights Seriously.* Cambridge, MA: Harvard University Press.

Easton, D. 1965. *A Systems Analysis Of Political Life.* London, UK: John Wiley & Sons Ltd.

Eckert, P. 2008. "Variation and the Indexical Field." *Journal of Sociolinguistics* 12 (4): 453–476.

Edward, D. 1995. "How the Court of Justice Works." *European Law Review* 20:539–558.

Eisenberg, T., J. Goerdt, B. Ostrom, and D. Rottman. 1995. "Litigation Outcomes in State and Federal Courts: A Statistical Portrait." *Seattle University Law Review* 19 (3): 433–453.

Eisenberg, T., and C. Lanvers. 2008. "Summary Judgment Rates Over Time, Across Case Categories, and Across Districts: An Empirical Study of Three Large Federal Districts." Cornell Law School Working Paper No. 08-22, Cornell Law School, Ithaca, NY. https://scholarship.law.cornell.edu/lsrp_papers/108/.

Epstein, D., and S. O'Halloran. 1996. "Divided Government and the Design of Administrative Procedures: A Formal Model and Empirical Test." *Journal of Politics* 58 (2): 373–397.

Epstein, L., A. D. Martin, K. M. Quinn, and J. A. Segal. 2007. "Ideological Drift among Supreme Court Justices: Who, When, and How Important?" *Northwestern University Law Review* 101 (4): 1483–1542.

Epstein, R., and R. E. Robertson. 2015. "The Search Engine Manipulation Effect (SEME) and its Possible Impact on the Outcomes of Elections." *Proceedings of the National Academy of Sciences* 112 (33): E4512–E4521.

Eren, O., and N. Mocan. 2016. "Emotional Judges and Unlucky Juveniles." *American Economic Journal: Applied Economics* 10 (3): 171–205.

Eun Kim, S., and J. Urpelainen. 2017. "Environmental Public Opinion in US States, 1973–2012." *Environmental Politics* 27 (1): 1–26.

Evans, M., W. McIntosh, J. Lin, and C. Cates. 2007. "Recounting the Courts? Applying Automated Content Analysis to Enhance Empirical Legal Research." *Journal of Empirical Legal Studies* 4 (4): 1007–1039.

Fabrigar, L. R., and D. T. Wegener. 2011. *Exploratory Factor Analysis.* Oxford, UK: Oxford University Press.

Fagan, J., and E. Ash. 2017. "New Policing, New Segregation? From Ferguson to New York." *Georgetown Law Journal* 106 (1): 25–102.

Farina, C. R., M. J. Newhart, C. Cardie, D. Cosley, and Cornell eRulemaking Initiative. 2011. "Rulemaking 2.0." *University of Miami Law Review* 65 (2): 395–448.

Farina, C. R., M. Newhart, and J. Heidt. 2012. "Rulemaking vs. Democracy: Judging and Nudging Public Participation that Counts." *Michigan Journal of Environmental & Administrative Law* 2 (1): 123–172.

Feldman, A. 2016. "Counting on Quality: The Effects of Merits Brief Quality on Supreme Court Decisions." *Denver Law Review* 94 (1): 43–70.

Ferguson, R. A. 1990. "The Judicial Opinion as Literary Genre." *Yale Journal of Law & Humanities* 2 (1): 201–219.

Fiorina, M. P., and R. G. Noll. 1978. "Voters, Bureaucrats and Legislators: A Rational Choice Perspective on the Growth of Bureaucracy." *Journal of Public Economics* 9 (2): 239–254.

Fischman, J. B. 2014. "Measuring Inconsistency, Indeterminacy, and Error in Adjudication." *American Law & Economics Review* 16 (1): 40–85.

———. 2015. "Do the Justices Vote like Policy Makers? Evidence from Scaling the Supreme Court with Interest Groups." *The Journal of Legal Studies* 44 (S1): S269–S293.

Fowler, J. H. 2006. "Connecting the Congress: A Study of Cosponsorship Networks." *Political Analysis* 14 (4): 456–487.

Francis, W. L. 1989. *The Legislative Committee Game: A Comparative Analysis of Fifty States.* Columbus, OH: Ohio State University Press.

François, T., and C. Fairon. 2012. "An 'AI Readability' Formula for French as a Foreign Language." In *Proceedings of the 2012 Joint Conference on Empirical Methods in Natural Language Processing and Computational Natural Language Learning,* 466–477. Stroudsburg, PA: Association for Computational Linguistics.

Frankenreiter, J. 2017. "The Politics of Citations at the ECJ—Policy Preferences of EU Member State Governments and the Citation Behavior of Members of the European Court of Justice." *Journal of Empirical Legal Studies* 14 (4): 813–85.

———. 2018. "Are Advocates General Political? An Empirical Analysis of the Voting Behavior of the Advocates General at the European Court of Justice." *Review of Law & Economics* 14 (1): 1–43.

Friedman, B. 2009. *The Will of the People: How Public Opinion Has Influenced the Supreme Court and Shaped the Meaning of the Constitution.* New York, NY: Macmillan.

Friedman, J. H. 2000. "Greedy Function Approximation: A Gradient Boosting Machine." *Annals of Statistics* 29 (5): 1189–1232.

Fujimura, J. H. 1988. "The Molecular Biological Bandwagon in Cancer Research: Where Social Worlds Meet." *Social Problems* 35 (3): 261–283.

Furlong, S. R., and C. M. Kerwin. 2005. "Interest Group Participation in Rule Making: A Decade of Change." *Journal of Public Administration Research and Theory* 15 (3): 353–370.

Gabel, M., M. Malecki, C. Carrubba, and J. Fjestul. 2003. "The Politics of Decision-Making in the European Court of Justice: The System of Chambers and Distribution of Cases for Decision." Paper originally presented at the Meetings of the European Consortium for Political Research, September 2003, Marburg, Germany.

Gailmard, S., and J. W. Patty. 2007. "Slackers and Zealots: Civil Service, Policy Discretion, and Bureaucratic Expertise." *American Journal of Political Science* 51 (4): 873–889.

Galanter, M. 2004. "The Vanishing Trial: An Examination of Trials and Related Matters in Federal and State Courts." *Journal of Empirical Legal Studies* 1 (3): 459–570.

Ganchev, T., N. Fakotakis, and G. Kokkinakis. 2005. "Comparative Evaluation of Various MFCC Implementations on the Speaker Verification Task." In *Proceedings of the SPECOM-2005,* 191–194. Moscow, Russia: Moscow State Linguistic University.

Ganglmair, B., and M. Wardlaw. 2017. "Complexity, Standardization, and the Design of Loan Agreements." Working paper, Social Science Research Network (SSRN). https://ssrn.com/abstract=2952567.

Garg, N., L. Schiebinger, D. Jurafsky, and J. Zou. 2018. "Word Embeddings Quantify 100 Years of Gender and Ethnic Stereotypes." *Proceedings of the National Academy of Sciences* 115 (16): E3635–E3644.

Garner, B. A. 2013. *Legal Writing in Plain English: A Text with Exercises.* Chicago, IL: University of Chicago Press.

Gelman, A. 2011. "Causality and Statistical Learning." *American Journal of Sociology* 117 (3): 955–966.

Gelman, A., and G. Imbens. 2013. "Why Ask Why? Forward Causal Inference and Reverse Causal Questions." NBER Working Paper No. 19614, National Bureau of Economic Research, Cambridge, MA. https://www.nber.org/papers/w19614.

Gentzkow, M., and J. M. Shapiro. 2010. "What Drives Media Slant? Evidence from US Daily Newspapers." *Econometrica* 78 (1): 35–71.

Gentzkow, M., J. M. Shapiro, and M. Taddy. 2016. "Measuring Polarization in High-Dimensional Data: Method and Application to Congressional Speech." NBER Working Paper No. 22423, National Bureau of Economic Research, Cambridge, MA. https://www.nber.org/papers/w22423.

Gerrish, S., and D. M. Blei. 2011. "Predicting Legislative Roll Calls from Text." In *Proceedings of the 28th International Conference on International Conference on Machine Learning,* 489–496. Madison, WI: Omnipress.

Gibson, J. L., and G. A. Caldeira. 2009. *Citizens, Courts, And Confirmations.* Princeton, NJ: Princeton University Press.

Gibson, J. L., G. A. Caldeira, and L. K. Spence. 2003. "Measuring Attitudes toward the United States Supreme Court." *American Journal of Political Science* 47 (2): 354–367.

Gibson, J. L., M. Lodge, and B. Woodson. 2014. "Losing, but Accepting: Legitimacy, Positivity Theory, and the Symbols of Judicial Authority." *Law & Society Review* 48 (4): 837–866.

Gibson, J. L., and M. J. Nelson. 2014. "The Legitimacy of the US Supreme Court: Conventional Wisdoms and Recent Challenges Thereto." *Annual Review of Law & Social Science* 10 (1): 201–219.

Gilmore, G. 1951. "Book Review: The Bramble Bush." *Yale Law Journal* 60 (7): 1251–1253.

Glaeser, E. L., and A. Shleifer. 2002. "Legal Origins." *Quarterly Journal of Economics* 117 (4): 1193–1229.

Glenn, H. P. 2006. "Comparative Legal Families and Comparative Legal Traditions." In *The Oxford Handbook of Comparative Law,* edited by M. Reimann and R. Zimmermann, 422–439. Oxford, UK: Oxford University Press.

Goel, S., J. M. Rao, and R. Shroff. 2016. "Precinct or Prejudice? Understanding Racial Disparities in New York City's Stop-and-Frisk Policy." *Annals of Applied Statistics* 10 (1): 365–394.

Goelzhauser, G., and D. M. Cann. 2014. "Judicial Independence and Opinion Clarity on State Supreme Courts." *State Politics & Policy Quarterly* 14 (2): 123–141.

Goffman, E. 1963. *Stigma: Notes on the Management of Spoiled Identity.* Upper Saddle River, NJ: Prentice-Hall.

Golden, M. M. 1998. "Interest Groups in the Rule-Making Process: Who Participates? Whose Voices Get Heard?" *Journal of Public Administration Research and Theory* 8 (2): 245–270.

Goodfellow, I., Y. Bengio, and A. Courville. 2016. *Deep Learning.* Cambridge, MA: MIT Press.

Goodnow, F. J. 1900. *Politics and Administration: A Study in Government.* New York, NY: The Macmillan Company.

Gordley, J. 2006. "Comparative Law and Legal History." In *The Oxford Handbook of Comparative Law,* edited by Mathias Reimann and Reinhard Zimmermann, 753–774. Oxford, UK: Oxford University Press.

Goutal, J. L. 1976. "Characteristics of Judicial Style in France, Britain and the USA." *American Journal of Comparative Law* 24 (1): 43–72.

Gray, V., and D. Lowery. 1995. "Interest Representation and Democratic Gridlock." *Legislative Studies Quarterly* 20 (4): 531–552.

Greiner, D. J., and C. Wolos Pattanayak. 2012. "Randomized Evaluation in Legal Assistance: What Difference Does Representation (Offer and Actual Use) Make?" *Yale Law Journal* 121 (8): 2118–2214.

Grimmer, J. 2015. "We Are All Social Scientists Now: How Big Data, Machine Learning, and Causal Inference Work Together." *PS: Political Science & Politics* 48 (1): 80–83.

Grimmer, J., and B. M. Stewart. 2013. "Text as Data: The Promise and Pitfalls of Automatic Content Analysis Methods for Political Texts." *Political Analysis* 21 (3): 267–297.

Grinstead, C. M., and J. L. Snell. 1997. *Introduction to Probability.* Providence, RI: American Mathematical Society.

Grogger, J. 2011. "Speech Patterns and Racial Wage Inequality." *Journal of Human Resources* 46 (1): 1–25.

Groseclose, T., and J. Milyo. 2005. "A Measure of Media Bias." *Quarterly Journal of Economics* 120 (4): 1191–1237.

Grus, J. 2015. *Data Science from Scratch: First Principles with Python.* Sebastopol, CA: O'Reilly Media.

Gryski, G. S., and G. Zuk. 2008. "A Multi-User Data Base on the Attributes of US Appeals Court Judges, 1801–2000." Unpublished database. http : / / artsandsciences.sc.edu/poli/juri/auburn_appct_codebook.pdf.

Gunning, R. 1969. "The Fog Index After Twenty Years." *Journal of Business Communication* 6 (2): 3–13.

Guthrie, C., J. J. Rachlinski, and A. J. Wistrich. 2000. "Inside the Judicial Mind." *Cornell Law Review* 86 (4): 777–830.

———. 2007. "Blinking on the Bench: How Judges Decide Cases." *Cornell Law Review* 93 (1): 1–44.

Hadfield, G. K., and B. R. Weingast. 2012. "What Is Law? A Coordination Model of the Characteristics of Legal Order." *Journal of Legal Analysis* 4 (2): 471–514.

Haire, S. B., D. R. Songer, and S. A. Lindquist. 2003. "Appellate Court Supervision in the Federal Judiciary: A Hierarchical Perspective." *Law & Society Review* 37 (1): 143–168.

Hall, M. A., and R. F. Wright. 2008. "Systematic Content Analysis of Judicial Opinions." *California Law Review* 96 (1): 63–122.

Hamm, K. E. 1980. "US State Legislative Committee Decisions: Similar Results in Different Settings." *Legislative Studies Quarterly* 5 (1): 31–54.

Hamm, K., R. Hedlund, and N. M. Miller. 2014. "State Legislatures." In *The Oxford Handbook of State and Local Government,* edited by D. Haider-Markel, 293–318. Oxford, UK: Oxford University Press.

Hannak, A., P. Sapiezynski, A. Molavi Kakhki, B. Krishnamurthy, D. Lazer, A. Mislove, and C. Wilson. 2013. "Measuring Personalization of Web Search." In *Proceedings of the 22nd International Conference on World Wide Web,* 527–538. New York, NY: ACM.

Hansen, B. E. 2001. "The New Econometrics of Structural Change: Dating Breaks in US Labor Productivity." *Journal of Economic Perspectives* 15 (4): 117–128.

Hansford, T. G., and C. Coe. 2019. "Linguistic Complexity, Information Processing, and Public Acceptance of Supreme Court Decisions." *Political Psychology* 40 (2): 395–412.

Harbridge, Laurel M. 2016. "Book Review: *Legislative Effectiveness in the United States Congress: The Lawmakers* by Craig Volden and Alan E. Wiseman." *Journal of Politics* 78 (1): e3–e4.

Hastie, T., R. Tibshirani, and J. Friedman. 2008. *The Elements of Statistical Learning: Data Mining, Inference, and Prediction.* 2nd. New York, NY: Springer.

Hedge, D. 1998. *Governance And The Changing American States.* New York, NY: Routledge.

Helbing, D. 2010. *Quantitative Sociodynamics: Stochastic Methods and Models of Social Interaction Processes.* New York, NY: Springer Science & Business Media.

Hicks, W., and D. Smith. 2009. "Do Parties Matter? Explaining Legislative Productivity in the American States." Presented at *The State of the Parties: 2008 and Beyond* Conference, October 2009, The Bliss Institute, University of Akron, OH.

Hill, M., W. Kelly, B. Lockhart, and R. Ness. 2013. "Determinants and Effects of Corporate Lobbying." *Financial Management* 42 (4): 931–957.

Ho, D. E., S. Sherman, and P. Wyman. 2018. "Do Checklists Make a Difference? A Natural Experiment from Food Safety Enforcement." *Journal of Empirical Legal Studies* 15 (2): 242–277.

Holland, P. W. 1986. "Statistics and Causal Inference." *Journal of the American Statistical Association* 81 (396): 945–960.

Holmes, O. W. 1897. "The Path of the Law." *Harvard Law Review* 10 (8): 457–478.

Hopkins, D. J., and G. King. 2010. "A Method of Automated Nonparametric Content Analysis for Social Science." *American Journal of Political Science* 54 (1): 229–247.

Hughes, J. M., N. J. Foti, D. C. Krakauer, and D. N. Rockmore. 2012. "Quantitative Patterns of Stylistic Influence in the Evolution of Literature." *Proceedings of the National Academy of Sciences* 109 (20): 7682–7686.

Huq, A. Z. 2018. "Democratic Erosion and the Courts: Comparative Perspectives." *New York University Law Review* 93 (1): 21–31.

Imai, K., L. Keele, and D. Tingley. 2010. "A General Approach to Causal Mediation Analysis." *Psychological Methods* 15 (4): 309–334.

Imbeau, L. M., F. Pétry, and M. Lamari. 2001. "Left–Right Party Ideology and Government Policies: A Meta-Analysis." *European Journal of Political Research* 40 (1): 1–29.

Imbens, G. W., and D. B. Rubin. 2015. *Causal Inference for Statistics, Social, and Biomedical Sciences: An Introduction.* New York, NY: Cambridge University Press.

Ioannidis, J. P. A. 2005. "Why Most Published Research Findings Are False." *PLoS Medicine* 2 (8): e124 (696–701).

Ioannidis, J., T. D. Stanley, and H. Doucouliagos. 2017. "The Power of Bias in Economics Research." *Economic Journal* 127 (605): F236–F265.

Isele, H. G. 1949. "Ein halbes Jahrhundert deutsches Bürgerliches Gesetzbuch." *Archiv für die Civilistische Praxis* 150 (1): 1–27.

Iyyer, M., P. Enns, J. L. Boyd-Graber, and P. Resnik. 2014. "Political Ideology Detection Using Recursive Neural Networks." In *Proceedings of the 52nd Annual Meeting of the Association for Computational Linguistics,* 1113–1122. Stroudsburg, PA: Association for Computational Linguistics.

Jackson, M. O. 2010. *Social and Economic Networks.* Princeton, NJ: Princeton University Press.

Jacques, P. J., R. E. Dunlap, and M. Freeman. 2008. "The Organisation of Denial: Conservative Think Tanks and Environmental Scepticism." *Environmental Politics* 17 (3): 349–385.

Jaiswal, D., H. Ochani, R. Vunikili, R. Deshmukh, D. L. Chen, and E. Ash. 2019 (forthcoming). "Analysis of Vocal Implicit Bias in SCOTUS Decisions Through Predictive Modelling." *Proceedings of Experimental Linguistics.*

James, G., D. Witten, T. Hastie, and R. Tibshirani. 2013. *An Introduction to Statistical Learning.* New York, NY: Springer.

Jelveh, Z., B. Kogut, and S. Naidu. 2015. "Political Language in Economics." Columbia Business School Research Paper No. 14-57, Columbia University, New York. https://ssrn.com/abstract=2535453.

Jensen, J., S. Naidu, E. Kaplan, L. Wilse-Samson, D. Gergen, M. Zuckerman, and A. Spirling. 2012. "Political Polarization and the Dynamics of Political Language: Evidence from 130 Years of Partisan Speech." *Brookings Papers on Economic Activity* 43 (2): 1–81.

Johnson, S. M. 2014. "The Changing Discourse of the Supreme Court." *New Hampshire Law Review* 12 (1): 29–68.

Johnston, C. D., and A. O. Ballard. 2016. "Economists and Public Opinion: Expert Consensus and Economic Policy Judgments." *The Journal of Politics* 78 (2): 443–456.

Jones, S. E. 2013. *The Emergence of the Digital Humanities.* New York, NY: Routledge.

Jurafsky, D., and J. H. Martin. 2000. *Speech and Language Processing: An Introduction to Natural Language Processing, Computational Linguistics, and Speech Recognition.* Upper Saddle River, NJ: Prentice Hall.

Kadiri, Y., T. Leble, Z. Pajor-Gyulai, E. Ash, and D. L. Chen. 2018. "Tone of Voice Predicts Political Attitudes: Evidence from US Supreme Court Oral Arguments." Working paper. https://users.nber.org/~dlchen/papers/Tone_of_Voice_Predicts_Political_Attitudes.pdf.

Kagan, E. 2001. "Presidential Administration." *Harvard Law Review* 114 (8): 2246–2320.

Kasparov, G. 2017. *Deep Thinking: Where Machine Intelligence Ends and Human Creativity Begins.* New York, NY: Public Affairs.

Katz, D. M., and M. J. Bommarito. 2014. "Measuring the Complexity of the Law: The United States Code." *Artificial Intelligence & Law* 22 (4): 337–374.

Katz, D. M., M. J. Bommarito, and J. Blackman. 2017. "A General Approach for Predicting the Behavior of the Supreme Court of the United States." *PLoS ONE* 12 (1): e0174698.

Kim, P. T., M. Schlanger, C. L. Boyd, and A. D. Martin. 2009. "How Should We Study District Judge Decision-Making?" *Washington University Journal of Law & Policy* 29:83–112.

Kincaid, J. P., J. A. Aagard, and J. W. O'Hara. 1980. *Development and Test of a Computer Readability Editing System (CRES).* (Navy) Training Analysis and Evaluation Group, Orlando, FL, TAEG Report 83.

Kincaid, J. P., R. P. Fishburne Jr, R. L. Rogers, and B. S. Chissom. 1975. *Derivation of New Readability Formulas for Navy Enlisted Personnel.* Naval Technical Training Command: Millington, TN, Research Branch Report 8-75.

Kirilenko, A., S. Mankad, and G. Michailidis. 2014. "Do US Regulators Listen to the Public? Testing the Regulatory Process with the RegRank Algorithm," 1–2. New York, NY: Association for Computing Machinery.

Kleinberg, J. 1999. "Authoritative Sources in a Hyperlinked Environment." *Journal of the Association for Computing Machinery* 46 (5): 604–632.

Kleinberg, J., H. Lakkaraju, J. Leskovec, J. Ludwig, and S. Mullainathan. 2018. "Human Decisions and Machine Predictions." *The Quarterly Journal of Economics* 133 (1): 237–293.

Kleinberg, J., J. Ludwig, S. Mullainathan, and Z. Obermeyer. 2015. "Prediction Policy Problems." *American Economic Review* 105 (5): 491–495.

Klerman, D. M., P. G. Mahoney, H. Spamann, and M. I. Weinstein. 2011. "Legal Origin or Colonial History?" *Journal of Legal Analysis* 3 (2): 379–409.

Klingenstein, S., T. Hitchcock, and S. DeDeo. 2014. "The Civilizing Process in London's Old Bailey." *Proceedings of the National Academy of Sciences* 111 (26): 9419–9424.

Klofstad, C. A., R. C. Anderson, and S. Peters. 2012. "Sounds Like a Winner: Voice Pitch Influences Perception of Leadership Capacity in Both Men and Women." *Proceedings of the Royal Society of London B: Biological Sciences* 279 (1738): 2698–2704.

Knill, C., M. Debus, and S. Heichel. 2010. "Do Parties Matter in Internationalised Policy Areas? The Impact of Political Parties on Environmental Policy Outputs in 18 OECD Countries, 1970–2000." *European Journal of Political Research* 49 (3): 301–336.

Komárek, J. 2009. "Review: Questioning Judicial Deliberations." *Oxford Journal of Legal Studies* 29 (4): 805–826.

Korfiatis, N., E. Garcia-Bariocanal, and S. Sanchez-Alonso. 2012. "Evaluating Content Quality and Helpfulness of Online Product Reviews: The Interplay of Review Helpfulness vs. Review Content." *Electronic Commerce Research & Applications* 11 (3): 205–217.

Korn, J. 2004. "Teaching Talking: Oral Communication Skills in a Law Course." *Journal of Legal Education* 54 (4): 588–596.

Kornhauser, L. A. 1992. "Modeling Collegial Courts I: Path-Dependence." *International Review of Law and Economics* 12 (2): 169–185.

———. 2012. "Judicial Organization and Administration." In *Procedural Law and Economics,* edited by C. W. Sanchirico, 27–44. Cheltenham, UK: Edward Elgar Publishing.

Kornilova, A., D. Argyle, and V. Eidelman. 2018. "Party Matters: Enhancing Legislative Embeddings with Author Attributes for Vote Prediction." In *Proceedings of the 56th Annual Meeting of the Association for Computational Linguistics,* 510–515. Stroudsburg, PA: Association for Computational Linguistics.

Kozinski, A. 1992. "What I Ate for Breakfast and Other Mysteries of Judicial Decision-Making." *Loyola of Los Angeles Law Review* 26 (4): 993–999.

Kozlowski, A. C., M. Taddy, and J. A. Evans. 2018. "The Geometry of Culture: Analyzing Meaning through Word Embeddings." Working paper, arXiv:1803.09288 [cs.CL]. https://arxiv.org/abs/1803.09288.

Krawiec, K. D. 2013. "Don't 'Screw Joe the Plummer': The Sausage-Making of Financial Reform." *Arizona Law Review* 55 (1): 53–103.

La Porta, R., F. Lopez-de-Silanes, and A. Shleifer. 2008. "The Economic Consequences of Legal Origins." *Journal of Economic Literature* 46 (2): 285–332.

La Porta, R., F. Lopez-de-Silanes, A. Shleifer, and R. W. Vishny. 1998. "Law and Finance." *Journal of Political Economy* 106 (6): 1113–1155.

———. 2000. "Agency Problems and Dividend Policies Around the World." *Journal of Finance* 55 (1): 1–33.

———. 2002. "Investor Protection and Corporate Valuation." *Journal of Finance* 57 (3): 1147–1170.

Lakkaraju, H., J. Kleinberg, J. Leskovec, J. Ludwig, and S. Mullainathan. 2017. "The Selective Labels Problem: Evaluating Algorithmic Predictions in the Presence of Unobservables." In *Proceedings of the 23rd ACM SIGKDD International Conference on Knowledge Discovery and Data Mining,* 275–284. New York, NY: Association for Computing Machinery.

LaLonde, R. J. 1986. "Evaluating the Econometric Evaluations of Training Programs with Experimental Data." *American Economic Review* 76 (4): 604–620.

Laqueur, H., and R. Copus. 2016. "Synthetic Crowdsourcing: A Machine-Learning Approach to the Problems of Inconsistency and Bias in Adjudication." Working paper, Social Science Research Network (SSRN). https://ssrn.com/abstract=2694326.

Lau, J. H., and T. Baldwin. 2016. "An Empirical Evaluation of Doc2Vec with Practical Insights into Document Embedding Generation." Working paper, arXiv:1607.05368 [cs.CL]. https://arxiv.org/abs/1607.05368.

Lauderdale, B. E., and T. S. Clark. 2014. "Scaling Politically Meaningful Dimensions Using Texts and Votes." *American Journal of Political Science* 58 (3): 754–771.

Law, D. S. 2016. "Constitutional Archetypes." *Texas Law Review* 95 (2): 153–243.

Le, Q., and T. Mikolov. 2014. "Distributed Representations of Sentences and Documents." In *Proceedings of the 31st International Conference on Machine Learning,* edited by E. P. Xing and T. Jebara, 32:1188–1196. Beijing, China: PMLR.

Lebovits, G. 2004. "Short Judicial Opinions: The Weight of Authority." *New York State Bar Association Journal* 76 (7): 60–64.

Lee, F. E. 2015. "How Party Polarization Affects Governance." *Annual Review of Political Science* 18:261–282.

Lee, J. A., and M. Verleysen. 2007. *Nonlinear Dimensionality Reduction.* New York, NY: Springer Science & Business Media.

Lee, T. R., and S. C. Mouritsen. 2017. "Judging Ordinary Meaning." *Yale Law Journal* 127 (4): 788–879.

Leibon, G., M. A. Livermore, R. Harder, A. Riddell, and D. Rockmore. 2018. "Bending the Law: Geometric Tools for Quantifying Influence in the Multinetwork of Legal Opinions." *Artificial Intelligence & Law* 26 (2): 145–167.

Levitt, S. D. 1997. "Using Electoral Cycles in Police Hiring to Estimate the Effect of Police on Crime." *The American Economic Review* 87 (3): 270–290.

Levy, O., and Y. Goldberg. 2014. "Dependency-Based Word Embeddings." In *Proceedings of the 52nd Annual Meeting of the Association for Computational Linguistics,* 302–308. Stroudsburg, PA: Association for Computational Linguistics.

Levy, O., Y. Goldberg, and I. Dagan. 2015. "Improving Distributional Similarity with Lessons Learned from Word Embeddings." *Transactions of the Association for Computational Linguistics* 3:211–225.

Lewis, D. E. 2003. *Presidents and the Politics of Agency Design: Political Insulation in the United States Government Bureaucracy, 1946–1997.* Palo Alto, CA: Stanford University Press.

———. 2007. "Testing Pendleton's Premise: Do Political Appointees Make Worse Bureaucrats?" *Journal of Politics* 69 (4): 1073–1088.

———. 2008. *The Politics of Presidential Appointments: Political Control and Bureaucratic Performance.* Princeton, NJ: Princeton University Press.

Linder, F., B. A. Desmarais, M. Burgess, and E. Giraudy. 2018. "Text as Policy: Measuring Policy Similarity Through Bill Text Reuse." Working paper, Social Science Research Network (SSRN). https://ssrn.com/abstract=2812607.

Little, L. E. 1998. "Hiding with Words: Obfuscation, Avoidance, and Federal Jurisdiction Opinions." *UCLA Law Review* 46 (1): 75–112.

Livermore, M. A. 2014. "Cost–Benefit Analysis and Agency Independence." *University of Chicago Law Review* 81 (2): 609–688.

———. 2018. "Will AI Eat ELS?" *Journal of Institutional & Theoretical Economics (JITE)* 174 (1): 214–219.

Livermore, M. A., F. Dadgosari, M. Guim, P. Beling, and D. N. Rockmore. 2018. "Law Search as Prediction." Virginia Public Law and Legal Theory Research Paper No. 2018-61. https://ssrn.com/abstract=3278398.

Livermore, M. A., V. Eidelman, and B. Grom. 2018. "Computationally Assisted Regulatory Participation." *Notre Dame Law Review* 93 (3): 977–1034.

Livermore, M. A., A. Riddell, and D. Rockmore. 2017. "The Supreme Court and the Judicial Genre." *Arizona Law Review* 59 (4): 837–901.

Llewellyn, K. 1951. *The Bramble Bush.* New York, NY: Oceana Publications.

Long, L. N., and W. F. Christensen. 2011. "Does the Readability of Your Brief Affect Your Chance of Winning an Appeal?" *Journal of Appellate Practice & Process* 12 (1): 145–162.

———. 2013. "When Justices (Subconsciously) Attack: The Theory of Argumentative Threat and the Supreme Court." *Oregon Law Review* 91 (3): 933–960.

Macey, J., and J. Mitts. 2014. "Finding Order in the Morass: The Three Real Justifications for Piercing the Corporate Veil." *Cornell Law Review* 100 (1): 99–156.

Maltzman, F., and P. J. Wahlbeck. 2004. "A Conditional Model of Opinion Assignment on the Supreme Court." *Political Research Quarterly* 57 (4): 551–563.

Mancini, G. F., and D. T. Keeling. 1995. "Language, Culture, and Politics in the Life of the European Court of Justice." *Columbia Journal of European Law* 1 (3): 397–413.

Manning, C. D., P. Raghavan, and H. Schütze. 2008. *Introduction to Information Retrieval.* Cambridge, UK: Cambridge University Press.

Margolin, D., Y.-R. Lin, and D. Lazer. 2013. "Why so Similar?: Identifying Semantic Organizing Processes in Large Textual Corpora." Working paper, Social Science Research Network (SSRN). https://ssrn.com/abstract=2353705.

Marlon, J., E. Fine, and A. Leiserowitz. 2017. *A Majority of Americans in Every State Say the US Should Participate in the Paris Climate Agreement.* New Haven, CT: Yale Program on Climate Change Communication.

Martin, A. D., and K. M. Quinn. 2002. "Dynamic Ideal Point Estimation via Markov Chain Monte Carlo for the US Supreme Court, 1953–1999." *Political Analysis* 10 (2): 134–153.

Mayew, W. J., and M. Venkatachalam. 2012. "The Power of Voice: Managerial Affective States and Future Firm Performance." *Journal of Finance* 67 (1): 1–43.

McAleer, P., A. Todorov, and P. Belin. 2014. "How Do You Say Hello? Personality Impressions from Brief Novel Voices." *PLoS ONE* 9 (3): e90779.

McArdle, A. 2006. "Teaching Writing in Clinical, Lawyering, and Legal Writing Courses: Negotiating Professional and Personal Voice." *Clinical Law Review* 12 (2): 501–539.

McAuliffe, K. 2012. "Language and Law in the European Union: The Multilingual Jurisprudence of the ECJ." In *The Oxford Handbook of Language and Law*, edited by L. M. Solan and P. M. Tiersma, 200–216. Oxford, UK: Oxford University Press.

McComb, M. F. 1949. "A Mandate from the Bar: Shorter and More Lucid Opinions." *American Bar Association Journal* 35 (5): 382–384.

McConnell, M. W., and R. A. Posner. 1989. "An Economic Approach to Issues of Religious Freedom." *The University of Chicago Law Review* 56 (1): 1–60.

McCubbins, M. D., R. G. Noll, and B. R. Weingast. 1987. "Administrative Procedures as Instruments of Political Control." *Journal of Law, Economics, & Organization* 3 (2): 243–277.

McFarland, D. A., K. Lewis, and A. Goldberg. 2016. "Sociology in the Era of Big Data: The Ascent of Forensic Social Science." *The American Sociologist* 47 (1): 12–35.

McKay, A., and S. W. Yackee. 2007. "Interest Group Competition on Federal Agency Rules." *American Politics Research* 35 (3): 336–357.

McLaughlin, G. H. 1969. "SMOG Grading—A New Readability Formula." *Journal of Reading* 12 (8): 639–646.

Medwed, D. S. 2007. "The Innocent Prisoner's Dilemma: Consequences of Failing to Admit Guilt at Parole Hearings." *Iowa Law Review* 93 (5): 491–557.

Mendelson, N. A. 2011. "Rulemaking, Democracy, and Torrents of E-Mail." *George Washington Law Review* 79 (5): 1343–1380.

Mikolov, T., K. Chen, G. Corrado, and J. Dean. 2013. "Efficient Estimation of Word Representations in Vector Space." Working paper, arXiv: 1301.3781 [cs.CL]. https://arxiv.org/abs/1301.3781.

Mikolov, T., I. Sutskever, K. Chen, G. S. Corrado, and J. Dean. 2013. "Distributed Representations of Words and Phrases and their Compositionality." In *Advances in Neural Information Processing Systems 26*, 3111–3119. Red Hook, NY: Curran Associates.

Mikva, A. J. 1987. "For Whom Judges Write." *Southern California Law Review* 61 (5): 1357–1371.

Miller, J. B., and A. Sanjurjo. 2018. "Surprised by the Gambler's and Hot Hand Fallacies? A Truth in the Law of Small Numbers." *Econometrica* 86 (6): 2019–2047.

Moravčík, M., M. Schmid, N. Burch, V. Lisý, D. Morrill, N. Bard, T. Davis, K. Waugh, M. Johanson, and M. Bowling. 2017. "DeepStack: Expert-Level Artificial Intelligence in Heads-Up No-Limit Poker." *Science* 356 (6337): 508–513.

Moretti, F. 2013. *Distant Reading*. London, UK: Verso.

Morgan, S. L., and C. Winship. 2014. *Counterfactuals and Causal Inference*. Cambridge, UK: Cambridge University Press.

Mosteller, F., and D. L. Wallace. 1963. "Inference in an Authorship Problem: A Comparative Study of Discrimination Methods Applied to the Authorship of the Disputed Federalist Papers." *Journal of the American Statistical Association* 58 (302): 275–309.

Mouritsen, S. C. 2010. "The Dictionary Is Not a Fortress: Definitional Fallacies and a Corpus-Based Approach to Plain Meaning." *BYU Law Review* 2010 (5): 1915–1980.

Mullainathan, S., and J. Spiess. 2017. "Machine Learning: An Applied Econometric Approach." *Journal of Economic Perspectives* 31 (2): 87–106.

Mullane, N. 2012. *Life After Murder: Five Men in Search of Redemption*. New York, NY: Public Affairs.

Munafò, M. R., B. A. Nosek, D. V. M. Bishop, K. S. Button, C. D. Chambers, N. P. du Sert, U. Simonsohn, E.-J. Wagenmakers, J. J. Ware, and J. P. A. Ioannidis. 2017. "A Manifesto for Reproducible Science." Article number 0021, *Nature Human Behaviour* 1 (1).

Mustard, D. B. 2001. "Racial, Ethnic, and Gender Disparities in Sentencing: Evidence from the US Federal Courts." *Journal of Law and Economics* 44 (1): 285–314.

Nakosteen, R., and M. Zimmer. 2014. "Approval of Social Security Disability Appeals: Analysis of Judges' Decisions." *Applied Economics* 46 (23): 2783–2791.

Nass, C., and K. M. Lee. 2001. "Does Computer-Synthesized Speech Manifest Personality? Experimental Tests of Recognition, Similarity–Attraction, and Consistency–Attraction." *Journal of Experimental Psychology: Applied* 7 (3): 171–181.

Nay, J. J. 2017. "Predicting and Understanding Law-Making with Word Vectors and an Ensemble Model." *PLoS ONE* 12 (5): e0176999.

Neuman Jr., R. C. 1999. "Organic Molecules and Chemical Bonding." In *Organic Chemistry*, 1–55. Self-published. https://people.chem.ucsb.edu/neuman/robert/orgchembyneuman/BookContents.html.

Neumark, D. 2016. "Experimental Research on Labor Market Discrimination." *Journal of Economic Literature* 56 (3): 799–866.

Newman, M. E. J., and M. Girvan. 2004. "Finding and Evaluating Community Structure in Networks." *Physical Review E* 69 (2): 026113.

Nguyen, V.-A., J. L. Boyd-Graber, P. Resnik, and K. Miler. 2015. "Tea Party in the House: A Hierarchical Ideal Point Topic Model and Its Application to Republican Legislators in the 112th Congress." In *Proceedings of the 53rd Annual Meeting of the Association for Computational Linguistics and the 7th International Joint Conference on Natural Language Processing,* 1438–1448. Stroudsburg, PA: Association for Computational Linguistics.

Niculescu-Mizil, A., and R. Caruana. 2005. "Predicting Good Probabilities with Supervised Learning." In *Proceedings of the 22nd International Conference on Machine Learning,* 625–632. New York, NY: Association for Computing Machinery.

Nielsen, L. B., R. L. Nelson, and R. Lancaster. 2010. "Individual Justice or Collective Legal Mobilization? Employment Discrimination Litigation in the Post Civil Rights United States." *Journal of Empirical Legal Studies* 7 (2): 175–201.

Nixon, D. C. 2004. "Separation of Powers and Appointee Ideology." *Journal of Law, Economics, & Organization* 20 (2): 438–457.

Noble, S. U. 2018. *Algorithms of Oppression: How Search Engines Reinforce Racism.* New York, NY: NYU Press.

Oldfather, C. M., J. P. Bockhorst, and B. P. Dimmer. 2012. "Triangulating Judicial Responsiveness: Automated Content Analysis, Judicial Opinions, and the Methodology of Legal Scholarship." *Florida Law Review* 64 (5): 1189–1242.

Open Science Collaboration. 2015. "Estimating the Reproducibility of Psychological Science." *Science* 349 (6251): aac4716.

Owens, R. J., and J. P. Wedeking. 2011. "Justices and Legal Clarity: Analyzing the Complexity of US Supreme Court Opinions." *Law & Society Review* 45 (4): 1027–1061.

Page, L., S. Brin, R. Motwani, and T. Winograd. 1999. "The PageRank Citation Ranking: Bringing Order to the Web." In *Proceedings of the 7th International World Wide Web Conference,* 161–172. Palo Alto, CA: Stanford InfoLab.

Pedregosa, F., G. Varoquaux, A. Gramfort, V. Michel, B. Thirion, O. Grisel, M. Blondel, et al. 2011. "Scikit-learn: Machine Learning in Python." *Journal of Machine Learning Research* 12:2825–2830.

Peppers, T. C. 2006. *Courtiers Of The Marble Palace: The Rise And Influence Of The Supreme Court Law Clerk.* Palo Alto, CA: Stanford Law & Politics.

Peppers, T. C., and C. Zorn. 2008. "Law Clerk Influence on Supreme Court Decision Making: An Empirical Assessment." *Depaul Law Review* 58 (1): 51–78.

Perju, V. 2009. "Reason and Authority in the European Court of Justice." *Virginia Journal of International Law* 49 (2): 307–377.

Perron, P. 2006. "Dealing with Structural Breaks." In *Palgrave Handbook Of Econometrics: Econometric Theory,* edited by T. C. Mills and K. Patterson, 278–352.

Persson, P. 2018. "Attention Manipulation and Information Overload." *Behavioural Public Policy* 2 (1): 78–106.

Petersilia, J. 2003. *When Prisoners Come Home: Parole and Prisoner Reentry.* Oxford, UK: Oxford University Press.

Poole, K. T. 2005. *Spatial Models of Parliamentary Voting.* Cambridge, UK: Cambridge University Press.

Poole, K. T., and H. L. Rosenthal. 1985. "A Spatial Model for Legislative Roll Call Analysis." *American Journal of Political Science* 29 (2): 357–384.

———. 2007. *Ideology and Congress.* New Brunswick, NJ: Transaction Publishers.

Pope, D. G., J. Price, and J. Wolfers. 2013. "Awareness Reduces Racial Bias." NBER Working Paper No. 19765, National Bureau of Economic Research, Cambridge, MA. https://www.nber.org/papers/w19765.

Posner, R. A. 1973. "An Economic Approach to Legal Procedure and Judicial Administration." *Journal of Legal Studies* 2 (2): 399–458.

———. 1995. "Judges' Writing Styles (And Do They Matter?)" *University of Chicago Law Review* 62 (4): 1421–1449.

———. 1996. "Pragmatic Adjudication." *Cardozo Law Review* 18 (1): 1–20.

———. 2002. "Legal Scholarship Today." *Harvard Law Review* 115 (5): 1314–1326.

———. 2004. "Judicial Behavior and Performance an Economic Approach." *Florida State University Law Review* 32 (4): 1259.

———. 2010. *How Judges Think.* Cambridge, MA: Harvard University Press.

Preacher, K. J., and A. F. Hayes. 2008. "Asymptotic and Resampling Strategies for Assessing and Comparing Indirect Effects in Multiple Mediator Models." *Behavior Research Methods* 40 (3): 879–891.

Prendergast, C. 2007. "The Motivation and Bias of Bureaucrats." *American Economic Review* 97 (1): 180–196.

Press, S. J. 2009. *Subjective and Objective Bayesian Statistics: Principles, Models, and Applications.* Hoboken, NJ: John Wiley & Sons.

Purnell, T., W. Idsardi, and J. Baugh. 1999. "Perceptual and Phonetic Experiments on American English Dialect Identification." *Journal of Language & Social Psychology* 18 (10): 10–30.

Quinn, K. M., B. L. Monroe, M. Colaresi, M. H. Crespin, and D. R. Radev. 2010. "How to Analyze Political Attention with Minimal Assumptions and Costs." *American Journal of Political Science* 54 (1): 209–228.

Rachlinski, J. J., S. L. Johnson, A. J. Wistrich, and C. Guthrie. 2009. "Does Unconscious Racial Bias Affect Trial Judges?" *Notre Dame Law Review* 84 (3): 1195–1246.

Rachlinski, J. J., A. J. Wistrich, and C. Guthrie. 2013. "Altering Attention in Adjudication." *UCLA Law Review* 60:1586–1618.

Rakoff, S. H., and R. Sarner. 1975. "Bill History Analysis: A Probability Model of the State Legislative Process." *Polity* 7 (3): 402–414.

Ramji-Nogales, J., A. I. Schoenholtz, and P. G. Schrag. 2007. "Refugee Roulette: Disparities in Asylum Adjudication." *Stanford Law Review* 60 (2): 295–411.

Rasmussen, H. 1986. *On Law and Policy in the European Court of Justice.* Dordrecht, Netherlands: Martinus Nijhoff Publishers.

Rauterberg, G., and E. Talley. 2017. "Contracting Out of the Fiduciary Duty of Loyalty: An Empirical Analysis of Corporate Opportunity Waivers." *Columbia Law Review* 117 (5): 1075–1152.

Reimann, M. 2008. "The Good, the Bad, and the Ugly: The Reform of the German Law of Obligations." *Tulane Law Review* 83 (4): 877–919.

Revesz, R. L. 1997. "Environmental Regulation, Ideology, and the DC Circuit." *Virginia Law Review* 83 (8): 1717–1772.

Rice, D. 2014. "The Impact of Supreme Court Activity on the Judicial Agenda." *Law & Society Review* 48:63–90.

Riddell, A. 2014. "How to Read 22,198 Journal Articles: Studying the History of German Studies with Topic Models." In *Distant Readings: Topologies of German Culture in the Long Nineteenth Century,* edited by M. Erlin and L. Tatlock, 91–114. Rochester, NY: Camden House.

Robins, J. M., A. Rotnitzky, and L. P. Zhao. 1994. "Estimation of Regression Coefficients When Some Regressors Are Not Always Observed." *Journal of the American Statistical Association* 89 (427): 846–866.

Rodriguez, D. B., and M. D. McCubbins. 2006. "The Judiciary and the Role of Law: A Positive Political Theory Perspective." In *The Oxford Handbook On Political Economy,* edited by B. Weingast and D. Wittman, 273–287. Oxford, UK: Oxford University Press.

Rodriguez, D. B., M. D. McCubbins, and B. R. Weingast. 2009. "The Rule of Law Unplugged." *Emory Law Journal* 59 (6): 1455–1495.

Roe, M. J. 2006. "Legal Origins, Politics, and Modern Stock Markets." *Harvard Law Review* 120 (2): 460–527.

Romantz, D. S. 2003. "The Truth About Cats and Dogs: Legal Writing Courses and the Law School Curriculum." *University of Kansas Law Review* 52 (1): 105–146.

Rosenbaum, P. R., and D. B. Rubin. 1983. "The Central Role of the Propensity Score in Observational Studies for Causal Effects." *Biometrika* 70 (1): 41–55.

Rosenthal, A. 1974. *Legislative Performance in the States: Explorations of Committee Behavior.* New York, NY: Free Press.

Rosenthal, J. S., and A. H. Yoon. 2011. "Detecting Multiple Authorship of United States Supreme Court Legal Decisions Using Function Words." *Annals of Applied Statistics* 5 (1): 283–308.

Rubin, D. B. 1974. "Estimating Causal Effects of Treatments in Randomized and Nonrandomized Studies." *Journal of Educational Psychology* 66 (5): 688–701.

———. 2008. "For Objective Causal Inference, Design Trumps Analysis." *Annals of Applied Statistics* 2 (3): 808–840.

Rubinstein, A. 2000. *Economics and Language: Five Essays.* Cambridge, UK: Cambridge University Press.

Rudolph, M., and D. Blei. 2017. "Dynamic Bernoulli Embeddings for Language Evolution." Working paper, arXiv:1703.08052 [stat.ML]. https://arxiv.org/abs/1703.08052.

Rudolph, M., F. Ruiz, S. Athey, and D. Blei. 2017. "Structured Embedding Models for Grouped Data." In *Advances in Neural Information Processing Systems,* 30:250–260. Red Hook, NY: Curran Associates.

Rudolph, M., F. Ruiz, S. Mandt, and D. Blei. 2016. "Exponential Family Embeddings." In *Advances in Neural Information Processing Systems,* 29:478–486. Red Hook, NY: Curran Associates.

Ruiz, F. J. R., S. Athey, and D. M. Blei. 2017. "SHOPPER: A Probabilistic Model of Consumer Choice with Substitutes and Complements." Working paper, arXiv: 1711.03560 [stat.ML]. https://arxiv.org/abs/1711.03560.

Rzhetsky, A., J. G. Foster, I. T. Foster, and J. A. Evans. 2015. "Choosing Experiments to Accelerate Collective Discovery." *Proceedings of the National Academy of Sciences* 112 (47): 14569–14574.

Salton, G., A. Wong, and C.-S. Yang. 1975. "A Vector Space Model for Automatic Indexing." *Communications of the Association for Computing Machinery* 18 (11): 613–620.

Sankari, S. 2013. *European Court of Justice Legal Reasoning in Context.* Amsterdam, Netherlands: Europa Law Publishing.

Sarosy, C. 2013. "Parole Denial Habeas Corpus Petitions: Why the California Supreme Court Needs to Provide More Clarity on the Scope of Judicial Review." *UCLA Law Review* 61 (4): 1134–1191.

Schanzenbach, M. 2005. "Racial and Sex Disparities in Prison Sentences: The Effect of District-Level Judicial Demographics." *Journal of Legal Studies* 34 (1): 57–92.

Schauer, F. 1995. "Opinions as Rules." *University of Chicago Law Review* 62 (4): 1455–1475.

Schauer, F., and V. J. Wise. 1996. "Legal Positivism as Legal Information." *Cornell Law Review* 82 (5): 1080.

Scheb, J. M., and W. Lyons. 2000. "The Myth of Legality and Public Evaluation of the Supreme Court." *Social Science Quarterly* 81 (4): 928–940.

Scherer, K. R. 1979. "Voice and Speech Correlates of Perceived Social Influence in Simulated Juries." In *The Social Psychology of Language,* edited by K. R. Scherer, H. Giles, and R. St. Clair, 88–120. London, UK: Blackwell.

Schultze, S. J. 2018. "The Price of Ignorance: The Constitutional Cost of Fees for Access to Electronic Public Court Records." *Georgetown Law Journal* 106 (4): 1197–1227.

Segal, J. A., and H. J. Spaeth. 1996. "The Influence of Stare Decisis on the Votes of United States Supreme Court Justices." *American Journal of Political Science* 40 (4): 971–1003.

———. 2002. *The Supreme Court and the Attitudinal Model Revisited.* Cambridge, UK: Cambridge University Press.

Senter, R. J., and E. A. Smith. 1967. *Automated Readability Index.* Technical report. Cincinnati University, OH.

Shalev-Shwartz, S., and S. Ben-David. 2014. *Understanding Machine Learning: From Theory to Algorithms.* Cambridge, UK: Cambridge University Press.

Shammas, C., M. Salmon, and M. Dahlin. 1987. *Inheritance in America from Colonial Times to the Present.* New Brunswick, NJ: Rutgers University Press.

Shayo, M., and A. Zussman. 2011. "Judicial Ingroup Bias in the Shadow of Terrorism." *Quarterly Journal of Economics* 126 (3): 1447–1484.

Shipan, C. R., and W. R. Lowry. 2001. "Environmental Policy and Party Divergence in Congress." *Political Research Quarterly* 54 (2): 245–263.

Shor, B., C. Berry, and N. McCarty. 2010. "A Bridge to Somewhere: Mapping State and Congressional Ideology on a Cross-Institutional Common Space." *Legislative Studies Quarterly* 35 (3): 417–448.

Shor, B., and N. McCarty. 2011. "The Ideological Mapping of American Legislatures." *American Political Science Review* 105 (3): 530–551.

Shwed, U., and P. S. Bearman. 2010. "The Temporal Structure of Scientific Consensus Formation." *American Sociological Review* 75 (6): 817–840.

Silberzahn, R., E. L. Uhlmann, D. P. Martin, P. Anselmi, F. Aust, E. Awtrey, Š. Bahník, et al. 2018. "Many Analysts, One Data Set: Making Transparent How Variations in Analytic Choices Affect Results." *Advances in Methods & Practices in Psychological Science* 1 (3): 337–356.

Silver, D., J. Schrittwieser, K. Simonyan, I. Antonoglou, A. Huang, A. Guez, T. Hubert, et al. 2017. "Mastering the Game of Go without Human Knowledge." *Nature* 550 (7676): 354–359.

Simon, D. 2012. *In Doubt: The Psychology of the Criminal Justice Process.* Cambridge, MA: Harvard University Press.

Skocpol, T., and V. Williamson. 2016. *The Tea Party and the Remaking of Republican Conservatism.* Oxford, UK: Oxford University Press.

Slapin, J. B., and S.-O. Proksch. 2008. "A Scaling Model for Estimating Time-Series Party Positions from Texts." *American Journal of Political Science* 52 (3): 705–722.

Smith, J. L. 2014. "Law, Fact, and the Threat of Reversal from Above." *American Politics Research* 42 (2): 226–256.

Songer, D. R., and S. Haire. 1992. "Integrating Alternative Approaches to the Study of Judicial Voting: Obscenity Cases in the US Courts of Appeals." *American Journal of Political Science* 36 (4): 963–982.

Spache, G. 1953. "A New Readability Formula for Primary-Grade Reading Materials." *The Elementary School Journal* 53 (7): 410–413.

Spaeth, H. J., L. Epstein, A. D. Martin, J. A. Segal, T. J. Ruger, and S. C. Benesh. 2016. *Supreme Court Database, Version 2016 Release 01.*

Spamann, H. 2009a. "Contemporary Legal Transplants: Legal Families and the Diffusion of (Corporate) Law." *BYU Law Review* 2009 (6): 1813–1878.

———. 2009b. "The Antidirector Rights Index Revisited." *Review of Financial Studies* 23 (2): 467–486.

———. 2015. "Empirical Comparative Law." *Annual Review of Law & Social Science* 11 (1): 131–153.

Spencer, S. B., and A. Feldman. 2018. "Words Count: The Empirical Relationship between Brief Clarity and Summary Judgment Decisions." *Legal Writing: Journal of the Legal Writing Institute* 22:61–108.

Splawa-Neyman, J. 1990. "On the Application of Probability Theory to Agricultural Experiments. Essay on Principles. Section 9." Translated by D. M. Dabrowska and T. P. Speed. *Statistical Science* 5 (4): 465–472.

Squire, P. 2007. "Measuring State Legislative Professionalism: The Squire Index Revisited." *State Politics & Policy Quarterly* 7 (2): 211–227.

Starr, S. B. 2014. "Evidence-Based Sentencing and the Scientific Rationalization of Discrimination." *Stanford Law Review* 66 (4): 803–72.

Steffensmeier, D., and S. Demuth. 2000. "Ethnicity and Sentencing Outcomes in US Federal Courts: Who is Punished More Harshly?" *American Sociological Review* 65 (5): 705–729.

Stiglitz, E. H. 2014. "Unaccountable Midnight Rulemaking? A Normatively Informative Assessment." *New York University Journal of Legislation & Public Policy* 17 (1): 137–192.

Stith, K., and J. A. Cabranes. 1998. *Fear of Judging: Sentencing Guidelines in the Federal Courts.* Chicago, IL: University of Chicago Press.

Sulam, I. 2014. *Editor in Chief: Opinion Authorship and Clerk Influence on the Supreme Court.* Unpublished manuscript.

Sunstein, C. R. 1996. *Legal Reasoning and Political Conflict.* New York, NY: Oxford University Press.

Sunstein, C. R., D. Schkade, and L. M. Ellman. 2004. "Ideological Voting on Federal Courts of Appeals: A Preliminary Investigation." *Virginia Law Review* 90 (1): 301–354.

Sunstein, C. R., D. Schkade, L. M. Ellman, and A. Sawicki. 2006. *Are Judges Political?: An Empirical Analysis of the Federal Judiciary.* Washington, DC: Brookings Institution Press.

Susskind, R. E. 1998. *The Future of Law.* New York, NY: Oxford University Press.

Sypnowich, C. 2014. "Law and Ideology." In *The Stanford Encyclopedia of Philosophy,* Winter 2014, edited by Edward N. Zalta. Palo Alto, CA: Metaphysics Research Lab, Stanford University.

Talbert, J. C., and M. Potoski. 2002. "Setting the Legislative Agenda: The Dimensional Structure of Bill Cosponsoring and Floor Voting." *Journal of Politics* 64 (3): 864–891.

Tanenhaus, J., M. Schick, M. Muraskin, and D. Rosen. 1963. "The Supreme Court's Certiorari Jurisdiction: Cue Theory." In *Judicial Decision-Making,* edited by G. Schubert, 113–115. Glencoe, IL: Free Press.

Thomas, M., B. Pang, and L. Lee. 2006. "Get out the Vote: Determining Support or Opposition from Congressional Floor-Debate Transcripts." In *Proceedings of the 2006 Conference on Empirical Methods in Natural Language Processing (EMNLP 2006),* 327–335. Stroudsburg, PA: Association for Computational Linguistics.

Thomson, R. J., and M. T. Zingraff. 1981. "Detecting Sentencing Disparity: Some Problems and Evidence." *American Journal of Sociology* 86 (4): 869–880.

Tigue, C. C., D. J. Borak, J. J. M. O'Connor, C. Schandl, and D. R. Feinberg. 2012. "Voice Pitch Influences Voting Behavior." *Evolution & Human Behavior* 33 (3): 210–216.

Tiller, E. H., and F. B. Cross. 1999. "A Modest Proposal for Improving American Justice." *Columbia Law Review* 99 (1): 215–234.

Todorov, A., A. N. Mandisodza, A. Goren, and C. C. Hall. 2005. "Inferences of Competence from Faces Predict Election Outcomes." *Science* 308 (5728): 1623–1626.

Turing, A. M. 1950. "Computing Machinery and Intelligence." *Mind* 59 (236): 433–460.

Turney, P. D., and P. Pantel. 2010. "From Frequency to Meaning: Vector Space Models of Semantics." *Journal of Artificial Intelligence Research* 37 (1): 141–188.

Tversky, A., and D. Kahneman. 1974. "Judgment under Uncertainty: Heuristics and Biases." *Science* 185 (4157): 1124–1131.

Van der Laan, M. J., E. C. Polley, and A. E. Hubbard. 2007. "Super Learner." *Statistical Applications in Genetics & Molecular Biology* 6 (1): 1–23.

Van der Laan, M. J., and S. Rose. 2011. *Targeted Learning: Causal Inference for Observational and Experimental Data.* New York, NY: Springer Science & Business Media.

van der Maaten, L., and G. Hinton. 2008. "Visualizing Data using t-SNE." *Journal of Machine Learning Research* 9:2579–2605.

Vehtari, A., A. Gelman, and J. Gabry. 2017. "Practical Bayesian Model Evaluation Using Leave-One-out Cross-Validation and WAIC." *Statistics & Computing* 27 (5): 1413–1432.

Vilhena, D. A., J. G. Foster, M. Rosvall, J. D. West, J. Evans, and C. T. Bergstrom. 2014. "Finding Cultural Holes: How Structure and Culture Diverge in Networks of Scholarly Communication." *Sociological Science* 1:221–239.

von Luxburg, U. 2007. "A Tutorial on Spectral Clustering." *Statistics & Computing* 17 (4): 395–416.

Wahlbeck, P. J., J. F. Spriggs II, and L. Sigelman. 2002. "Ghostwriters on the Court? A Stylistic Analysis of US Supreme Court Opinion Drafts." *American Politics Research* 30 (2): 166–192.

Wald, P. M. 1995. "The Rhetoric of Results and the Results of Rhetoric: Judicial Writings." *The University of Chicago Law Review* 62 (4): 1371–1419.

Waldron, J. 2008. "The Concept and the Rule of Law." *Georgia Law Review* 43 (1): 1–63.

Waltl, B., J. Landthaler, and F. Matthes. 2016. "Differentiation and Empirical Analysis of Reference Types in Legal Documents." In *Legal Knowledge and Information Systems,* 294:211–214. Frontiers in Artificial Intelligence and Applications. Amsterdam, Netherlands: IOS Press.

Wang, S. I., and C. D. Manning. 2012. "Baselines and Bigrams: Simple, Good Sentiment and Topic Classification." In *Proceedings of the 50th Annual Meeting of the Association for Computational Linguistics,* 2:90–94. Stroudsburg, PA: Association for Computational Linguistics.

Weiler, J. H. H. 2001. "Epilogue: The Judicial Après Nice." In *The European Court of Justice,* edited by G. de Búrca and J. H. H. Weiler, 215–226. New York, NY: Oxford University Press.

Weinshall-Margel, K., and J. Shapard. 2011. "Overlooked Factors in the Analysis of Parole Decisions." *Proceedings of the National Academy of Sciences* 108 (42): E833–E833.

Weisberg, R., D. A. Mukamal, and J. D. Segall. 2011. *Life in Limbo: An Examination of Parole Release for Prisoners Serving Life Sentences with the Possibility of Parole in California.* Palo Alto, CA: Stanford Criminal Justice Center.

Westreich, D., J. Lessler, and M. J. Funk. 2010. "Propensity Score Estimation: Machine Learning and Classification Methods as Alternatives to Logistic Regression." *Journal of Clinical Epidemiology* 63 (8): 826–833.

Wetter, J. G. 1960. *The Styles of Appellate Judicial Opinions.* Leiden, Netherlands: A. W. Sythoff.

Williamson, J. M. L., and A. G. Martin. 2010. "Analysis of Patient Information Leaflets Provided by a District General Hospital by the Flesch and Flesch–Kincaid Method." *International Journal of Clinical Practice* 64 (13): 1824–1831.

Wold, S., A. Ruhe, H. Wold, and W. J. Dunn, III. 1984. "The Collinearity Problem in Linear Regression. The Partial Least Squares (PLS) Approach to Generalized Inverses." *SIAM Journal on Scientific and Statistical Computing* 5 (3): 735–743.

Wood, B. D., and R. W. Waterman. 1991. "The Dynamics of Political Control of the Bureaucracy." *American Political Science Review* 85 (3): 801–828.

Wydick, R. C. 1994. *Plain English for Lawyers.* 3rd ed. Durham, NC: Carolina Academic Press.

Yackee, S. W. 2005. "Sweet-Talking the Fourth Branch: The Influence of Interest Group Comments on Federal Agency Rulemaking." *Journal of Public Administration Research & Theory* 16 (1): 103–124.

———. 2015. "Participant Voice in the Bureaucratic Policymaking Process." *Journal of Public Administration Research & Theory* 25 (2): 427–449.

Yamamoto, E. K. 1990. "Efficiency's Threat to the Value of Accessible Courts for Minorities." *Harvard Civil Rights—Civil Liberties Law Review* 25 (2): 341–429.

Yano, T., N. A. Smith, and J. D. Wilkerson. 2012. "Textual Predictors of Bill Survival in Congressional Committees." In *Proceedings of the 2012 Conference of the North American Chapter of the Association for Computational Linguistics: Human Language Technologies,* 793–802. Stroudsburg, PA: Association for Computational Linguistics.

Yates, J., D. M. Cann, and B. D. Boyea. 2013. "Judicial Ideology and the Selection of Disputes for US Supreme Court Adjudication." *Journal of Empirical Legal Study* 10 (4): 847–865.

Yoshino, K. 2006. *The Hidden Assault on Our Civil Rights.* New York, NY: Random House.

Young, A. 2017. "Consistency without Inference: Instrumental Variables in Practical Application." Working paper, London School of Economics, London, UK. https://personal.lse.ac.uk/YoungA/ConsistencywithoutInference.pdf.

Zhang, A. H., J. Liu, and N. Garoupa. 2018. "Judging in Europe: Do Legal Traditions Matter?" *Journal of Competition Law & Economics* 14 (1): 144–178.

Zweigert, K., and H. Kötz. 1998. *An Introduction to Comparative Law.* Translated by
    T. Weir. Oxford, UK: Clarendon Press.

MICHAEL A. LIVERMORE is Professor of Law at the University of Virginia.

DANIEL N. ROCKMORE is the William H. Neukom 1964 Professor of Computational Science and Professor of Mathematics and Computer Science at Dartmouth College. He is a longtime member of the external faculty of the Santa Fe Institute.

SFI PRESS

THE SANTA FE INSTITUTE PRESS

The SFI Press endeavors to communicate the best of complexity science and to capture a sense of the diversity, range, breadth, excitement, and ambition of research at the Santa Fe Institute. To provide a distillation of discussions, debates, and meetings across a range of influential and nascent topics.

To change the way we think.

SEMINAR SERIES

New findings emerging from the Institute's ongoing working groups and research projects, for an audience of interdisciplinary scholars and practitioners.

ARCHIVE SERIES

Fresh editions of classic texts from the complexity canon, spanning the Institute's thirty years of advancing the field.

COMPASS SERIES

Provoking, exploratory volumes aiming to build complexity literacy in the humanities, industry, and the curious public.

For forthcoming titles, inquiries, or news about the Press, contact us at
SFIPRESS@SANTAFE.EDU

ABOUT THE SANTA FE INSTITUTE

The Santa Fe Institute is the world headquarters for complexity science, operated as an independent, nonprofit research and education center located in Santa Fe, New Mexico. Our researchers endeavor to understand and unify the underlying, shared patterns in complex physical, biological, social, cultural, technological, and even possible astrobiological worlds. Our global research network of scholars spans borders, departments, and disciplines, bringing together curious minds steeped in rigorous logical, mathematical, and computational reasoning. As we reveal the unseen mechanisms and processes that shape these evolving worlds, we seek to use this understanding to promote the well-being of humankind and of life on earth.

COLOPHON

The body copy for this book was set in EB Garamond, a typeface designed by Georg Duffner after the Egenolff-Berner type specimen of 1592. Headings are in Kurier, a typeface created by Janusz M. Nowacki, based on typefaces by the Polish typographer Małgorzata Budyta. Additional type is set in Cochin, a typeface based on the engravings of Nicolas Cochin, for whom the typeface is named.

The SFI Press complexity glyphs used throughout this book were designed by Brian Crandall Williams.

SANTA FE INSTITUTE
COMPLEXITY
GLYPHS

ZERO

ONE

TWO

THREE

FOUR

FIVE

SIX

SEVEN

EIGHT

NINE

-A-

-B- -C- -D-

-E- -F- -G-

-H- -I- -J-

-K- -L- -M-

-N- -O- -P-

-Q- -R- -S-

-T- -U- -V-

-W- -X- -Y-

-Z-

www.ingramcontent.com/pod-product-compliance
Lightning Source LLC
Chambersburg PA
CBHW031505180326
41458CB00044B/6700/J